Management in Engineering

MANAGEMENT IN ENGINEERING

Principles and Practice

Second Edition

Gail Freeman-Bell
James Balkwill

PRENTICE HALL
London   New York   Toronto   Sydney   Tokyo   Singapore
Madrid   Mexico City   Munich

First published 1996 by
Prentice Hall Europe
Campus 400, Maylands Avenue
Hemel Hempstead
Hertfordshire, HP2 7EZ
A division of
Simon & Schuster International Group

© Prentice Hall Europe 1996

Typeset in 10/12pt Plantin
by Mathematical Composition Setters Ltd, Salisbury, UK

Printed and bound in Great Britain by Redwood Books,
Trowbridge, Wiltshire

Library of Congress Cataloging-in-Publication Data

Freeman-Bell, Gail.
    Management in engineering: principles and practice/Gail Freeman
-Bell, James Balkwill. – 2nd ed.
        p.   cm.
    Includes bibliographical references and index.
    ISBN 0-13-233933-1 (pbk.   : alk. paper)
        1. Engineering–Management.   I. Balkwill, James.   II. Title.
    TA190.F74   1996
    620'.0068–dc20
                                                    95–44646
                                                    CIP

British Library Cataloguing in Publication Data

A catalogue record for this book is available
from the British Library

ISBN 0-13-233933-1

---

  2 3 4 5   00 99 98 97

To Jack

Elizabeth, Two-bean, George, Mint

and (with much sadness since the first edition)
to the late Willie

Contents

Part 2 The management of engineering   71

Foreword

There is always a question about whether a perception is actually formed from elements of reality or, in fact, becomes a self-fulfilling prophecy.

My earliest mental image of an engineer was that of a rather dishevelled, introverted individual characterised by the protrusion of a slide-rule from the top pocket of his crumpled jacket. The chances are that, these days, the younger generation of engineers may not even recognise a slide-rule, let alone carry one.

One of the longstanding complaints of engineers from all generations is the lack of public recognition that they enjoy. Overseas, our engineers are highly regarded but in the United Kingdom tend to be viewed as subordinate to other professions such as medicine and the law—a prophet is not without honour, except in his own country?

I believe that this can be due, in part, to the public perception of the engineer fuelled by the engineer's own dedication to a specialism. For too long, engineers have tended to concentrate their efforts within the—often narrow—band which is their sphere of expertise, to the exclusion of involvement in the general enterprise. Personnel, commercial and finance issues have been left to the departments considered appropriate, with the result that management and business issues have been settled without an engineering input. Furthermore, engineers working in a company environment have often felt their creativity constrained by the specialists from other company areas. Representatives from personnel, finance, sales and marketing each have their own, particular styles which can hinder effective team play.

The need for broadly based engineers who can communicate intelligently and knowledgeably on the 'home ground' of all company functions has long been recognised. Considerable effort has been expended in providing the means for this educational process. Gone are the days when an engineer would be promoted beyond his level of competence as a manager purely because he was a good engineer.

We are now beginning to witness the advent of the engineer who not only has an understanding of the alternative disciplines but whose thought processes actually emulate those of the indigenous experts. These skills which the engineers embrace, in addition to their core skills, are those which equip them to be effective and successful managers. It is this level of overall understanding and competence which has provided the environment demanded for effective teamworking.

There are three major factors upon which every manager must focus:

- Operational processes.
- Management style.
- Organisational structure.

All other business topics will rest comfortably beneath these generic titles. The declared priority is crucial.

This book considers all the business elements that an engineer who desires to manage effectively will require and I am glad to see it. Understanding the elements described will provide a sound foundation from which engineers will be able to establish a knowledge of how those elements are applied through discussions with the fact-holders from within their own organisations. It is imperative that the readers are not satisfied with the theory alone, for therein lies sterility, but actually seek this practical information at an operational level.

Nothing, apart from bankruptcy, comes easily. The tools for engineers in management are known and available—use them, tailor them to your circumstances and style, and manage well!

John Towers
Group Managing Director
Rover Group Limited

Preface to the second edition

The first edition of *Management in Engineering* was published in 1993. It appears to have been well received both by those who learnt from it and those who taught with it. Management in engineering is a rapidly changing field although, naturally, some management concepts are as constant as human nature itself. In this second edition we have placed considerable importance on the opinions of the book's many users. We believe we have retained all that was good about the first edition but we have extended the scope of the text considerably. There are new chapters on product development, operations management and quality management. We have also made changes to other parts of the text to reflect changes that have taken place, to update material and to allow for the different emphasis now placed by industrialists on certain topics.

Commercial success comes from being able to meet the needs of customers at a profit. For engineers this involves the creative application of engineering principles. As they become more senior it also means becoming an effective manager. This book addresses the management needs of engineers and is primarily written as an undergraduate text. The learning objectives are:

- To be able to apply management topics as tools with which to achieve engineering solutions
- To show how effective management is achieved in engineering organizations

These objectives are met primarily by the way in which the content has been determined, structured and presented. In addition, we have included case studies that illustrate the techniques described, as well as many worked examples and revision questions.

The book is totally biased towards engineering and the examples are drawn from engineering industry. All the chapters are self-contained which allows them to be used on their own. However, the book leads the reader through all the appropriate management topics in a logical progression, culminating in a case study of a large engineering project which shows how all the topics in the book come together in practice.

Acknowledgements

Many different people have played an important role in the creation of this book and it is appropriate to spend a few lines acknowledging them. We wish to thank the companies that have allowed us to produce case study material, in particular Oxford Lasers Ltd. We also wish to thank those engineers who have allowed us to use their personal experiences: Hilary Briggs, Martin Twycross, Matthew Hobbs, Simon Tu, Rhys Lewis, Richard Benfield, John Boaler, and Steff Inns. We thank Mr John Towers for preparing the Foreword. There are many others, who have helped and advised us during the process of writing, in particular the many people who have used the text and provided valued reviews of it. We acknowledge their help and support.

Our editor, Chris Glennie, deserves a special mention this time, not because of his editing skills although he certainly has them, rather for being so good at paying the bill at Le Manoir aux Quat' Saisons where we had all those legendary publishing lunches. We remember his promise of a lunch centred around a bottle of Chateau De La Tour 1961 at the Manoir with the manuscript for the third edition and look forward to it very much!

Finally, in the first edition we saluted the wines of Messieurs Alain Voge (Saint Peray) and Pierre Ricard (Vacqueyras) without which, writing would have been impossible. Sadly, supplies of the best years of these wines have dried and the weather in the region since the first edition has not been good. Worthy of recommendation now are the 'new' wines of the Languedoc – the varietals, of which 'Domaine Saint Hilaire-Viognier' will take a lot of beating. Additionally, Les Trois Clochers 1990 of Réunis à Unidor (St Laurent des Vignes Dordogne), makes excellent drinking at the moment, but not for many, my father bought the last three cases a year ago! We are also grateful to the following, whose responses to a questionaire helped shape the content of the book.

Dr David Pollard
University of Abertay, Dundee
Dundee Business School

Ms Ann Hollings
Senior Lecturer Human Resource Management
Staffordshire University
Department of Management Studies
Business School

James Brown
Lecturer
Accrington and Rossendale College
Division of Technology and Construction

Mr Rajaratnam Rajahmaheswaran
Aylesbury College
Department of Engineering Technology

Mr Alan Walters
Bell College of Technology
Department of Management and Business Administration

Dr Robert Matthew
University of Bradford
Department of Civil Engineering

Alan Peirce
Senior Lecturer
Brunel: The University of West London
Department of Manufacturing and Engineering Systems

Professor James Powell
Director of the Graduate School
Salford University
Graduate School

Professor A P Ambler
Brunel: The University of West London
Department of Electrical Engineering

Mr David Briggs
Senior Lecturer
Manchester Metropolitan University
Department of Mechanical Engineering, Design and Manufacture

Mrs Margaret Johnson
Principle Lecturer
Cheltenham & Gloucester College of Higher Education
Centre for Management

Eur Ing Dennis A Snow
De Montfort University
Department of Mechanical and Manufacture

Roger Wilson
Head of Engineering
Dudley College of Technology
Faculty of Science and Technology

Mr George B Fyfe
University of Abertay, Dundee
Department of Electronic and Electrical Engineering

Robert B Tait
Senior Lecturer
Farnborough College of Technology
School of Engineering

John W Evans
Head of Programme
Fife College of Further and Higher Education
Department of Management

Miss Carol J Urquhart
Fife College
Department of Accounting

Ms Linda Major
University of Glamorgan
Department of Business Studies
Gwent Tertiary College

Michael Bretherick
Head of Academic Services
Hartlepool College of Further Education

Moyra Fowler
Lecturer
University of Hertfordshire
Division of Manufacturing Systems Engineering

Mr Andrew M D Armitage
Highbury College of Technology
Faculty of Technology

Ms Michelle Fraser
University of Humberside
School of Applied Science & Technology

Mr Dennis Baker
Senior Lecturer in Engineering
Isle of Man College
Department of Engineering

Mr Roger Armstrong
University of Central Lancashire
Department of Management Development
Lancaster Business School

Barry S Stanway
Synergy Dynamics
Corporate & Educational Development Planning Specialists

Mr Patrick Gunnigle
University of Limerick
Department of Management

Professor Robert M Parkin
Loughborough University of Technology
Department of Mechanical Engineering

William T Low
Napier University
Department of Management Studies

P Cameron-MacDonald
Napier University
Department of MMSE

Derek J Williams
Head of Technology
Northampton College
Department of Science and Technology

Mike Miles
Senior Lecturer
University of Plymouth
School of Manufacturing Materials and Mechanical Engineering

Eur Ing Norman Edward
Robert Gordon University
School of Mechanical and Off Shore Engineering

Mr Bob Dickenson
Shrewsbury College of Arts & Technology
School of Business

Mr John W Cross
Southampton Institute of Higher Education
Southampton Business School

Dr S K Banerjee
University of Strathclyde
Department of Design, Manufacture and Engineering Management

Jacqueline E Gavaghan
Lecturer
Tralee Regional Technical College
Department of Business

Mr Michael J Deacon
Tresham Institute
Department of Engineering and Construction

Dr Steve New
University of Manchester Institute of Science and Technology
School of Management

1

Engineers as managers

The most important skill of a creative mind is the ability to ask good questions. Before we embark on the body of this textbook there is a question that should be asked, and we must provide an answer to it before we can sensibly proceed with the study of 'Management in Engineering'. The question is 'why?' Why are we bothering, as engineers, to study management and what relevance does it have? This answer to this question, however, rests on another one. 'What do engineers actually do?' If they have a need for management then the book is worthy of study, if they don't then it should be thrown away now.

Many dictionaries define an engineer as 'one who drives engines' or 'one who designs and builds engines'. Few professional engineers are engine drivers and only a tiny fraction design and build engines. Popular understanding is often worse, with many people thinking that the words 'engineer' and 'mechanic' are interchangeable. They are not. The word 'engineer' has two roots, one means 'the creative one' or 'the one who uses ingenuity'. The other means 'to bring about' or 'to make manifest'. These are much closer to the truth, but they still do not fully explain what an engineer does and whether engineers have a need for management.

The engineering profession is like other professions, such as law or medicine, in that it has a regulating body controlling qualifications, there is a code of ethics, professional status is distinct and the law regulates the practice of engineering. It is therefore not surprising that the period of study prior to achieving membership of a professional engineering body is a lengthy and testing one. In the UK it usually takes a minimum of four years after graduation, and is often much longer. To examine whether engineers need to study management or not we will look at the two distinct periods of an engineer's life: training and professional practice.

During the training period required to become a qualified professional engineer the student is continually under pressure to acquire technical knowledge and skills. A mechanical engineer's training places emphasis on thermodynamics, stress analysis, and mechanical manufacturing processes. Electronics and electrical engineers will be putting their efforts into analog and digital circuit design, data processing and electronics manufacturing techniques. The emphasis is on technical excellence, on the application of advanced theory and the ability to describe the world with mathematics and then to manipulate it to meet desired needs. Management is a curious thing to study at this time. Whenever was a student of applied mathematics helped by knowing how to hold a performance-related appraisal? For engineering undergraduates to study 'management' may seem not only unreasonable in terms of the extra effort required but also unconnected with undergraduate engineering.

But what about life after qualifying? Of course, most undergraduates are aware that life after qualification may involve different skills and that management may yet prove to be useful. What reasons are there for thinking this?

Perhaps it is worth opening the dictionary again to see what it has to say about 'management'. The *Oxford Concise Dictionary* (Clarendon Press, 1990) gives a number of definitions of management. These include 'the process or an instance of managing or being managed'. Certainly this doesn't tell us very much, as we don't understand the process of managing. Another definition is 'trickery; deceit', not a definition we would use! The best definition for the term as it is used in engineering is 'the professional administration of business activities'.

How does this affect practising engineers, and what are business activities? In fact, business activities are anything that relate to the organization and operation of a business. Engineers certainly play a major role in the organization and operation of an engineering business. More than this, it is engineering that defines the very structure of a company. A jet engine manufacturer is constrained by the limits imposed by the design and manufacture of the jets. Engineers wrestle with the design of the engines, striving to get more power, greater reliability, more profit, and greater safety. It is engineers who state how the product will actually be when it exists. All other aspects of the company's management must derive from this, not the other way round. The question to ask now is 'what sort of jobs do engineering organizations need their engineers to do?' The following are extracted from advertisements that have appeared in professional engineering journals:

Principal Mechanical Engineer ... provide technical and commercial direction within the engineering field. Matching business acumen with a detailed understanding of Thermal and Acoustics...

Design Engineers ... With responsibility for design, manufacturing, specification, installation, commissioning and the provision of operational support, you will be encouraged to pursue the creative application of new technology.

Design/Development Engineer ... The successful applicant will be engaged in product concepts generation, detail design and analysis, development testing and client liaison.

In the jobs advertised there is no explicit reference to 'the professional administration of business activities', but let us look more carefully at each job in turn.

In the first job there is reference to commercial direction and business acumen, as well as to technical issues. 'Direction' implies that this job involves getting others to do something, showing them the way. For this job then one could assume that in addition to the technical knowledge required you have to know something about marketing, finance, and directing others.

The second job looks much more technical, except where it says 'with responsibility for...'. There are many technical issues involved in this job but it is clearly the intention that the successful applicant will be responsible for others who will assist in the work. There is simply too much for one person to do. The third job is similar.

Clearly, the technical issues so familiar to undergraduate engineers are important here but so is something else. In all these jobs, the advertisers are looking for specific technical knowledge and something more. The phrase 'the professional administration of business activities' may not trip off the tongue but it does describe the 'something else', that the advertisers of these jobs want. The 'something else' is, of course, management skill. These advertisements show that there is need for a combination of skills, both technical and management, and that the two skills go together.

How can an engineer's training period best provide for post-qualification life, both in technical skills and managerial ones? Clearly it is important that while in the qualifying period an engineer should gain a sound technical knowledge and learn enough of the physical principles so that when a new situation is encountered it may be analyzed. The engineer will then be able to identify the knowledge needed to solve the problem. The academic environment is ideal for this and most engineering courses are characterized by intense study.

But what of the management? What important management principles may be learnt at this time? It is clearly not possible to study and experience management in undergraduate life in the way that it will be experienced in post-graduate life, so what is required is a combination of technical and managerial courses. The undergraduate must be sufficiently well prepared for the business environment so that the transition is not traumatic or unexpected but is beneficial and the graduate can quickly absorb the new environment.

From a management training point of view it is clear that in order to prepare for life in engineering business, a training is required that provides the important management skills as they are needed by engineers. The training should make clear how the topics relate specifically to engineering. Management topics that do not help engineers should be omitted. Those of use to engineers, but often omitted from other management courses, should be included. The material should be written for engineers by engineers, and it should explain how an individual's engineering abilities can be greatly extended by good management, for that is the essence of management. This book provides the material for such a training.

We need a structure for such a large undertaking and this book has been carefully constructed to present the important aspects of engineering management in a logical and readable format. To begin with, we study the environment of engineering practice and then examine the needs to which this environment gives rise. These needs are studied as separate issues since this makes their understanding easier. Of course, in reality the issues are not separate and occur together. For this reason, we have cross-referenced material and used engineering examples to show its relevance. More importantly, we have provided a large case study of a real engineering project which includes a description of the technical side of the project. This case study brings together the separate themes of the book, shows the essence of engineering, and illustrates vividly the equal importance of management and technical skills. It serves as an excellent illustration of how the important principles explained in the book are used in practice.

We will now explain in a little more detail the contents of each chapter. In Part 1 we examine what organizations are and what they actually do. In Chapter 2 we start examining the nature of organizations, explaining that organizations have a purpose, a legal identity, and a need to survive in the long term. This provides the context for our work as engineers. Once we have this picture of an engineering organization we can examine the various tasks that such an engineering organization will need to execute. Chapter 3 is called 'Functions of organizations' and is a detailed explanation of the tasks that must be managed if the objective of the organization and purpose of its existence are to be met.

After completing Chapters 2 and 3 the reader has a complete picture of the engineering environment. We have explained why the organization exists, what it is trying to achieve and the tasks, or 'functions', that it must accomplish in order to be successful.

From this point on the nature of the book is different. It now explains the management tools that are required in order for the company's personnel to be successful in their aims. These tools come in two forms. First, there are a set of tools used to manage *engineering* and second, there are tools used to manage *engineers*. Part 2 (Chapters 4–8) deals with engineering and Part 3 (Chapters 9–12) with engineers.

In Chapter 4 we consider the management of money. In this chapter we look at the role of money in the operation of a company and specifically at the financial issues that involve engineers. The management of money is a large issue in its own right and some people never work with anything else. As engineers we have a great responsibility and it is important that we understand issues such as product costing and financial control. Chapter 4 provides an explanation of the financial issues that are important to engineers.

In Chapter 5 we deal with the development of new products. Over a period of time customers change, technology changes, expectations change and so too do products. Unless organizations continue to develop new products they soon find themselves with products that people don't want and insufficient funds to develop ones that people do. In this chapter we examine the process of product development and its management. This is a subject of particular interest to engineers since it involves management of the design process where the technical side of engineering is so important.

In Chapter 6 we examine the central function of an engineering organization, manufacturing. The chapter explains the organization of manufacturing, how it is planned and controlled, how materials are managed and how the whole operation is arranged in a profitable and effective way. Again this is an area of special interest to engineers since many spend entire careers in this field.

In Chapter 7 we consider a tool used to manage engineering that pervades the whole organization. Quality starts with suppliers and finishes with customers, and on its journey it passes through every area of the organization. The chapter starts with an investigation of the quality standard ISO 9000, explaining the requirements that this makes on engineering organizations. We then explain Total Quality Management and investigate some quality management tools and techniques that are essential for engineers. Lastly, we examine the factors that shape the use of quality in engineering organizations, particularly the cost associated with implementing a quality system.

Engineers often undertake work in project groups, either as team members or team managers. Chapter 8 looks specifically at the management of projects, their definition, planning and monitoring. It describes techniques that can be used for planning and considers the aspects that have to be investigated when deciding whether or not to proceed with a project, such as the cost. There is also a case study which shows how one company tackled a project that had a significant impact on its business and profitability.

Part 3, the management of engineers, starts with Chapter 9, 'Personnel management'. If you visit a company on a Sunday you will more than likely find the whole place at a standstill. There is nothing wrong with the organization. All the things that are there during the week are still there on a Sunday with one exception. The people are missing. No other single resource has such a dramatic effect on an organization. In Chapter 9 we look at the specific issues associated with employing and managing people. The contents of Chapter 9 give the reader a formidable and comprehensive battery of personnel skills that will serve well for life in professional practice.

The emphasis in Chapter 10 is still on personnel management but it is a very specific aspect, that of team working. Successful team working brings a staggering capacity for output. In Chapter 10 we study how this comes about in order that we may be able to control and enhance the success of teams to assist the organization in achieving its goals. Engineers are not the only professionals to recognize the benefits of successful team working and much of the material in this chapter is of use to other professions.

Engineers must be creative and able to work on more than one thing at a time. However, if engineers are to be able to orchestrate technical work and deliver results on time then they must be well organized. Personal organization is not an attribute of our genetic coding; we are not born tidy or messy people. We can give ourselves good personal organization by working at it. Chapter 11 tells you how. Clearly those who have good personal organization will be of more interest to the originators of the advertisements above than those who do not.

Engineers are processors of information, they gather it from a wide range of sources. If engineers are to be competent and enjoy the challenge of creative engineering, they must be good at gathering information and then disseminating it. It is clear that one skill required by all the above aspects of engineering and management has not been addressed. It is not possible to be successful in any of these areas if one is not able to communicate. Of course, we can all communicate at some level but professional engineers, along with other professionals, have the most demanding needs for clarity and excellence in communication. Chapter 12 explains this need and provides a training in communications that forms the basis of a separate undergraduate course in many engineering programmes.

Finally, in Part 4 all the different skills are brought together in a case study which clearly shows the balance of technical and managerial skills that a practising engineer actually uses. In the case study the boundaries between the separate management subjects have gone. In practice, they are never there. The technical and managerial issues are solved simultaneously and engineers use many of the skills at once. Just as a good piece of music is composed of different instruments playing together, each making exactly the ideal contribution for the given instant, so is good management practised. The project described in this case study lasted for two years and produced a prestigious new product for the company concerned. Naturally, this means that any useful description of it will take some time. For this reason the case study is provided in sections each of which deal with one or two management issues at a time. The case study is followed by a debrief in which the material is discussed. Comments are made about the project and how management theories may or may not be applicable to each section.

It is in Chapter 13 in which the very spirit of engineering management is encapsulated. All the issues explained in the previous chapters are brought together and used in conjunction with the commensurate technical skills of the qualified engineer. It is certainly important to dissect and examine the individual tools of good management as we have in the previous chapters, but it is also important to see how they fit back together again and understand how the skills are used in practice.

This book therefore expounds a complete understanding of engineering management. It first makes clear the environment in which engineering management is practised. It then dissects each of the important tools of management that an engineer needs. Finally, it brings them all back together explaining, with a real example, how successful management by engineers may be

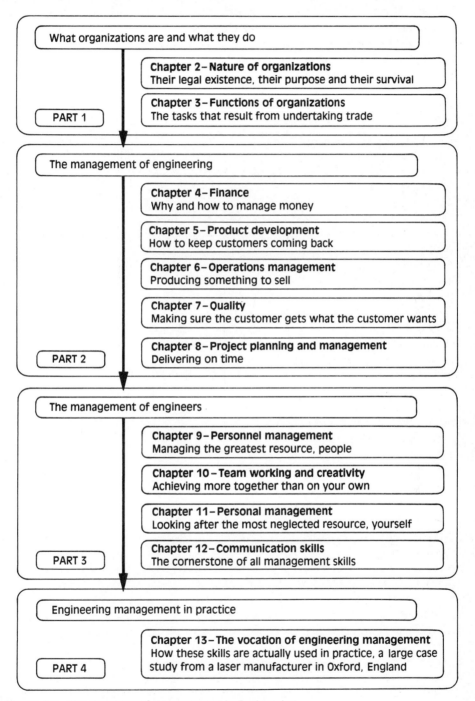

What organizations are and what they do

Chapter 2 – Nature of organizations
Their legal existence, their purpose and their survival

Chapter 3 – Functions of organizations
The tasks that result from undertaking trade

PART 1

The management of engineering

Chapter 4 – Finance
Why and how to manage money

Chapter 5 – Product development
How to keep customers coming back

Chapter 6 – Operations management
Producing something to sell

Chapter 7 – Quality
Making sure the customer gets what the customer wants

Chapter 8 – Project planning and management
Delivering on time

PART 2

The management of engineers

Chapter 9 – Personnel management
Managing the greatest resource, people

Chapter 10 – Team working and creativity
Achieving more together than on your own

Chapter 11 – Personal management
Looking after the most neglected resource, yourself

Chapter 12 – Communication skills
The cornerstone of all management skills

PART 3

Engineering management in practice

Chapter 13 – The vocation of engineering management
How these skills are actually used in practice, a large case study from a laser manufacturer in Oxford, England

PART 4

Figure 1.1   The structure of *Management in Engineering*

achieved. Figure 1.1 shows this structure diagrammatically and includes a very short summary of the contents of each chapter.

In the first paragraph of this introduction we asked 'why are we bothering, as engineers, to study management?' As you can see, it is impossible to be a professional engineer without this knowledge.

Part 1

WHAT ORGANIZATIONS
ARE AND WHAT THEY DO

2

Nature of organizations

Overview

In this chapter we look at the nature of organizations and we start by considering their fundamental aims. It often seems that organizations exist simply to make money, but in reality this is rarely the case, though, of course, money plays an important part. In order to determine the optimum activities for the employees of a business to undertake, there must be a clear statement of the aims and objectives of the whole organization. In this chapter we look at the role of corporate objectives and the way in which they define the overall aim of the company.

Once an organization has a purpose and undertakes activities to achieve those aims, it will require a legal identity. The law regulates the existence of trading organizations in order to protect the rights of individuals. No organization can trade legally without conforming to the requirements of the law, and so in our investigation of the nature of organizations we must develop some understanding of the forms of business that the law recognizes. We will therefore consider the legal forms of organizations in the UK, including a review of some of the legislative and other factors that affect the way in which organizations are able to transact business.

Once organizations exist with an aim and a legal identity they naturally try to survive. In order to survive, organizations must provide something that their customers want. There are many strategies that may be used to achieve this and we will briefly examine the most important ones. Finance has a special role to play in survival and we shall consider this as well.

Finally, we will look at an engineering company and the way that it has been established and operates.

2.1 Introduction

Many qualified engineers work in the engineering industry in manufacturing companies, some work in design consultancies and others in public organizations. There are also many people who qualify as engineers who take a different career route, moving away from engineering but using their numerical and managerial skills in areas such as finance, computing and management services, or administration. All the engineers entering these areas will need the management knowledge and skills described in this book. They will also all work in organizations.

Organizations vary enormously from the two- or three-person design consultancy, where everyone is a partner, to a multinational conglomerate where an engineer could be working in a division with 200 or 300 other engineers. However, if the engineer is to manage a project, finance or personnel effectively an understanding of the organization's overall aims is necessary to put the management requirement into context. The engineer must know what the goals of the organization are, and should be able to assess the contribution of the engineering activity to meeting these goals. This is particularly important for engineers who, as agents of change, are responsible for implementing new technology or working practices.

The operation of a company will be constrained by the way in which it has been formed. For example, the methods by which it can raise capital for expansion will be greatly affected by its legal form. While it is true that engineers do not often come into contact with the fund-raising process, it is important that they have some understanding of the legal implications of the form of a company and can understand something of the reasons for choosing one form over another.

In this chapter we will examine how businesses set their overall goals and why this is an important thing to do. We shall then consider the way in which businesses can be legally established in the UK. We also look at how organizations develop strategies to ensure that they will survive in the long term.

The learning objectives for this chapter are:

1. To understand how organizations set out their overall aims
2. To illustrate how organizations ensure that all employees are engaged in tasks that ensure that these aims are achieved
3. To look at the types of business that can legally be formed in the private sector in the UK
4. To understand the tasks an organization must undertake if it is to ensure its long-term survival.

2.2 The aim of organizations (corporate objectives)

2.2.1 The need for corporate objectives

It is common for newly qualified engineers to join organizations and start practising the skills they have learnt. As their competence advances, the contribution they are able to make becomes too large for them to achieve alone. Before long, the successful engineer earns managerial responsibility and is provided with constant assistance from subordinates. This process continues and senior managers often have subordinates who themselves have subordinates and so on, many times over.

With such large numbers of people, simply deciding what everybody should be doing can be a major problem. There are always many more constructive tasks available than it is practical to complete. Consequently, successfully managing such numbers involves choosing the most effective options for each person. In short, the professional manager will want to know not only that everybody is doing something constructive, but also that it is the most worthwhile task that they could be doing. It is very much in the interests of the organization to ensure this. People engaged in less than optimal tasks represent time wasted, since they could contribute more effectively by doing something else. For instance, a design engineer clearly knows that any material or labour savings, that do not compromise performance, are desirable. However, unless the engineer knows how much the organization stands to benefit from a project no rational choice can be made between this work and any alternatives. The question is not can I do something constructive, instead it is which of the many things would be best to do.

It is also of benefit to the individual to be certain that the work in progress is the most beneficial alternative for the company. Such knowledge is very motivational since it makes clear the value of the work. We are all motivated by doing useful work and demotivated by pointless labour. In Section 11.2.1 we explain how objectives are needed from a personal, psychological point of view. People need to know what they are supposed to be doing in order to be happy, and one cannot be fulfilled while having no direction to one's efforts. Having an important objective for which to aim answers this need directly and understanding how our objective fits into and assists the aims of our organization is part of being convinced that what we are doing is worthwhile. Successful organizations therefore ensure that everyone is not only engaged in constructive tasks but can also see the benefit of the contribution they are making and so are motivated to achieve their aims.

How can these important aims be achieved? In order for everyone to know what tasks are best and for all managers to be able correctly to identify optimal projects for those who report to them, everybody must have access to some underpinning theme and to which all efforts are directed. Such an aim is clearly going to be of great importance to the organization and its selection must be a matter for senior management, since, if it is chosen poorly, the whole organization will be directing efforts into activities that provide less than the best return. Once a good underpinning theme has been chosen, a method must be identified by which everybody in the organization can have access to, and be guided by, this central aim.

We will now introduce the two areas of management that engineering organizations use to achieve this organization-wide coordination of effort. The first step is choosing the underpinning theme, and this is dealt with in the next section. After this, Section 2.2.3 explains how the efforts of every member of the organization can be guided by the central aim.

2.2.2 The mission statement

Imagine a company, perhaps a small pump manufacturer. The business employs, say, thirty people. What is the purpose of the business? Does it exist to push back the frontiers of pump design? Probably not, although the company might conceivably be engaged in doing so. Does it exist to provide wealth for its employees? Certainly, the owners of the business may be anxious to provide fair pay for their workforce, but this is unlikely to be the reason the company was founded. Does the company exist to make money? Many companies produce less profit than would be provided by investing the assets elsewhere but still they carry on trading; money alone is therefore not enough to explain a company's existence.

There may be any number of different reasons for which the company exists and each reason brings its own strategy. For example, if the pump manufacturer does exist to lead the world in pump technology, then much research and development would be appropriate. If, on the other hand, the purpose of the company is long-term survival then perhaps moderate research and development coupled with the establishment of alternative business lines would be best. If the aim is to meet all the pump needs of a particular market sector then forging special links with that market and excluding competitors from it would be best.

From this demonstration of how the purpose of the company affects its management we can conclude that before any rational business decisions can be made, the company must decide once and for all, at the top, what it is trying to achieve. You cannot steer a course without knowing where you are heading.

Well-managed companies have such a statement, made at the highest level, explaining what they are trying to achieve, and this is known as a 'mission statement'. This endeavours to be a lasting summary of the whole spirit of the company and to explain succinctly what the company is about. It is this statement that sets the direction and thrust of the organization. The statement is an objective and, as such, should have the properties of a well-written objective that are described in Section 11.3. That section explains the mechanics of good objective writing and the properties that well-written objectives must have. It explains why they must have these properties and it should be read before preparing any objectives or mission statements oneself.

A salutary lesson in the importance of choosing the right mission statement may be taken from the example of the American railway companies early this century. One such company might have had a mission statement such as:

> To provide a quarter of all the railway miles sold in the USA over the next decade.

This statement seems a perfectly sensible goal; the organization is engaged in making money from the sale of rail journeys. We cannot tell from such a statement whether it is a realistic target for the particular company but it certainly does sound like a worthy objective for the organization. In fact, many companies at the time used exactly such mission statements and yet they ultimately went out of business. In the first years of this century, at a place called Kittyhawk, in the USA, two brothers spent their days making flying machines. A few years later one of them actually flew in one such machine and history was made as the Wright brothers became the first humans to fly in a machine weighing more than the air it displaced. What relevance had this to the mission statement of the rail companies? In their view, it had none and the rail companies turned a blind eye towards this new development. After only thirty years, people had put the aeroplane to all kinds of uses and cargo flights were routine. The new machines carried freight with unchallenged speed. The railways were relegated to transporting goods to ports where they were loaded onto ships to cross the Atlantic but the advent of transatlantic flight took even this from them. The error of the rail companies was very simple: they chose the wrong mission statement. Imagine how different things would have been if the rail companies had chosen the following mission statement:

> To provide a quarter of all the transport miles sold in the USA over the next decade.

There is only one word that has been changed in this new mission statement: 'railway' has become 'transport'. A rail company with this mission statement would have turned a keen eye towards developing new transport technologies, and such a company would have been the first to add air transport to its portfolio of products. A rail company that was successful in this goal would have ended up owning the airlines instead of going out of business because of them, all because just one word was incorrectly chosen.

2.2.3 Managing by objectives

Once a mission statement has been selected and agreed upon, the management of the organization must communicate it to all members within the organization.

This, however, means far more than simply making the words of the statement available. To arrive at tasks for all members of the organization that contribute to this central task clearly involves a great deal of subdivision and is far more than one person, or even a small team, is likely to be able to do. To process starts with an examination of the implications of the mission statement and from these, subordinate tasks can be identified. Each of these tasks may itself have component tasks and may be further subdivided. Consider the example of the pump manufacturer. The mission statement might be of the form:

> To return a profit of 14% of turnover by supplying centrifugal water pumps to European customers over the next calendar year.

From this central aim the senior management can use their technical knowledge and skills to produce supporting tasks. By considering the sales price and the profit margin for all the different models of pump, target numbers for each model to be sold can be prepared. The organization will have to manufacture the pumps and they will have to be sold and shipped. Using their knowledge of the business and the company, the directors can now start to make objectives for each department. Indeed, even the very presence of the departments is governed by the mission statement. The production department of the pump manufacturer exists to meet the manufacturing needs of the mission statement, not the other way round.

The sales department might be given the objective of logging fifty orders a month. The manufacturing department might be given the objective of meeting every order within twenty-two days with an average cost of manufacture of less than 65% of sales price. And so on for the other departments. All these departmental objectives are chosen to support the mission statement. If the mission had been different, the departmental objectives would also have been different.

Once this first level of objectives have been agreed, individuals are made responsible for achieving them. Normally in a company it will be the directors who take on objectives at this level.

When departmental objectives have been finalized, the objectives for the groups of individuals within the departments can be set. In the example above the manufacturing manager might know that the current material and labour cost is too high to meet the departmental objective and so sets up group of engineers to solve the problem. The importance of the departmental objectives can now be seen. For instance, the manufacturing department may feel that the amount of subcontract labour is too high or that commonality of parts between models should be increased. The departmental objectives are used to decide which tasks are actually important and only those which support the departmental objectives, and therefore the mission statement, are selected.

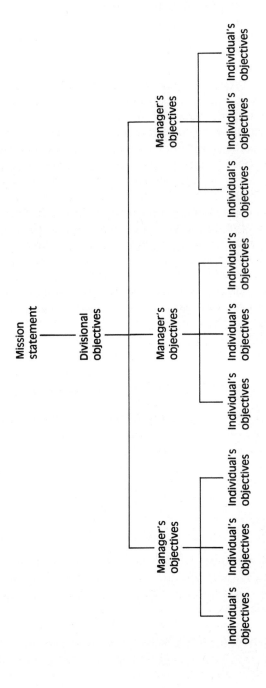

Figure 2.1 Management by objectives

The number of times objectives are subdivided depends only on the size of the organization. The important thing is that by this progressive subdivision of the main task the whole organization is directed to achieve the same overall goal. No effort is wasted on undesirable directions and no contradictory programmes of activity occur. The objective setting extends as far as necessary throughout the organization. Through the use of appraisals it is common to find management by objectives extending to every single member of staff in the organization. This tree structure of management by objectives is shown in Figure 2.1.

Finally, we have seen the need for objectives, we have seen the great effect mission statements can have on organizations and we have seen how objectives may be used to manage the activities of everybody in an organization. People tend to think of objectives, both at the corporate and the personal level, as being things that don't change. This is wrong: times do change and business opportunities change with them. For this reason, the senior managers of the organization will meet regularly to review progress towards each of their objectives and to reassess the value and sensibility of the mission statement and its consequent objectives. Corporate objectives are not immovable and they need maintaining to keep them appropriate. It is clearly not advantageous for organizations to stick doggedly to objectives which have become outdated and so miss opportunities. Conversely, there is no benefit in constantly changing objectives in response to every external factor. The most testing assessment of a management team is how well they maintain their corporate objective in order that they steer the organization between these two pitfalls.

2.3 The legal establishment of organizations

In this section we will examine five ways in which businesses in the private sector in the UK can be established. These are sole traders, partnerships, cooperatives, and public and private companies. We will also consider the use of franchises. These types of organizations will exist in countries outside the U.K. but the legal forms and legislation that affect them will vary from country to country.

2.3.1 Sole traders

The establishment of a business with a single owner, a sole trader, is the simplest form of business and is therefore very common, particularly in the high street. Sole traders, for example, run small shops and newsagents or work as electricians, plumbers, builders and consultants. This form of business is also the simplest to dissolve.

If you want to establish a business as a sole trader it is quite simple: you just have to start trading. The business has no separate legal identity from the owner which means that all debts, liabilities, and profits belong to the owner. You do not have to use your own name for the business, although this is usual, but if you use a different name you have to be careful not to use another company's trade name. Operating as a sole trader doesn't mean that you have to work on your own, you can still employ other people. Some features of sole trader businesses are as follows:

1. The accounts do not have to be disclosed publicly, and are therefore not available to competitors.
2. The owner makes all the decisions, and therefore has total control of the business. This can be a great advantage, particularly because of the speed with which decisions can be made and implemented. It can also be a disadvantage as the owner's competence as a manger and businessperson can restrict the business. In addition, the sole trader may have no-one with a vested interest in the business, with whom to discuss ideas.
3. A sole trader can offer a personal service which may be valued more highly by some customers.
4. All the profits and the debts belong to the owner. This can result in hardship if a business does badly as many sole traders use their homes as security for their business. In any case, if debts are excessive it can lead to the forced sale of personal possessions; liability for debts is unlimited.
5. Often the only form of security a sole trader will have is their own home and it can therefore be very difficult to get extra capital for development or expansion of the business.
6. Working for yourself, particularly with all the administration and book-keeping required to run a business, can mean that you work very long hours. It may also be some time before you are able to pay yourself an appropriate salary. Similarly, taking holidays can be a problem unless there is someone that can be trusted to run the business while you are away.
7. It can be a high-cost enterprise because the sole trader rarely benefits from economies of scale. However, some businesses overcome this problem by forming a sort of collective. This is particularly common in retail trade where a collective can purchase larger quantities from wholesalers and manufacturers, thus getting a better price.

2.3.2 Partnerships

Partnerships exist when there are a number of people involved who are part-owners of the business. This type of business can be set up without any formality but it is usually advisable to draw up a deed of partnership which covers things such as the way in which the profit will be split between the partners.

Under English law the maximum number of partners that is allowed is twenty. There are two exceptions to the limit. In banking the maximum number is ten and in certain professional partnerships where companies are not allowed to be formed by law, such as for solicitors and accountants, there may be more than twenty.

In a partnership each partner is jointly liable for any contract made by any other partner, irrespective of prior knowledge. This means that if, for example, one partner agrees to a contract to supply a piece of equipment within a certain time and another partner does not agree to this then the contract is still binding on all the partners. If a partner leaves the business then all previous contacts have to be told of the situation and made aware of the fact that the person will not be liable for future contracts. The former partner's name should also be removed from the business stationery.

The main features of operating as a partnership are as follows:

1.   The liability and decision making are shared. This can be an advantage when the skills and experience of the partners complement each other and the management of the business is carried out effectively. However, problems can ensue because liability as a partner exists even if the partner did not know of, or agree to, the contracts made. It may also be more difficult to reach decisions since they will all require consultation and agreement. This problem can be alleviated by having very clear lines of demarcation for the various aspects of management and operation of the business.
2.   There is less formality and expense than is involved when forming a company.
3.   Accounts do not have to be publicly disclosed to competitors.
4.   Because you have a number of people with resources that can be drawn upon it may be possible to raise large amounts of capital initially. However, extra capital may be difficult to obtain due to limited additional security, and many people use their homes as security for a loan.
5.   The partnership could be suddenly dissolved if one partner dies or goes bankrupt, unless this event is covered in the deed of partnership.
6.   Holidays and illness can be covered by the other partners.

2.3.3 Cooperatives

Cooperatives involve a voluntary association of people, called members, who operate an enterprise collectively. The aim of a cooperative may not necessarily be to make a profit. Cooperatives are run on the basis of one member, one vote irrespective of how much each member has invested. The managers are also democratically elected on this basis. In the UK limited liability for the members is achieved by registering with the Register of Friendly Societies.

The main advantages of cooperative operation are that they give support to the members of the cooperative, there is little asset stripping (where assets are sold off to provide profits), and because of the membership you should have good industrial relations.

Five principles are involved and form the main features of running a cooperative:

1. There is open and voluntary membership for all who work in the organization.
2. There is one member, one vote, the management being elected by the membership.
3. Interest paid on capital is limited.
4. The business is conducted for the mutual benefit of the members and all profits are shared. In particular, people are paid on the basis of what they need rather than on the basis of 'market rate for the job'.
5. The business must be socially aware and act responsibly towards other businesses, customers, suppliers and the local community; they are not necessarily profit-making.

2.3.4 Companies

A company has a separate legal identity from its owners and will continue to operate even if the owners and managers change. It is owned by shareholders who each have a liability which is limited to the nominal value of their shares. This nominal value is the amount originally promised in return for the allotment of shares. The nominal value is usually different from the price of shares being traded on the stock market. The trading price reflects the value placed on the company by investors. In the UK companies, both public and private, are registered with the Registrar of Companies. We will now look at how UK companies are established, and consider the main features of operating a company.

The creation of a company
The process of creating a company is called incorporation. The process requires that two documents are drawn up: these are the Memorandum of Association and the Articles of Association. The Memorandum relates to the external affairs of the company giving information such as the company name, its registered office, the purpose (objects) of the company, and its share capital. The Articles relate to the internal operation of the company and include such things as the rights of the shareholders and the powers of its directors.

These two documents and others defining the amount of capital and number of directors are submitted, with a fee, to the Registrar of Companies, who is then able to issue a certificate of incorporation. Following incorporation, the company details are made available for public inspection and a private company can start to trade immediately. A public company may not trade until the Registrar is satisfied that the necessary share capital has been allotted.

The regulation of UK companies is achieved through the Companies Act 1981. This Act states that the company must regularly prepare and publish information regarding its activities. The company is also allowed to carry out only the activities defined in the Objects clause of its Memorandum of Association. However, the Objects clause can be amended in accordance with the Act.

The annual report

Ownership of limited liability companies belongs to the shareholders and they delegate the operation of the company to the directors. It is the directors who are responsible for producing, in the correct form, annual reports and accounts as required by law. The annual report must contain both the balance sheet and the profit and loss account, and these are discussed in detail in Chapter 4.

Large companies also have to include in the annual report other items of financial and general information such as a Chairman's Review and the auditors' report. A statement of source of funds and the way in which they are used is also included, giving a full record of the financial transactions undertaken during the accounting period.

Closing a company

If a company experiences financial difficulties and cannot pay its debts it may be closed by court order. The way in which a company is wound up is called liquidation, and this can be done voluntarily or compulsorily. Voluntary liquidation occurs when the company agrees to sell its assets and pay off any debts. Compulsory liquidation is when the company is given no choice and the court appoints an official receiver, who will be put into the company to sell off the assets and pay off debts. In some cases the receiver may be able to sell the company as a going concern, which means that it continues to trade.

The main features of operating as a company are as follows:

1. Shareholders have limited liability, which is limited to the nominal value of their shares.
2. The company has separate legal identity from its owners and managers.
3. Extra capital can be raised in public companies by issuing more shares. Shares in public companies can be freely transferred without consulting other shareholders.
4. Once formed, a company has perpetual succession since the company management is separate from its ownership.
5. The accounts have to be disclosed publicly, and therefore it is possible to closely monitor what is happening in a competing company.
6. Company activities are limited by the Objects clause.
7. The company is more closely controlled and regulated by outsiders than sole traders or partnerships, because of the Companies Act 1981.

Public companies

Public companies tend to be larger than private companies and can be distinguished from private companies in that their names end in the letters 'plc' (public limited company). In order to become public the company must have a minimum defined amount of capital. It must also have at least one quarter of each share actually paid for. It can offer shares to the public and transfer them freely, subject to rulings by the Monopolies Commission. This share dealing is done through the Stock Exchange. The minimum membership of a public company is two and the maximum is defined by the number of shares. The company must have at least two directors and its accounts have to be disclosed to the public.

Private companies

Private companies can be distinguished because their names must contain the word 'Limited', abbreviated to 'Ltd'. The company is owned by shareholders but, unlike the public company, the shares cannot be offered to the public, unless this is done under agreements made with the Inland Revenue. In this case shares can be offered to company employees. Private companies do not have to disclose as much information in their accounts as public companies. They also require only one director but must have a company secretary.

2.3.5 Franchising

Franchising is an increasingly popular way of setting up a business. A person wanting to go into business will pay a sum of money to another organization for a franchise. The franchise is a licence to sell the franchisor's product or service and this form of business is very common in the service industry. In return for the franchise fee and other on-going payments the business receives the support of the franchisor, including a brand name or label under which they are licensed to trade and full support for marketing and training. In return for this support the franchisee is exclusively bound to the franchisor for the period of the franchise. In addition to paying the franchise fee the franchisee has to find premises and equipment in the house style. Sole traders, partners, and companies can operate as franchises.

The main features of operating a franchise are as follows:

1. There is limited outlay and risk due to the support of the franchisor. However, there may be problems if the support provided is not adequate.
2. The product has been tested, marketed, and there is a well-known image. In addition, the franchisor may provide large-scale advertising.
3. The franchisee may not be able to manage the product in the best way for their own business because it is not in the franchisor's best interest.

2.4 Strategies for survival

In Section 2.1 we saw how organizations define what they are actually trying to achieve and in the previous section we met the legal identities that an organization can take. In this section we will examine how an organization tries to guarantee its survival.

The simplest way to understand how companies may do this is to think of the most basic requirements of continued company survival and see how these may be achieved. To survive, a company must have customers who continue to purchase its products. This is the fundamental prerequisite of survival. Any planned activity that ensures that this continues to be true is therefore a survival strategy.

For example, a company may choose to guarantee its survival by having an effective product development mechanism that always ensures that its products can be developed more rapidly than its competition. Alternatively, it might have a manufacturing-based strategy in which it ensures that it is constantly in possession of the most cost-effective manufacturing machines and so, even though its product areas may change greatly, it will continue to survive. In yet another approach a company may choose to have a marketing-based survival strategy in which an effective marketing mechanism ensures that the company is always in an area of business with guaranteed customer spending power.

Underpinning all these, however, is the need to ensure that financial control is maintained. No matter what the area of business or the survival strategy, the company must have some way of determining how much money is raised from sales and then directing it into activities that support its survival. This is not, however, a strategy for survival, it is a necessary but not sufficient condition for it. A company must clearly have such a system, but merely having it does not guarantee survival. Finance is dealt with in detail in Chapter 4. We shall now concentrate on the strategies mentioned above.

2.4.1 Strategic marketing

Marketing ensures that a match is found between the customers' needs and the attributes of the product, at a profit. It examines how future markets and products may be brought together to the benefit of the company. When the senior management of a company take a long-term view and examine the future over a period that will span the introduction of many new products they are undertaking a process usually called strategic marketing. This is a very important process since it shapes the company's activities over many years in the future. For example, a company making hi-fi will have many product developments in progress based on current technology and customers but it needs to look further than this. In the long term totally different technologies may emerge and change their products beyond all recognition. Customer expectations may change and products become unfashionable. The manufacturers of carbon paper used to have a very stable market niche, now they have no market at all. Product developments of the kind that reduced the price of the carbon paper or made it less messy did nothing to save such companies. To survive, these companies would have had to develop photocopiers or move into a completely different business area. How can companies be good at seeing such pitfalls coming? The answer is by being eternally vigilant to the world about them, reacting to it and perhaps even shaping it. This is the stuff of strategic marketing, a tool used to look into the commercial future.

Marketing is, by its nature, forward-looking and so is concerned with determining the needs of customers for future products or modifications. For this reason, marketing should have a formal input into the product development process. It must also be able to comment on the ideas of the designers in relation to their commercial viability. It is able to do this because a significant factor of market research is determining what the basis is for customers' buying decisions. Thus it can build a picture of features that are essential to sell the product, and those that are desirable and might improve sales. These features may not be restricted to particular product features but may relate to, for example, availability of spare parts, geographical location of distributors, or speed of delivery.

Strategic marketing should also make significant input to business planning and corporate strategy by providing forecasts of demand in terms of products and quality which are used to produce budgets and prepare cash-flow forecasts, as well as allowing decisions on company expansion, or contraction, to be made. It should provide the data on which the company can make decisions about how it is to develop. Figure 2.2 shows the long-term options open to companies wishing to improve performance.

Market penetration involves maintaining sales by attracting more customers to your product rather than the competition, or by getting the customers of your competitors to buy your product. You might achieve increased market penetration through lower pricing or by increasing awareness of your product through advertising.

Product development involves targeting new or modified products to the same groups of customers to which you previously sold. This is where marketing

	Present products	New products
Present markets	Market penetration	Product development
New markets	Market development	Diversification

Figure 2.2 Strategic marketing options

will make a direct input to the design and development process. In some businesses, particularly those greatly affected by changing technology such as personal computer manufacture, a certain rate of product development is normal. Product development is also used to refresh the market by providing something that has a slightly better performance, better appearance, or is easier to use, thus prompting customers to make a further purchase and upgrade their existing product.

Through market development a company aims to improve sales of its current product range by selling to groups of people who have not previously purchased, or who have but have had a different use for the product. An example of this is the way in which baby shampoo is sold to adults who wash their hair frequently and thus need a mild shampoo. A further example is the way in which yoghurt, previously a healthy dessert, is promoted as a low-fat healthy alternative to ice-cream and is marketed as a snack food.

Product diversification is the riskiest strategy for any company as it means putting a new product into an unknown market. Again, the marketing input to design is crucial and has to be supported by extensive market research.

We shall now look briefly at some other activities that are important to the development and success of a product which are often undertaken by strategic marketing personnel.

Promotion

Product promotion can take many forms and usually involves marketing personnel. Choosing specific market segments and promoting products within them is a strategic issue. The style of promotion used depends very much on the product in question, and often organizations have products that require different promotion strategies. Advertising a custom storage shelving system for warehouses would require different advertising and promotion from that for plastic bins that might be used for storing components in a small laboratory, yet both these products might be made by one organization. Promotion means more than just advertisements on television. Advertisements can be placed in newspapers, journals and magazines. It can mean conferences, shows, trade fairs, hosting visits, assisting others, sponsorship, discount schemes, games, competitions, customer visits, exhibitions or publishing papers. Even lodging patents or fighting court cases can be seen as promotion. In strategic marketing anything that affects the customer's perception of the organization or the product can be manipulated to promote the product.

Image and identity

An important part of the strategic marketing process is the act of establishing and maintaining a corporate image and identity. An organization selling cameras, for example, may wish its involvement with film sales to be made public in order to promote its image as a comprehensive supplier of photographic equipment. In reality, being a retailer of film offers little or no advantage to the camera-manufacturing process and the association is made simply to foster an image in the customer's eyes. Image and identity can be created through many means, and almost any product attribute that is visible to the customer can be used in this way. Certain car manufacturers, for example, use a particular combination of lines in the bonnet or the radiator grill to make an image and give identity. The purpose is to reinforce in the customer's eyes that this is the product of a particular manufacturer. Choosing how and in what way the image of the organization is promoted is a strategic issue since it affects customers' long-term view of the organization and therefore their purchasing decisions.

In addition to these activities, marketing will also have to develop strategies, for both current and new products, in terms of pricing, promotion and distribution. The pricing policies will require input from the engineers who will have to indicate costs, or give estimates of future costs, for manufacture of the product. Strategic marketing will also have to consider the additional costs of distribution and promotion, as well as considering what is happening in the marketplace. Promotion raises awareness of the product with potential customers and therefore includes the way it is packaged, the way it is advertised, and any promotional literature that is used to describe the product. Distribution describes the way in which product is transferred to your customer and so will involve an examination of the sorts of outlets that are required, their geographical location, and the way in which the product is sold.

2.4.2 Simultaneous Engineering

The advantage given to an organization that can get new products into the marketplace at the chosen time is enormous. It is not so much a question of getting into the marketplace first, many organizations trade very successfully by choosing to be second or even deliberately late. Such organizations benefit from all the market development money poured in by the leaders and have the advantage of being able to choose products to develop that have already been tested in the marketplace. Products developed later are more likely to be accurate in meeting the customers' needs and so will sell well. Naturally, however, there is less market share available by this time and entering an established market brings its own difficulties. The issue is not being first or even last, the important point is that when an organization makes a decision to research a new product it is able to move from initial market investigation to marketplace sales in the minimum possible time. Simultaneous Engineering (SE) is a means by which this may be achieved and so can be used as a strategy for continued survival.

SE simply means that, as much as is possible, the tasks that have to be completed in order to launch a new product are performed in parallel. At its most basic, SE is simply good project management and the techniques described in Chapter 8 are of importance. At its most sophisticated SE is a marketing weapon that will persuade customers to buy from the organization because they believe that the organization's product development process is so good and so quick that they will be receiving a better and more up-to-date product because of it. This customer perception can be a very powerful influence in marketplaces that change quickly. For example, engineers buying computer equipment will often prefer to buy from organizations whose product development is fast and proven, since they are more likely to benefit from a coherent product development strategy. Software subsequently released is likely to work on older systems and the benefits of tomorrow's research and development can be used, when it is available, on today's equipment.

The two tasks that are overlapped to the greatest degree are the preparation of the marketing specification and the technical aspects of the product development process itself. In the idealized product development process the marketing specification is completed before the technical work starts. There are clearly economies that can be made, however. For example, certain technical requirements may be identified very early in the marketing process. Once a technical aspect has been identified work may start on it, and even though it is not known exactly what form it will take in the finished product, work can progress. The advantages of such an approach are clearly that it makes the whole process shorter. However, the major disadvantage is that it is likely to be more expensive. This is because unnecessary technical work may take place before the technical requirements are accurately known. Another disadvantage is that the work is more difficult to manage and consequently it is more expensive. Difficulties in management may come from trying to plan future blocks of work on outcomes that are not yet known and from the demotivating effects of personnel being involved in work that is not ultimately used.

2.4.3 Manufacturing strategies

Some organizations employ manufacturing strategies to ensure that they survive. These strategies all aim to develop some aspect of the manufacturing capability of the company in such a way that it becomes more than the function responsible for production of the finished goods and something unique and commercially potent in its own right. For a company that has a good market share this may be achieved by investing in equipment that can supply the entire market and then aggressively selling at the narrowest profit margin. The low price offered by such an organization means that others are soon forced out of the market, and before long the organization has the overwhelming majority of the market to itself. From this position it can then put up prices to a level that makes it just not worth while for the competition to invest in the market. The organization then owns a manufacturing function that guarantees survival as long as there is a market for its products. This is a much stronger position than in a strongly competitive market. Examples of this manufacturing strategy are found in the micro-chip manufacturing sector where only a few companies produce all the world's chips.

Alternatively, in the brewing industry just three companies supply virtually all the malted barley for the whole of the UK.

Another manufacturing strategy that may be employed is continually to invest in manufacturing equipment that offers the best production rates. For example, a machining company may always buy the latest machines, disposing of the old ones long before they have reached the end of their working life. Such a business offers its customers the very best in machine tool technology and, although its customers may change considerably over the years, the organization will be guaranteed business since there will always be customers who need to make use of these machines and for whom purchase is not desirable.

Yet another way in which manufacturing may be used to develop a strategy is the skill level of the workforce. For example, some companies offer contract assembly. An electronics company that does this may have a workforce that is particularly skilled in the manufacture and test of PCBs for the music industry. Such a company is able to offer manufacture more cheaply and with a greater level of quality than some companies can achieve in-house. A company that trades by designing new circuits for the music industry is not necessarily going to be good at manufacture, whereas the company specializing in manufacture has no diversion for the manufacturing function such as design or product development and so can develop a manufacturing system that is extremely effective. For example, some test equipment is very expensive and manufacturers may not be able to justify the expense, whereas a contract manufacturer performing the assembly of PCBs for many customers could. For a company that offers this kind of manufacturing excellence it is again true that the development of manufacturing has become the strategy which guarantees continued survival. Indeed, for a company such as this, the product is not the PCBs themselves, rather the service of manufacturing them.

2.5 Case study – JP Engineering

This case study shows the development of a typical small engineering company and the sort of strategic decisions that have to be made when running a company so that it may continue to prosper.

JP Engineering is a subcontract machining company based in Oxfordshire. Its business is to produce parts to other people's drawings using a variety of machine tools, including CNC. The business was formed six years ago by John and Phil. At that time they were working together for a company but decided that they wanted to work for themselves. They are both fully skilled machine operators, having served apprenticeships.

When they decided to go into business John and Phil wanted to stay with an area that they knew well, i.e. machining. However, their goal was to become the best subcontract machining company in the region in terms of quality and reliability. They did not necessarily want to be the biggest. JP Engineering was established as a partnership, with a partnership agreement signed by both partners.

One of the first jobs for the partners was to raise capital to start the business and in order to do this they had to produce a business plan describing how they thought the company would trade and grow. Most importantly, the plan gave a projection of expected incomes and outgoings. That first plan made a number of bold statements. For the first six months of trading the business would employ only John and Phil, and would then increase by one employee. The partners projected that after 12 months the business would employ a total of four or five people. They planned the purchase of their first CNC machine after 24 months and made an estimate of turnover for the first year. With their business plan the partners were able to secure funding, although some of this was secured on their homes.

During those early days John and Phil worked as machinists building up a client base of local companies. The job of running the company was done at weekends or in the evening. However, their goals were realized and the business went well. In the first six months they employed a further two machinists, an inspector, and a part-time secretary, and also bought a CNC machine. The first year's turnover was two and a half times greater than their target.

Over the six years the business has continued to grow and has survived a recession, although this forced the partners to make four people redundant. The business now employs 15 people in total, has moved to premises four times greater than the original ones, and turnover has increased almost fivefold over the first-year level.

The changes in the business operation have been subtle. John and Phil are still working many hours but now 75% of their time is spent running the business. The management tasks, such as scheduling the machine shop, are divided evenly between the partners. There has also been an increase in non-production staff as the growth has led to the arrival of someone responsible for purchasing and sales. The partners' goal to be best has led them down the road of BS 5750 Quality systems (now known as BS EN ISO 9000, see Chapter 7).

The business is still run as a partnership although this is regularly reviewed. As soon as it becomes feasible and beneficial to change status to a limited company then this will be done. The small, local client base is maintained and the partners aim to balance the work from customers to avoid the risks of having too much reliance on a single source. The relationship with clients is informal and the partners feel this is important to their business operation. It is also a necessity when trying to match the needs of the client with the business objectives, trading delivery dates so that quality of the product is not jeopardized.

At this time there are no plans for further expansion of the premises but the partners do plan to update machinery which will lead to improvements in productivity and quality when linked to plans to introduce shift-working and to increase the number of machinists by one or two.

2.6 Summary

In Section 2.2 we saw the benefits that can be achieved by companies who use corporate objectives to manage their business. Corporate objectives guide the whole organization and from a fundamental platform from which all other decisions are deduced.

In Section 2.3 we saw that in the UK there are five legal ways to establish a business. The method chosen will have implications for the way the organization can transact business, the means by which it can be financed, and the amount of information that has to be publicly divulged. The five forms of business are:

Sole trader
Partnership
Cooperative
Public company
Private company.

In Section 2.4 we saw how organizations may try to ensure their continued survival. These strategies may be based on the product development process, the manufacturing function or strategic marketing.

Finally, in Section 2.5 we looked at an engineering company, at the way in which it was established and at the changes that have taken place as it has grown.

2.7 Revision questions

1. Write down a mission statement for the following:

   (a) A private company that you know
   (b) The American space agency NASA
   (c) The department responsible for defence
   (d) A public transport system
   (e) The next five years of your life

2. Write your own summary of why corporate objectives are important. Include the benefits they bring and any disadvantages or extra work you can think of.

3. Investigate your organization to discover its mission statement. How does this mission statement relate to you? Write about the effects of the mission statement listing areas where the behaviour of the institution seems to agree with the mission statement and where it disagrees. (If your institution does not have a mission statement compose one yourself.)

4. What might cause the mission statement of an organization to be revised? How often should this be done? What are the dangers of revising it and of not revising it?

5. What are the advantages of operating as a partnership rather than as a sole trader?

6. As an engineer what possibilities exist for you to operate a franchise? What are the disadvantages of franchising rather than developing your own product?

7. Compare the operation of a partnership and a limited company.

8. Describe a strategy for survival that might be employed by a computer manufacturer, a domestic appliance manufacturer and a design consultancy.

9. Why is an effective financial administration described as a necessary but not sufficient criterion for survival?

10. Explain the terms mission statement, strategic marketing, survival strategy and MBO.

Further reading

Argenti, J. (1986), *Systematic Corporate Planning*, Van Nostrand Reinhold (UK). A comprehensive book about corporate planning, explaining why it is needed and how it may be achieved. The book includes chapters on corporate objectives, expected performance and GAP analysis. The second half of the book deals with planning and executing corporate plans.

Dale, E. and Michelon, L. C. (1986), *Modern Management Method*, Penguin Books. A readable and thorough guide to many issues in management, including managing and communicating by objectives, managerial decision making, decision making and payoff tables, critical path analysis and PERT. Based on the wide experience of two famous management consultant authors.

Handy, C. B. (1985), *Understanding Organizations*, 3rd edition, Penguin Business.

Kadar, A., Hoyle, K. and Whitehead, G. (1987), *Business Law Made Simple*, Heinemann.

Ohmae, K. (1983), *The Mind of the Strategist*, Penguin Group. In his native Japan, Kenichi Ohmae has been described as 'Mr Strategy'. His abilities as an adviser to top management have earned him considerable respect as an authority on his subject. In this book Ohmae places strategic thinking at the centre of success. The book is in three parts. In part one the art of strategic thinking is expounded, in part two the techniques for building successful strategies are examined, and in part three the realities of modern strategic planning are explained.

3

Functions of organizations

Overview

In order to design a device an engineer must have an understanding of the use to which it will be put. The requirement for the device gives the context. Before we can manage people or resources we need to know in what context we are working and understand why management is required. Knowing how businesses come together and operate gives that context. Unless we have an understanding of the whole it makes it very hard to put the detail in perspective.

In Chapter 2 we looked at the way in which companies can be formed and set objectives. In this chapter we examine the functions that must exist in business, considering how they can be organized and managed. We also present some case studies that show how these functions interact in real companies and the role that engineers can have.

3.1 Introduction

In any manufacturing organization there will be various functions that have to be carried out in order for the business to operate. To illustrate this we will consider the example of the manufacture of electric kettles.

We will start with the assumption that research has been done to ensure that the product is viable and that the company is established and has adequate money to finance the operation. Bearing this in mind, we will consider how the business will operate.

In order to make a profitable business out of manufacturing kettles we will have to consider a number of important areas; the minimum that we would have to consider is shown in Figure 3.1. We need to procure (buy) some raw materials, we need to transform those raw materials into a saleable product, and then we need to sell the product at a high enough price to provide a profit. So, we have three main functions of business: buying, making, and selling.

As part of the business we have identified that parts will have to be bought in and kettles will have to be manufactured. Therefore there needs to be somewhere to keep all these items that can be used both as a means of safe storage and as an

Figure 3.1   The kettle business

aid to control the receipt and use of parts. This is usually achieved by having a stores system, which is often part of the purchasing function's remit.

By saying that we can manufacture we have assumed that we know what the kettle looks like, that the parts that go into it have been defined and that manufacturing processes have been specified. All these activities are carried out by some form of product development function which will include design and perhaps research. If expansion, growth, and development of the company are to be considered then how the company and its products will develop over a period of time in order to achieve a desired market share must be determined. Companies can achieve this by using research which is then allied to design activity. In the kettle business the research activity might involve trying to develop a new type of heat exchanger mechanism. The design activity would then take the new mechanism and consider the design for manufacturability, producing drawings and parts lists. It would also ensure compatibility with other parts in the kettle.

In order to sustain itself the business must be profitable, and for this to happen we need to ensure that the cost of buying the raw materials and making a kettle is less than the amount that we can sell it for. This introduces two further functions: finance and marketing. The finance function will be involved in calculating the costs to make the kettles, including buying the raw materials. Among many other activities, it will also be involved in making sure that there is enough cash to keep the business operating while waiting for customers to pay their bills. Marketing means ensuring that a match is found between customers' needs and the product's attributes, at a profit, and will be involved in finding out what the customers want and what the competition is offering.

This leads us to conclude that six areas have to be addressed in order to set up and operate a manufacturing company. However, there is also a requirement to ensure that the products made are always of a sufficient standard to meet the customers' expectations and that they conform to appropriate product and safety standards. Companies can address this area by having some form of quality function which is used as part of the management of the other functions to ensure that standards and customer requirements are met. In the kettle business 'quality' would involve all aspects of the business, including ensuring that the design process addresses the appropriate safety standards and that there are effective measures to ensure that people in the company know what the standards are and that they have the ability to meet them.

Finally people will be needed to work in the company in order to carry out all the tasks required. Their deployment and welfare, initial recruitment and selection as well as payment and conditions of employment will have to be considered. This is the role of the personnel function.

So we can conclude that in order for a company to manufacture electric kettles, or indeed any product, it needs to address several different areas. These areas (functions) are as follows:

1. The buying and safe storage of materials (purchasing and stores)
2. The manufacture of products (operations)
3. The selling of products (sales)
4. The pricing of the product and its competition (marketing)
5. The financing of the business and the cost of manufacture (finance)
6. The design and development of the product for manufacture (product development)
7. Consistent product standards (quality)
8. The people employed by the business (personnel).

In this chapter we shall consider each of the functions listed above. We will look at the way in which they might be organized in an engineering company and consider some of the activities that have to be carried out. Towards the end of the chapter we will examine some examples of the way in which engineers are employed in engineering companies and how they interact with each of the functions described.

The learning objectives for this chapter are as follows:

1. To understand how an engineering company operates
2. To consider the way in which engineers are required to interact with all departments in a company
3. To look at some examples of how engineers work in engineering companies.

3.2 Purchasing

A company will have to purchase goods and services to feed all aspects of its operation, from special parts required for research to paper products for the washroom. In all purchasing activities it is necessary to clarify what is required, obtain supplies and ensure that any business objectives are met. Meeting business objectives may relate to quality requirements or budget controls. It will certainly include making sure that the customer for the good or service gets what is required at the time that it is required.

The purchasing function is complicated by the fact that within a company there will be a diversity of goods and services to be purchased and different types of customers requiring different levels of service from the purchasing function. In a very large organization it may be appropriate to have a large purchasing department which handles all requirements and has a number of buyers each with responsibility for a particular type of good or service. In a small company it may be more appropriate to allow different departments to do their own purchasing. The organization of purchasing will also be affected by the quality system, if the company has one; the controls imposed by a parent company, if a subsidiary; and the business relationship with suppliers (for example, if the company has a direct computer link to its suppliers).

In this section we will look at the job of the purchasing function, the way in which purchasing can be organized, and some of the activities carried out by a purchasing department.

3.2.1 The role of the purchasing function

In an engineering company the majority of purchases will be used in some way to form the final product and this can lead to purchased goods accounting for as much as half of the total expenditure. Purchasing might involve buying parts for production, negotiating contracts for temporary staff, assessing suppliers, or negotiating discounts for bulk purchases. It always involves having to ensure the supply of goods of the right quality, in the right quantity, at the right time and at the best price for the level of service provided.

The purchasing function, however, is much more than just buying, it exists to meet the needs of its customers inside the organization whether in design, manufacturing, personnel, or finance. The needs of such a disparate group will be very different and the purchasing function must address them in the most appropriate and effective way.

3.2.2 Organization of the purchasing function

Often in large companies a centralized purchasing department is established which has control over all aspects of procurement, stockholding and supply. The department may also be known as materials management. In smaller companies these activities are more likely to be distributed, with different departments having responsibility for buying the goods and services that they require. However, there should still be an overall central purchasing policy which, as a minimum, allows cash flow control and will define the procedures by which suppliers will be selected and then paid.

When a company employs many buyers in a centralized department they may be given particular areas to cover. For example, there may be buyers with specific responsibilities for procurement of food, furniture, stationery, production parts, and so on. One advantage of this type of operation is that each buyer can develop very specialized knowledge of the market and provide even greater economies. Other advantages of centralization include being able to make bulk purchases with consequent price saving, minimizing the cost of placing orders, and avoiding duplication of effort. Against this, if a decentralized purchasing department is used instead it can mean that a buyer is able to react more quickly to a problem and that there is likely to be a better understanding of the local needs. Figure 3.2 shows how the purchasing function might be organized in a medium-sized engineering company.

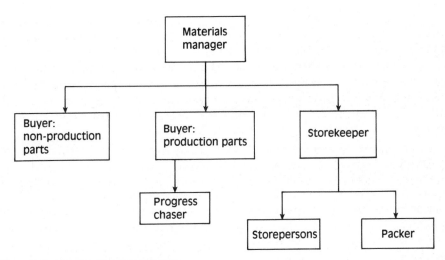

Figure 3.2 Purchasing organization

Some companies allow certain departments or groups the freedom to place purchase orders with a minimum of centralized control. This is typically found in research and product development activities where it is important that the designers and engineers can communicate with suppliers easily and when they require goods on an *ad-hoc* and quick turn-around basis. This can work very effectively, particularly if the alternative of using a central buyer would delay the orders or would be very bureaucratic. However, these problems do not have to occur if the purchasing function is properly organized and there are appropriate ways of dealing with different sorts of orders. There is a very large potential problem with this *ad-hoc* system and that relates to the level of control exerted by central management. This has therefore to be accompanied by strict budget controls and a normalization of purchasing procedures as a device moves from research prototyping to manufacture. There may also be a need to control the suppliers that can be used because of quality requirements or their ability to supply to production demands.

3.2.3 Activities in purchasing

The most important activities that will be undertaken within the purchasing function will be establishing purchasing policy, sourcing, buying, and stock control. In addition, the purchasing function must work with other departments to ensure that their needs are understood and met, and that quality is maintained.

Establishing purchasing policy

A purchasing policy will define the rules by which the purchasing function will operate. It may define the criteria on which suppliers are selected or the geographical location of suppliers. It may also define cash limits or specify the limits on the amount of business that can be put to any one supplier.

Purchasing policies will be developed over a period of time but one of the main management issues involved in purchasing will be the implementation of the policy and monitoring to check effectiveness. This will be most important, and most difficult to achieve, in companies that operate a decentralized purchasing system.

Sourcing

Sourcing is about finding suppliers that can meet the company requirements for goods and services. This will invariably require close consultation with the department requiring the goods or service so that the buyer fully understands the requirements. It will then involve contacting potential suppliers and collating information that can be used to make a decision about which supplier best meets the needs. Sourcing can take up a lot of time and effort and it can be more cost effective to let a specialist in a particular area carry it out. For example, in the software field it might be more economically viable to have a software consultant to undertake sourcing rather than someone who has only a general knowledge of the field and the terms that are likely to apply.

Many companies also carry out second sourcing in which an alternative supplier for a particular good or service is found. This ensures that there is some security of supply for the company in the event of a supplier not being able to meet its requirements and ensures that a competitive price is secured at all times.

The criteria for selection of suppliers should be defined in terms of quality, location, delivery reliability, financial stability of suppliers, and any other factors that are appropriate to the company objectives. In companies operating quality systems there will also be a need to ensure that any supplier can provide assurance that the required quality standards can be met. This may involve the purchasing department in carrying out supplier-appraisal visits and will always involve monitoring of the performance of suppliers for goods supplied, delivery, and quality of after-sales service.

Buying

There are several ways of buying and the methods chosen will depend on the company and on the particular types of parts, items, or services being bought. For companies operating a fairly stable production system buying production parts will be very much a case of buying from a list indicating the parts required and where they can be obtained. More specialist buying will require more detailed knowledge about the actual goods required and an understanding of the commercial implications for any contracts agreed.

Many companies hold stock of frequently used items in Stores. For these items new stock is purchased when a specified reorder level is reached. The quantity of new parts will also be specified. This will be the reorder quantity and will have been calculated taking into account purchase price, discounts, and stockholding costs. Companies buying for stock may use a call-off schedule with a supplier. In this type of buying one order is placed and it covers all the parts that will be required for a period although they may not all be needed at the same time. The company then 'calls off' the parts at the times when it does want them. The advantage to the company is that administration is reduced significantly and yet the parts are delivered as needed; there is no need to hold stock of one large order.

A vital aspect of purchasing for manufacturing is the ability to predict lead time. This is the time required to get the goods into Stores and available for manufacture, following the placing of an order. The lead time offered by suppliers will affect the lead time that a company is able to offer its customers as well as the amount of stock that a company must hold. The effect on stockholding can be quite significant because of the company's need to keep enough parts to cover fluctuations in demand and any problems with supply. Because of the importance of lead time many companies will specify lead times for goods, and the ability to meet lead time should be monitored as part of any supplier assessment scheme. Lead times may also be specified by the manufacturing system that is used by the organization, as in a Just-in-Time system, for example (see Chapter 6).

It is important, particularly for production parts, that the buyers have good communication links with manufacturing so that they can act quickly in cases of urgency or advise on any difficulties with supply. The criteria used for buying should be defined in the purchasing policy. Part of the buying role includes progress chasing. This is following up an order to ensure that the goods are going to be delivered on time and in the right quantity. Some large companies will employ people whose only job is to keep in touch with suppliers' progress on orders.

Stock control

Stock control concerns the procurement, safe storage, and release of materials. It necessarily involves establishing and maintaining a database of materials used in the organization, and it will often include the operation of a Store or warehouse which is used for the receiving, holding and release of goods. Chapter 6 contains sections on purchasing and stock control which describe the purchasing activities in more detail, and shows the data used to make purchasing decisions.

3.3 Operations

In using the term 'operations' we mean the central business activity of the company. In a manufacturing company therefore operations are the activities that transform raw materials into finished product. The majority of engineering companies are involved with manufacturing to a greater or lesser extent. Even in those companies that do not manufacture, designers have to be aware of the constraints imposed on product development by manufacturing processes and organization.

In this section we shall look in overview at some of the activities that are carried out within the operations function and examine how manufacturing is organized in order to meet company objectives. Chapter 6 describes the operational aspects of this function in detail.

3.3.1 Management activities in the operations area

There are many management tasks that have to be undertaken within the operations area. A few of these are described below.

Management of people

In a manufacturing company a large proportion of the workforce will be directly employed within the operations function. There will be people from many different backgrounds within the same department. Whatever the number of people involved and their skill level and position in the organization, they all have to be managed in a way that ensures that the objectives of the operations function are met. In this section we will look at some of the issues that relate specifically to the operations function. However, more generally, personnel management is dealt with in detail in Chapter 9.

The factors that will affect the way in which an Operations Manager will manage the operations workforce include the number of people in the operations function, how the workforce is organized, the level of automation in the operations area, and the personnel policy of the company. The activities involved in managing the personnel of the operations function include general personnel management, manpower planning, and training.

Manpower planning in the operations function is strategically important to the company as it will have a direct effect on output. It involves considering the skill levels of people available and those required to achieve the company's output targets, and then aims to match the available capacity to requirements. For example, if a worker can produce 2 units of output per hour then available capacity can be calculated by considering for how many hours that worker is available. When this is done for the whole workforce it is possible to calculate any differences between what is available and what is needed. This allows decisions to be made about recruitment, redundancy and working hours including shift-work and overtime, and takes into account planning for sickness, holidays and training.

Training is a particularly important activity in a manufacturing environment because of the variety of skills that are needed and the fact that people need to be retrained in new technologies. Decisions regarding training requirements may be made following employee-appraisal interviews or manpower planning. Training can then take place on- or off-site as appropriate. The effects of training have to be carefully considered by the Operations Manager because of the effect it will have on output, particularly if using experienced people to provide on-the-job training to less experienced people.

The other aspect of training that has to be considered in this environment is that of external controls that may be applicable. For example, if a company is providing a graduate training scheme that has been developed in conjunction with one of the engineering institutions or is operating a nationally recognized apprentice-training scheme there will be specific training and review requirements. There may also be requirements for attendance at external training courses or for external progress monitoring.

Management of machines and processes

Manufacturing requires the use of many different machines and processes and there are many aspects of machine management that have to be considered. The major factors influencing the management of machines and processes are the type of production being used, the level of automation, and the complexity of the processes. The actual processes used will depend upon the products being made and the way in which they have been designed, the numbers of product being produced, and the skills available in-house. For example, if a product is to be made in large volumes then it should be designed to be produced using processes that allow high-volume manufacture. The management activities relating to the processes will include making decisions about the types of production used, which processes are used, and monitoring and control of processes.

Monitoring and controlling processes are particularly important. Standards of workmanship and product specifications must be set and then the process controlled to ensure that design requirements can be met. Many monitoring techniques have been developed, statistical process control being an enduring favourite. All such techniques aim to inform of impending difficulties before the process has drifted out of specification.

One of the main activities in the management of machines will be selection of equipment to be used. This is not a continuous task as such capital investment decisions should not usually be undertaken on a regular basis. However, it is important to make the right decisions. Factors that will have to be considered in selection will relate to the processes used in the company, the desired capacity of the company, the return that the company would expect from any machinery purchased, as well as any plans that the company has for expansion or development and any technological changes that affect the company's processes.

Production planning and control

Production planning involves the planning of people, machines and materials in order to ensure that the output targets set for the operations area can be met. The factors that affect this are the lead times for manufacturing and purchasing, the production system used and its level of complexity, the size of the operations function, and the objectives set for the operations area.

The activities included in production planning will vary from long-term forecasting, as required for business planning, to daily or weekly shopfloor scheduling. However, before production can be planned there needs to be data on which the plans can be based. The data needed falls into two categories: (1) operational data, including stock data, and (2) product information. In each of these categories there will be data that is set up when the system is established and data that is developed during operation and which has to be continually monitored and fed back to the user. Thus, one of the main management activities within the operations area will be the collection, monitoring, and control of production data.

3.3.2 Organization of manufacturing

Manufacturing takes raw materials, parts and components and transforms them into finished product. This can involve a few or many processes and can vary from a simple machining process to a complex assembly and test procedure. The manufacturing activity will reflect the size of the organization and the diversity of products and processes.

When there are a number of products using similar processes manufacturing is commonly organized according to function. This means that all processes or machines of a similar type are grouped together. All products that then need these processes will be scheduled into that process area. Where the products are very diverse, or a great deal of customization is required, the organization is more likely to reflect the needs of individual products, with the processes required for a product grouped together.

The method of organization of the actual production facility will affect the ways in which it may be controlled and in which scheduling takes place. This is dealt with in detail in Chapter 6. However, in addition to the production facility there will be other aspects of the operations function which may or may not be grouped with manufacturing. These will include production planning, progress chasing, sometimes stores and stock control, and industrial engineering.

3.4 Marketing and sales

In many companies, particularly small ones, there is no physical distinction between the marketing function and any other, although it is likely that the various aspects of marketing are being addressed since businesses can only sustain themselves if they can produce what customers want. In addition, it is likely that they will only be able to attract funds from external sources, such as banks, if they can indicate how the company will develop over the next few years.

Some companies will organize themselves so that the functions of both sales and marketing are covered by one department. It tends to be only the large companies that have a separate marketing department. For those companies who operate without a separate marketing department there are several organizations who specialize in carrying out market research on a contract basis.

In many organizations, particularly those formed during the 1960s and 1970s, it is common to find a separate marketing department. In such organizations the task of understanding, examining and suggesting ways to exploit markets was given over to a separate and distinct function. The advantages of being able to specialize and of fitting in with the existing structure were seen to outweigh the disadvantages of separation and communication weaknesses. The marketing department interacted with the other functions as necessary and by having a director of marketing on the board the important issues raised by the function found their way into company strategy.

More recently, organizations have come to the conclusion that the communication disadvantage that a separate function places on the product development process is too great to ignore and the separate function has been replaced with a change in attitude within the product development process itself. In such an organization it is common to find the engineers undertaking the product development process directly involved in marketing issues. They meet customers, they analyse the needs of the different markets and, as part of the product development process itself, perform marketing activities.

It does not matter, in an absolute way, what method of meeting the organization's marketing needs is employed. The important point is that the organization is provided with acceptable market intelligence in order to meet its strategic goals. In this section we shall look at the way in which market information is accessed, using market research, and study how customers can be defined so that their needs can be met in the most effective way. Finally, we shall consider the role of the sales function.

3.4.1 Market research

Market research is the name given to any activity undertaken by the organization with the express purpose of gaining information about potential customers. There are four key elements to market research:

1. Analysis of current activities
2. Market intelligence
3. Market analysis
4. Product evaluation.

The analysis of current activities will include monitoring current sales, and trends in sales, and so will give a picture of who is buying what, and where. Market intelligence involves knowing what is happening in the market and would include analysing new developments, in particular knowing what the competition are planning. Market analysis is an examination of the consumer reaction to new pricing policies and new advertising campaigns and an assessment of the market potential of new products by contacting and gathering information from customers and potential customers. Evaluation of product after launch is done to determine the effectiveness of the strategies adopted for that product and makes apparent any changes that may be required for the future.

Market research provides the base for the decisions affecting marketing which then focus on the product and on how it is promoted, distributed and priced. During a market evaluation, marketing personnel will try to describe the market as accurately as they can. Graphical representations make understanding easier and it is common to find diagrammatic representations of the market used to convey the description. An example is Figure 3.3, which shows a pneumatic press market divided by power and production rate.

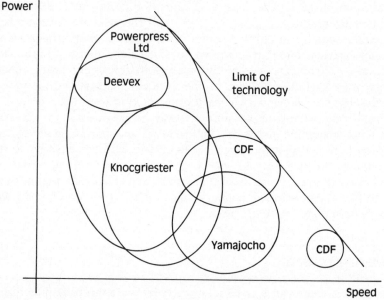

Figure 3.3   The structure of the pneumatic press market 1994

3.4.2 Customers and markets

Customers come in many different forms. They are the most important part of the business process since it is from them that all the wealth that supports the business must come. Every penny paid out in salaries must be obtained from a customer somewhere choosing the product developed by the organization instead of someone else's.

Some businesses have just one type of customer. A medical instrument company, for example, may have evolved to supply only to the National Health Service whereas a metal-forming company may supply simultaneously to customers in the food industry as well as the car industry. Each position has its own strengths and weaknesses. In this section we shall examine the different types of customer and their position in the market.

End user

The 'End User Customer' is the name given to those customers of the organization that purchase a product and then actually use it for its intended purpose. In the end of course every product is intended for an end user but often products are not sold directly to them. For example, consider an electronic chip manufacturer. The manufacturer will almost certainly sell the product to distributors and wholesale stockists who then sell the product on, either to smaller retail outlets or to manufacturers of products that include some quantities of the product. By one route or another, the chips may end up in a multitude of different manufacturers' products. The chain does not end here, for the products of a manufacturer may go through several more stages before finding their way to the end user. It is quite common for basic components to pass through up to ten different hands before reaching the end user. From this observation we can see that, in contrast to popular belief, the end user may represent a very small portion of an organization's markets, if indeed any sales are made to end users. Organizations that do sell directly to end users have the advantage that their normal trading brings them into contact with their customers. Those that do not have this contact have much further to go when gathering information about their customers.

Original Equipment Manufacture

The 'Original Equipment Manufacture' market, or OEM as it is usually called, is the name given to a particular type of customer. As the term implies, OEM customers purchase products from the organization and then incorporate them into products of their own which are then sold to on, perhaps to an end user or even another OEM. For example, a company might be manufacturing magnetrons for generating low-power microwaves. Several other organizations may then purchase these magnetrons and incorporate them into microwave ovens. These are then sold directly to consumers. In such a situation the company manufacturing the magnetrons is said to have several OEM customers or an OEM market.

Such customers are very desirable since they represent continuous business for the organization. OEMs tend to want deliveries scheduled months, or even years, in advance. In is common for OEMs to make contracts of supply with the organization and these may run for several years, stipulating maximum and minimum volumes of supply. Naturally, such agreements involve much negotiation since each side is intent on getting the best deal. The cooperation between a supplier and an OEM customer can be very great, each passing information to the other in the interests of the collaboration continuing to be successful. However, OEM customers have their dangers too, particularly if an organization has only one OEM customer. In such a situation the customer may ultimately wish to take over the supply of the product themselves, in order to reduce costs (perhaps by acquisition of the supplier or by alternative supply), or they may try to design the need for the product out of the equipment they supply. In either case, an organization with only one OEM customer is vulnerable to its commercial fortunes. The most desirable situation is therefore to supply several OEM customers, all of whom are in competition, selling into a broad market. In such a situation, the organization is not tied to the fortunes of just one OEM, and can be sure that the collection of competing OEMs will exploit their market as comprehensively as they can and in so doing provide the biggest market for the organization's product.

Retail customers

The 'Retail Market' or 'retail customers' simply means selling directly to individual customers. We are most familiar with it as the market in which we buy products such as hi-fi, DIY goods, and food, as consumers. Retail is used to describe the final point of sale of products that require extensive distribution and it is this that distinguishes the 'retail market' from the 'end user'. It is normal for a manufacturer to supply distributors who buy and sell in bulk at 'trade prices'. These suppliers then sell in bulk to a retailer who carries smaller quantities and sells individually and at a higher 'retail price'. In this way products requiring extensive distribution can find their way to the customers.

Displaced customers

Displaced customers are those that have been separated from the person responsible for the purchase. In such a case, although the product must meet the needs of the actual customer, it must also meet the needs of the purchaser, and this may not always be possible if the opinions regarding the important product attributes are not the same. For example, it is common for scientific equipment purchased for research to be used by post-graduate and research staff. However, the people who sign the purchase orders for such equipment are rarely the senior scientists in charge of the research. Most commonly they are the purchasing committee of the institution. In such a situation the users of the equipment may regard certain labour-saving features as worthwhile while those who make the purchase do not. A situation can occur where the organization cannot please both parties; the price and attributes that are acceptable to one group are not to the other. Competition and lack of understanding between the parties can make selling into such a situation difficult.

Another example of displaced customers is children. It is not the children that buy toys, it is their parents. However, the parents do not play with the toys.

Toy manufacturers are keenly aware of this and appeal to the needs of the parents as much as the children. Children are not, for example, interested in the safety of a toy as much as the enjoyment that it brings. The parents may take a very different view. For example, Christmas advertisements for toys show parents pleased with the happiness and convivial family atmosphere brought about by owning the toy. They are even screened after many of the children who will actually play with the toy have gone to bed! It is only by being clear about who the actual customer is that intelligent business decisions about product design and promotion can be made. Organizations that do not have this information therefore stand every chance of developing the wrong products and selling them ineffectively with obvious consequences.

Segmentation

Segmentation is the name given to the process of dividing large numbers of customers into smaller groups. Each group has its own distinct characteristics. For example, a manufacturer producing cars would not simply describe itself as selling into the car market. The marketplace for cars is enormous and may usefully be subdivided into smaller groups of customers, for example into hatchback customers, saloon car customers, sports or luxury car customers. A brief look at the models offered by a car manufacturer will soon show the extent to which manufacturers 'segment' these markets even further in order to exploit them. A typical family saloon car is available with a range of engine sizes from very basic and underpowered through to high-performance models that few customers are ever likely to be able to drive to the limits of their performance. After this the number of accessories offered is wide, as is the choice of trims, instrumentation, colour and warranty agreements. When these variants are considered together there are an alarming number of different combinations available. This can be borne out by investigating any large car park. Although there may be many examples of a given generic type of car, the number of examples of identical specifications is extremely low.

Why is this enormous range offered? The answer is that by segmenting the customers for cars into smaller groups and catering more accurately for their needs the organization can obtain a greater number of customers than would otherwise be possible. Of course, being able to supply so many different variants of a basic design makes for a very complex manufacturing environment. However, it is an extremely efficient way of producing what are, in effect, a large number of different products. In their efforts to avoid any potential customer not purchasing a car from their organization because they do not provide one that is ideal, the product development functions of the car manufacturers have led their manufacturing departments to provide enormous choice. Car production has changed a great deal under the influence of customers since the days of Henry Ford saying 'you can have any colour you like as long as it's black'. Few would doubt that the developments are desirable.

At the limit, this 'segmentation' of customer groups becomes one-off manufacturing. The ideal solution to the needs of a customer in the motor industry is unlikely to be met by one-off manufacturing since the initial expense and maintenance cost of an individually designed and manufactured motor car would be prohibitively expensive. It costs in the region of £1–5 billion to develop a new car model. By segmenting the market and taking advantage of the engineering capacity to mass produce and by introducing almost limitless number of combinations, customers are getting the advantage of a car that is almost individually tailored to their needs and yet are benefiting from the economics and serviceability of mass-produced goods.

Segmentation then forms groups out of large and unmanageable masses of customers. These groups may be accurately characterized and products then developed which more accurately meet their needs.

Market refreshment

After a customer base has been segmented and an organization has sold extensively into the different segments a situation may arise when demand starts to fall because all the customers already have the best product available to meet their needs. It is clearly of interest to the organization to identify ways of stimulating the customer base to buy products again. In order for these customers to buy again something new must be provided which satisfies a need not previously met. Market refreshment is the process by which this may be achieved.

The basis on which the process depends is that customers who buy a product do so, not because of an absolute need for a given level of performance but because of a relative need for performance. This concept can easily be illustrated. For example, someone who buys a camera will regard themselves as being of some particular level of competence in the use of cameras. If they regard themselves as a person of average photographic abilities they will conclude that a camera of average performance is appropriate for their needs. This will lead them to prefer to buy such a camera. This purchaser will do so with complete disregard for the absolute performance of the camera. One never hears such a person say that the average camera of forty years ago had fewer features and so that is the camera they require, yet one does hear them expound the benefits of their purchase as being suited to their needs. This can only be because such consumers arrive at their needs by considering their position in the marketplace in relative terms. This observation is borne out by observing the features provided with top-of-the-range cameras. Such cameras have far more features and abilities than are actually required by many of their purchasers, yet they sell, simply because of the customers' perception of themselves as being 'professional' photographers who need such a camera.

This creates a commercially very interesting possibility, namely, that if a new product of superior specification can be found to move into top position everybody in the market will have to buy a new product in order to stay in the same position. Those who regard themselves as being at the top will have to buy the new model, those who feel centrally placed will have to own the next model up and so on. When executed successfully, this market refreshment can stimulate everyone in the market to upgrade and buy again. Clearly, a second-hand market may develop in such a situation but the organization will certainly cash in on the new demand and can expect to sell many of the existing models throughout the range as people move up. A good example of this is the computer industry in which customers have to upgrade when a new generation of processor is released to avoid moving downwards in relative position in the marketplace.

Naturally, there are limits on such a process. One cannot simply go on bringing out new products with increasingly higher specifications and expect everybody to trade up every time. Eventually the gap between what is actually needed and what is on offer becomes too large for the customer to bear and the technique fails. When applied successfully the technique can have a powerful effect on the profitability of the organization and it is most likely to work where the dimension of the product used to refresh the market is a technologically limited one. An example of this might be the power-to-weight ratio of a jet engine. As technology advances and it becomes possible for a manufacturer to launch a new engine offering a higher power-to-weight ratio, so those customers who wish always to own the state-of-the-art engine, such as the military, will be forced to replace some of their existing engines with the new models. Much of the market may be refreshed by each new advance and the amount of profit from each development may therefore be substantially larger than the profit obtainable on the sales of the new product alone. It is for this reason that organizations with this opportunity are careful to ensure that the new product will not only sell well in its own right but also includes features that will refresh the market as effectively as possible.

An example of a customer group that cannot be refreshed might be a paint manufacturer. People do not go out and buy a new colour every time it comes out. Instead, they choose one they like and stay with it until its condition deteriorates sufficiently for them to become customers again when they re-evaluate the ranges of paints on sale.

3.4.3 Sales

Selling involves ensuring that buyers are found for the products that are available and is concerned with current activities. It may also include ensuring that customer needs are translated into production targets and with providing data for marketing and product development. In this section we will examine the organization of the sales function and consider some of the activities that have to be carried out.

Sales involves getting the product to the customer. In a sales department there will be three main types of activity: (1) selling, (2) providing customer contact, and (3) translating customer requirements into production orders.

There are several ways in which a sales department can be organized. Some companies organize their sales teams according to the particular products that they sell, some by geographical area, others by the type of selling that is used. The types of people in a sales department can be varied but in the engineering world, particularly in high-technology companies, you may meet the 'Sales Engineer'. This person may or may not be an engineer by training and qualification, but the term usually implies that he or she has a greater in-depth technical knowledge of the product than a 'salesperson'.

Selling

The type of selling selected by an organization will depend upon the products made and the market into which they are sold. A common method is selling by catalogue in which the company makes a number of standard products and then aims to sell them to customers off-the-shelf. In this case the specification for the product is known, the price is set and the delivery lead time will be known. Sales personnel need only a superficial knowledge of the product but will require good sales skills, as products sold in this way usually have a lot of competition.

Another method is that of selling by tender or quotation. This is commonly used when companies are providing a more specialist product or service, and is very common when goods are capital equipment purchases. In this case the sales personnel will need extensive knowledge of the product and will have to work closely with the other departments in the company. The procedure for selling in this way is as follows. The company wanting to buy the goods (the purchaser) sends out a tender or quotation request to potential suppliers. The suppliers may be chosen because they are known to make the goods required, or a more general tender request can be issued through an advertisement in an appropriate journal. If a company wishes to tender for the goods it then has to follow the procedure given in the tender request. This will usually mean that a bid is prepared and is sent to the customer for evaluation. The bid will provide full details of the goods to be supplied. This might mean that the customer's specification is accepted or that some modifications are made and the supplier includes a full specification of what they can supply. In addition to the product specification the bid will contain all the contractual terms to which the supplier is prepared to submit, for example cost, delivery, after-sales service and support, and warranty arrangements. All bids are sent to the customer and evaluated. If the customer accepts a bid then a legal contract is made. It is therefore imperative that in this type of selling the sales personnel understand the implications of any claims made in the bid.

Providing customer contact

The sales department will provide the first-line contact for customers who want queries answered and problems resolved. This has two advantages. First, it means that the customer has a contact, often a personal one, and will find it easier to build a relationship with the company. Second, it ensures that people in other departments are not interrupted by customers when they may not be in a position to give the attention a customer deserves.

Translating customer requirements into production orders

The sales procedures must ensure that when a customer places an order with the company, they receive what they want. In a company that sells off-the-shelf products this might be as simple as placing a request with the Finished Goods Stores. However, if the product is not in stock, or is manufactured to order, then the task is to ensure that the customer requirements are fully understood by the manufacturing department. This may mean filling in a production order form, or keying requirements into a sales order processing package that fronts an MRP package.

During this activity it may be necessary for the sales personnel to refer back to the customer, particularly if there are any problems with delivery or specification; at a minimum, the order should be acknowledged.

3.5 Finance

The finance function is concerned with money, bringing it into the company, looking at how it is used, and paying it out. Only by effective control of these activities can a company stay in business. The aim of the finance function is to ensure that this control provides the organization with the financial ability to meet its corporate goals.

In this section we will look at the finance function as it would be organized through a finance department and consider some of the activities that have to be carried out. Specific finance and accounting techniques will be dealt with separately in Chapter 4.

3.5.1 Organization of the finance department

Because of its importance the responsibility for the finance function in medium-sized and large organizations will usually lie with a director of the company, and will then be operated through a finance department. In a small company the owner or an executive director will have the responsibility for the task but may use the services of an external adviser.

The finance department will be responsible for the management of all the company's financial activities. In addition to accountants, the department may employ ledger clerks, payroll clerks and others with specific responsibilities for certain tasks such as dealing with taxation. It may even have responsibility for the purchasing function. The size of the department will depend upon the size of the company and the amount of financial management required. The finance personnel, with the exception of the finance director who will be responsible to the board, often have no line management responsibility outside the department but they may be responsible for setting and monitoring financial policy, such as defining spending limits and procedures for committing expenditure, which will then have to be used by managers in other areas. Because of this, the finance department must have good communications with all the other departments in the company.

In addition to internal communications, personnel in the department may have to deal with external organizations such as the Inland Revenue, banks, customers, and suppliers.

3.5.2 Activities of the finance department

We will now look at some of the specific activities carried out by the finance department and the requirement to report on these activities.

Bringing money into the company

The finance department is responsible for capital planning. This means that it must look at plans put forward by the directors and departments in order to determine the financial requirements that allow these plans to be achieved. It will then have to prepare proposals indicating the implications if the plans are to be followed and, if they are approved, will have to find and manage the funding. This may involve, for example, looking at taking loans, using reserves, or increasing shareholding.

In addition to this strategic activity the finance department will be responsible for the accounting activities of the company, including invoicing for payment of goods and services supplied to customers.

Using money

In order to operate efficiently a company needs to ensure that its money is being used effectively. To this end, it needs to know how much its products cost, which in turn allows sales prices to be set and cash flow requirements to be known. The finance department will be responsible for determining this information on the basis of data supplied by the operations departments. The analysis of the production costs will provide information about profit margin and estimates of turnover, and will allow budgets and spending plans to be prepared.

Cash flow forecasting on the basis of production data and planning is an important task. It involves defining what the company wants to achieve over a period of time and then analysing what this means in terms of cash that has to be paid out. For example, a company may be planning to make a number of units of product A. The product takes 3 months to make and will sell for £30 000 per unit. However, before you can get any money in from the sale of the product you need to spend money to buy materials, pay staff, purchase services and so on. The finance department's job here will be to look at when the sums of money have to be paid out and when sums of money will come into the company. It will then have to make decisions about how to deal with any problems such as there being a shortfall of cash coming in when bills have to be paid.

Paying money out

Money has to be paid out of the company for the goods and services that it uses in order to make a product. Paying for labour in the form of salaries is an important part of the finance department's function and allied to this will be the collection of taxes, national insurance and pensions. The department may also be directly involved in setting pay scales and determining bonuses.

Payment for goods and services purchased from other companies will be made on the basis of invoices received. As when sending out invoices for payment, it is important to balance payment of goods in order to control cash flow. A problem many small companies have is that of their customers not paying invoices on time, which causes serious problems if the companies have had to pay *their* invoices on time in order to secure supplies. This can put companies out of business because they are no longer able to finance their activities.

Reporting on financial activities

A company will require financial information for internal use. There will need to be information on budgets, sales figures, capital spend, cost of overheads, and cash flow. This will be used to determine the financial constraints within which the company, and its departments, have to operate. The finance department will be responsible for collecting the data needed to generate this information and for putting it into a format that can be understood and used by managers throughout the company.

In addition to the internal information, all private and public companies are required by law to produce financial statements. For the majority of companies these take the form of balance sheets and profit and loss accounts. The provision of this information is a prime responsibility of any finance department. There may also be requirements for financial information to be provided for other external sources, for example if a company is applying for a loan.

The role of the finance department is very wide, as can be seen above. In all its activities it is imperative that the information that is collected and disseminated has integrity, as the whole company relies on it to stay in business.

3.6 Product development

In the example at the beginning of this chapter we saw that a company wishing to make money by manufacturing kettles would have to develop the design of the kettle as time goes by. When a company improves the design of its products it is undertaking 'product development'. The aim of this process is to produce products that customers will continue to want to buy. However, there is much more to the successful exploitation of new products than simply launching bigger and better versions. The process of product development is complex and is examined in depth in Chapter 5. Our purpose in this section is to set the product development process in context and examine how it is organized and what it must achieve.

3.6.1 Activities of the product development function

The purpose of the product development function is to ensure that the organization has products in production that have customers who will buy them. If this can be achieved then the organization is guaranteed a market share and the only remaining problem is ensuring that the products reach the customers.

Customers play the central role in business life. It is customer choice that dictates whether the organization's products sell or not. All the revenue earned by an organization originates from its customers and so every wage packet and salary cheque ultimately comes from individual customers looking at the products on offer and choosing those of the organization over those of the competition.

In order to ensure that this situation is sustained there are various activities that must be undertaken, and these are briefly examined below.

Identifying customer needs

Customers choose to buy products in answer to some need. They want the product to do something for them. It therefore stands to reason that the engineers involved in the product development process must have an excellent understanding of what it is that the customers actually want. Often these are not the same as the needs that the design engineers think the customers have. The identification of needs is therefore taken very seriously in organizations with well-managed product development processes. Those involved meet customers face to face, spend time working with them and watching what they do and how it is done.

In a celebrated piece of product development undertaken by the Royal College of Art for the National Health Service, nearly one and a quarter million observations were taken of how hospital beds are actually used. The resulting design was a great success and is still in use today. It is testament to that fact that without understanding exactly what your customers want, you cannot hope to supply it. Conversely, the Sinclair C5 is often cited as an example of a product for which the design team had lost contact with what the customers wanted. Interest in what could be achieved technically and a desire to keep the project secret in order to have a dramatic launch meant that the designers were not in possession of data relating to real customer needs. Their imagined need for environmentally friendly, low-effort, personal, urban transport blinded them to the requirement for safety and particularly to be visible at windscreen height.

Product design specifications

Once the needs and desires that customers have of a new product are identified, a specification is prepared. This allows the engineers to define what the product must offer. The specification is not a solution. It does not, for example, contain details about how the product may be designed. Instead, it specifies what the product must do, what attributes it must have and what it is that makes this product different from that of the competition.

Technical design

Naturally, the development of new products requires a technical input from design engineers. New generations of microprocessors through to new models of battery charger all require technical input. This stage of product development usually involves engineers using their understanding of mathematical subjects together with their practical minds to synthesize new and imaginative solutions to the design problems posed by the specification. Some people regard this phase as actually being 'engineering'.

Preparation of documentation

Once a design in conceptualized, it must be expressed. Indeed, during the design process, development engineers will wish to express the design and communicate with each other. They do so using drawings. In addition to the drawings, there will be many other documents that need preparation. This need comes about from the requirement for production personnel to ultimately take over the manufacture of the products. Routine production must replace one-off prototypes. The knowledge and skill of the development engineers must be contained in the documentation so that production may proceed without the need for detailed discussion of each part. Such a documentation set is not easily prepared and good development engineers are as much good communicators as they are good designers.

Production implementation

Once the drawings and documentation are prepared, the product may be manufactured. When a new product is introduced to manufacturing there will almost inevitably be problems. Production methods may not operate exactly as imagined. Economies of scale may mean that some components should be changed. Jigs, fixtures, test schedules and assembly instructions all have to be used before they can be perfected. It is therefore common to find that one activity of the product development department is to assist in the introduction of new products to manufacturing and to support any technical problems that they may encounter.

3.6.2 Organization of the product development function

Marketing structure

Often companies have a marketing department to deal with the identification of new markets and products. Usually the marketing department is responsible for developing a draft product development specification which is then handed over to an engineering department in which the specification is transformed into a working, profitable product that can be manufactured to a specific price. In industries where sensitivity to customer changes is very great one finds the marketing function greatly increased in size. For example, in the car industry, minute differences between customer preference and product attributes select between success and failure. Consequently a large portion of profits must be diverted into marketing to ensure that the products on offer meet customer expectations.

Design structure

Another approach to the product development structure is to have the product designers involved in meeting customers and preparing the product development specification from the start. The advantage of this is that those people who actually do the designing are given a better understanding of what is required. The disadvantage is that people who are good at technical design are not necessarily most effective when dealing with customers and working with them to solve their problems. The changing marketplace of the early 1990s has led many organizations to conclude that the advantages of good contact with customers and consequent production of designs that more accurately meet their needs outweigh the disadvantages.

It is not ultimately important how organizations actually produce designs that meet customers needs, what matters is that it happens. We shall investigate the whole area of the management of product development in Chapter 5.

3.6.3 Research

In its scientific sense, the word 'research' is used to describe enquiry into the fundamental workings of nature and is often a very intellectual process. In commercial engineering the word is used in a much broader sense and describes extending the technologies of the organization. This may involve fundamental research, as is the case with, for example, the microprocessor manufacturers who constantly research the fundamental operation of their miniature circuits. Alternatively, commercial research may involve bringing established technologies into an organization. For example, a manufacturer of small metal equipment cases may wish to expand into supplying plastic products. The use of injection moulding for this purpose is not new in principle but it may be new to the organization. Research will have to carried out to see what effects the change might have, how it will be managed and how the new products and manufacturing techniques will fit into the organization.

The process of research is often stereotyped as mad professors in isolated laboratories proposing impractical gadgets that have no real customers. In practice, research could hardly be less like the stereotype. Successful organizations have well-managed research laboratories peopled with intelligent scientists and engineers who maintain the technical position of the organization within their budgets. Research is, by its very nature, an investigation into the unknown. The characteristics that bring success at it are creativity, a fertile imagination, technical excellence, good personal management and an ability to work with others, all characteristics that attract engineers. It is clear that managing a research function will have some problems that are specific to this unique discipline. However, it is also clear that the function must be managed, as must all the functions, to ensure that the overall objectives of the organization are achieved. Research is usually a very expensive pastime and if it is not well managed it can rapidly consume an unacceptable amount of profit. There is usually no upper limit on how much money can be employed to solve a research problem. Good research managers are therefore those that produce the results required but consume only the resources they originally requested.

Communication issues and links with other departments are very important in research. Good communication is required because of the importance of establishing customer needs and determining manufacturing constraints. It is also of vital importance that the research personnel are able to communicate the importance of what they have designed and researched, and why. They will have to sell their ideas to others in the company, and this may not be an easy task. Typically, the work involves a vast number of choices of routes to follow and ideas to evaluate. Keeping track of the work and chronicling the decision process becomes a considerable task in its own right; for this reason, the quality of documentation is always under scrutiny in the research department. In addition, due to the nature of the work and the long timescales that can be involved, the company needs to ensure that it is able to carry on with a project if a member of the team leaves. New members will have to understand why certain decisions were taken or why something was done in a particular way.

In some ways the strategic management of research is similar to that of other functions. The organization has its own agreed mission statement and from this the executive management produce objectives for each department and successive layers of the organization until everybody has their own personal or departmental objectives and budgets. Clearly, objectives for research activities are much more difficult to set than for other, more predictable, activities but plans can still be prepared and Section 10.4 gives details of some techniques that can help. There are many issues involved in the successful management of research such as how much profit it should consume, how radical may be the organization's 'new products', and how the link between research and the customer is to be made. These are complex questions and there are specific textbooks to deal with each.

Research can take a great deal of money to finance with a risk on return, therefore the financing can often be difficult. Some companies carry out very little research, preferring to develop markets rather than products, others license products which means that they pay for the right to manufacture a product that has been developed and tested by someone else. Other companies fund research from profits or use external funding provided by government agencies or customers. Some spend as much as they can on research and others as little as possible.

Because of its very technical and innovative nature the people involved in research tend to be more highly qualified and specialized than in other areas. Also, in order to suit the nature of the work the organization in research tends to be much less formal and more relaxed than elsewhere. Personnel, project and financial management are all very important issues in research, as are policies on publishing results, intellectual property rights, and patenting.

In summary, research is the process of extending the organization's technological abilities. Excellence in technological provision gives an organization an advantage over its competitors and so is of special interest to organizations. The process of research is a creative one and, for this reason it is intrinsically difficult to plan. However, it is clearly important to the success of the organization that research is controlled and that the results that are required by the organization are identified with an acceptably small level of expenditure. The strategy employed by the organization to provide its research needs can take many forms, ranging from a complete and separate department through to a subcontracted project placed with another organization altogether. How research is achieved is not particularly important. What is important is that the technological needs of the organization are provided for.

3.7 Quality

Quality means many things to many people. However, in the engineering industry there is a clear understanding of the term provided by the definition from the international standard ISO 8402 Quality vocabulary:

> Quality: The totality of features and characteristics of a product or service that bear on its ability to satisfy stated or implied needs.

The ISO 8402 definition of quality is quite clear but many other definitions are also used, usually incorporating ideas relating to meeting customer requirements and cost. However, quality should not be confused with expense. For instance, high quality does not have to mean high cost. A cheap pen that does the job for which it was intended and is then thrown away can still be a quality pen if that was what the customer expected.

Traditionally, there are two approaches to providing quality goods: (1) detecting defects using inspection and quality control to prevent defective product being sold and (2) preventing defects using quality assurance and total quality management. In the latter approach management controls are put in place to prevent defects being introduced into the design or manufacture of the product. Whichever approach is used, engineers will necessarily be involved with quality.

In this section we will look briefly at some of the activities involved in operating a quality system and at the way in which the quality function can be organized. Specific quality management principles and techniques are described in detail in Chapter 7.

3.7.1 Quality systems

The definition of a quality system, from ISO 8402, is:

> The organizational structure, responsibilities, procedures, processes and resources for implementing quality management.

Many companies have some form of quality system which ensures that the company's policies relating to quality assurance are implemented, used, and monitored. Many of these follow the requirements for quality systems given in the international standard ISO 9000 Quality systems. Total Quality Management (TQM) is an extension of this standard. The philosophy behind TQM is that quality should pervade all aspects of the business of the company whereas the standard is limited to certain sections of the business. In addition, there is much more emphasis in the TQM philosophy on employee and customer involvement.

NOTE that throughout this text the quality system standard will be referred to as ISO 9000, using its international designation. The UK designation is BS EN ISO 9000, and it was previously known in the UK as BS 5750. See Chapter 7 for further information.

The aim of any quality system is to provide a systematic approach to design and manufacture in order to prevent failures and to provide objective, documented evidence that an agreed quality level has been achieved. If the quality system meets the requirements of ISO 9000 it can be certified as such by an independent third party approved by the United Kingdom Accreditation Service (UKAS). This reduces the requirement for assessment by individual customers and the need for customers to define their own quality system requirements.

Certification may also mean that a company can address new markets and thus increase sales. Many markets require very high standards of reliability and traceability – for example, the nuclear market, the defence industry and the medical market. The customers in these markets require manufacturers to operate proven and effective quality systems, and certification provides assurance that this is the case. In addition, as more companies are achieving either a total quality system or certification to a quality system standard they are requiring their suppliers to have similarly rigorous procedures to ensure the quality standards of their goods and services. There are many other advantages of having an effective quality system some of which relate to the aspects of the system aimed at preventing defects.

3.7.2 Management activities in the quality function

It follows from the definition of quality system that the activities required to make a quality system should permeate all aspects of the business and these specific activities will be described in Chapter 7. However, for all the quality system activities there will be some general requirements which will be, usually, centrally administered. These relate to documentation, training, and auditing.

Documentation

ISO 9000 requires that the quality system is documented. However, even organizations that do not follow ISO 9000 generally document their quality activities. This approach ensures that everyone is clear about what is to be done and by whom. A number of levels of documentation will exist, the top level being the quality manual. The quality manual is the operating document for the system, defining what is to be achieved and indicating how this is to be done. It may or may not contain the detailed operating procedures.

Procedures define the tasks to be carried out in order to meet the requirements of the quality system. They give the job title of the person responsible for carrying out the task and of the person that has authority for making decisions about the work being undertaken. They clearly define the way in which the task is to be undertaken and the records that are to be kept in order to verify that the task has been carried out correctly.

Procedures should be prepared by the people who are going to use them, or are carrying out the work to be described. However, procedure writing must be very precise and unambiguous to ensure that the procedures can be followed by anyone that has a need. In addition, where compliance to ISO 9000 is required the procedures must ensure that the requirements defined by the standard are set.

The role of the quality function will be to aid in preparation of documentation, reviewing its appropriateness and assessing its effectiveness, and then to be responsible for its issue to relevant personnel and its control.

Training

The operation of a quality system will lead to a training requirement for personnel who work within the system. This training may be of a general form, on the subject of quality, or it may relate to specific aspects of the quality system operation, or the interpretation of ISO 9000. The quality function would be responsible for the provision of this training and the monitoring of its effectiveness.

Quality systems auditing

Auditing is a systematic and objective assessment of the quality system to ensure that the procedures are being followed and that they are effective for the achievement of quality. In order to maintain impartiality and ensure that the audits are effective they must be carried out by people who are trained in auditing techniques and who are independent of the service or area being audited. It is usual for this task to be performed by the quality function. There may be additional auditing of suppliers to be done. As part of a supplier appraisal scheme, this would also fall to the quality function. Auditing of the quality function must be done by someone external to it.

3.7.3 Organization of the quality function

Quality control is all the activities that are carried out to ensure conformance of parts or products to specification, and so will include both inspection and test activities. The way in which a company deals with quality control is usually based on how 'quality' was introduced into the company. In many companies there are inspection and test departments which may be a part of the operations department, part of the quality department, or a department in its own right reporting to a Chief Inspector who in turn reports to a senior manager or director. However, increasingly more inspection and test is done at source by operators, thus giving them 'ownership' of their work. It is important that authority to stop defective material and parts is also given to the operators working in this way, this authority always being part of an inspector's job.

Due to the modern views of quality management, particularly because of the concept of 'Total Quality', quality as a function is now more commonly found spread throughout the organization. However, many companies do retain a 'quality department' which has responsibility for implementing, coordinating and monitoring procedures to ensure that agreed quality levels are maintained and for carrying out audits. In this case the quality department is similar to the personnel department in that it does not have direct authority but it will need to integrate with all aspects of the business. It will also have to liaise with outside organizations such as accreditation bodies, customers and suppliers.

Management of the quality function, as with the personnel department, is complicated by this lack of direct authority but many companies alleviate some of the problems by giving a director, or a senior manager, responsibility for the function. Indeed, if the company operates in accordance with ISO 9000 this is required. It is of equal importance that the responsibility and authority of everyone working within the quality system is clearly defined so that they too can carry out the tasks assigned to them.

The Quality Manager is the person with defined responsibility and authority for ensuring that the requirements of the quality system standard are implemented and maintained. This person should have senior manager status, and should report directly to the Board of Directors in all matters concerning quality, irrespective of other duties.

3.8 Personnel

Aspects of the personnel function will be carried out in all departments throughout the company. Because of the need for every manager to manage personnel, this topic is covered in detail in Chapter 9. However, in this section we will look at how the personnel function can be organized and consider some of the activities that have to be carried out, because we are considering all the functions required to operate a business.

3.8.1 Organization of personnel

Anyone performing a management role will be involved in the management of personnel. However, a personnel department will normally act as central coordinator for this activity and will both prepare and monitor personnel policy. In many small companies the personnel department will be one person, perhaps in a part-time post.

3.8.2 Activities of a personnel department

An important activity of a personnel department is a strategic one, defining the personnel policy for the company. This policy will define the procedures and rules that will apply when managers in different departments deal with personnel issues. The personnel department may deal with some issues but generally it will have no line management responsibility outside the department. Its prime tasks will be to establish personnel policy, provide advice and training to people that will carry out the policy, and monitor the effectiveness of the policy to ensure fairness for all employees. Some of the specific personnel issues that will have to be dealt with are described below.

Manpower planning

Manpower planning matches the company's requirements for people and skills with those that are available. It should highlight deficiencies and surpluses in advance of them making an impact on the business. Manpower planning will be particularly important in the manufacturing area but will have to be addressed throughout the company. The personnel policy should show how manpower planning is to be performed, and possibly when. It should also indicate any constraints within which managers will have to work.

There are two levels of manpower planning, the highest level being the overall company requirements to meet corporate objectives. These would be identified in the business plan and would not normally be very detailed. However, at a lower level the line managers in each department would have to plan in detail their staff requirements in order to meet department objectives. For example, the business plan might say that in order to meet expansion plans the manufacturing area will have to grow by three people over the year. However, the manufacturing manager will have to specify what those people are to do and exactly when they are required. The manager will also have to plan in more detail to allow for holidays and sickness among personnel in the department.

Employee appraisal

Employee appraisal is used to assess performance of individual employees, following discussion, and then allows decisions to be made about the employee's future progress in the company. If an employee-appraisal system is used the personnel policy should define the way in which the appraisal is carried out, what records are made, and how often appraisal should take place. The personnel department would be responsible for ensuring that the managers carrying out appraisal are trained to do the job, and for monitoring the application of the appraisal system to ensure that all employees were being treated fairly and that consistency is maintained throughout the organization.

Recruitment and selection

This activity involves identifying requirements, following manpower planning, and then preparing job descriptions, advertising, and carrying out selection interviewing in conjunction with managers. It should also include induction of new staff. The personnel policy should define the recruitment and selection procedures to be used. The personnel department would normally provide appropriate training and advice to managers actually involved in recruitment. In addition, it will play an administrative support role in distributing application forms, arranging interviews, sending out contracts, and advertising the post. The personnel department would also be required to provide advice on the legal aspects of recruitment and selection.

Training

This activity involves the provision of training following from the needs identified during appraisal. It should also include monitoring and control of training provision to ensure that training needs have been met.

Health and safety

There are specific legal requirements for health and safety policies in companies and many will administer and monitor these policies through their personnel departments. There is also a requirement for first aid facilities, first aiders and an accident log, and these may be administered by the personnel department.

Welfare

The attention an employer pays to the welfare of its employees will affect levels of morale and motivation. The personnel department will be concerned with all policies to maintain and improve employee welfare. A counselling service for employees is often provided through the personnel department.

Consultation and negotiation

This may involve negotiating with unions, staff associations, or individuals, on areas of company remuneration and working conditions, as well as in cases of dispute. The personnel department will usually be responsible for preparing contracts of employment and for defining payment systems, both of which may require extensive negotiation with employee representatives.

Dismissal

There are legal regulations governing dismissal and all companies should have a policy on this. The personnel department has to formulate the policies, monitor them, and may be required to undertake dismissal proceedings.

Records

A company needs to keep records for each of its employees, and these will be maintained by the personnel department. These would include application forms, references, appraisal forms, pay and contract details, and any records of grievance or disciplinary proceedings. All this information will have to be held securely and access limited, these issues being addressed in the personnel policy.

3.9 Company operation and the role of engineers

Companies have to carry out many different functions in order to conduct business. We have looked at the major business functions in the last seven sections. However, it would be wrong to give the impression that each of these functions exists in isolation. For any company to transact business successfully there must not only be good communication between the various functions but there must be some meshing of the functions. Similarly, people cannot work in isolation and the smaller the company you work for, the greater the need to work closely with others in different functional areas.

To illustrate how companies really do function and to show the roles that can be taken by engineer we have prepared three case studies. The first case is about Matthew Hobbs, who works as a development engineer in a company employing 60 people. The second case concerns Martin Twycross, who, although an electrical engineer, is employed as an Area Sales Manager in a company employing 550 people. Finally, we have the case of Simon Tu, who is a quality engineer for a company in Singapore employing 1200 people.

3.9.1 Case study: Matthew Hobbs

Matthew Hobbs works as a development engineer at Warwick Power Washers Ltd, Oxford, UK. This is a small company employing approximately 60 people, and it has been trading for 26 years. The main business of the company is the design and manufacture of power washers, which are used by both domestic and industrial customers; they also sell the consumables, such as detergent, for the washers. Some of the larger industrial machines are made to individual customer specification, and these will often take the form of making modifications to a product made previously.

The company has a technical department which is responsible for the design and development of all products; this is where Matthew works, reporting to the

Chief Engineer. He graduated in mechanical engineering and has worked in the company for a year, since completing his studies.

The technical department consists of four people including the Chief Engineer, and Matthew is able to work in a semi-autonomous way, being able to organize his own time and priorities within the constraints imposed by the department's objectives. In designing new products Matthew is involved with the complete design from customer specification through to getting the product into manufacture.

The on-site sales team of the company has four or five staff, but selling is also done through a network of distributors and by service engineers. This sales team has prime responsibility for customer contact. However, because of the nature of the business the technical department also liaises with customers. Matthew, and the other members of the technical department, are sometimes contacted by customers or dealers who make a direct input to the design process and also answer their technical enquiries. Sales make the major input to the design process by passing on customer requirements in the form of a design specification, giving general customer feedback on product performance, indicating general market trends, and advising on new EC safety guidelines. They also take part in the design review which takes place after a prototype has been built by the department.

The company has a buyer who is responsible for buying all the parts for manufacture. In the design process Matthew is responsible for ensuring that, wherever possible, standard parts are used. However, if this is not feasible then he has to define a purchase specification and, sometimes, source a supplier; the buyer has responsibility for negotiating contract terms after the sourcing has been done. In addition, when new subcontractors for machining processes are sourced Matthew may visit them to consider their manufacturing capability for possible future requirements. He will also gain technical information which can be used to improve product design. Very occasionally Matthew may visit to check on the quality of goods and service, although this is usually undertaken by the company's quality engineer.

Following the design process Matthew is responsible for ensuring that the new product moves smoothly into manufacture. This involves preparing drawings, designing jigs and fixtures when required, and producing assembly instructions; bills of material are prepared by the department's technical administrator. Matthew will show the manufacturing supervisor how to build the new product. When the product is in manufacture he may be called on whenever there are production problems; he retains technical responsibility for the product.

3.9.2 Case study: Martin Twycross

Martin Twycross graduated with a BSc in Engineering, specializing in electrical engineering. For the first four years after graduation he was in the British Army, for the last four years he has been working in the Sales and Marketing Department of Research Machines.

Research Machines is based in Oxford, UK. It was started 21 years ago and now employs 550 people in Oxfordshire. Initially the company supplied electronic components, and now produces a range of industry-standard personal computers (PCs) and local area networks (LANs). It is a main supplier of PCs to the education sector in the UK.

Martin joined the company as a salesperson and following a series of promotions is now an Area Sales Manager; he has been in this post for 3 months. It has never been a company requirement for people in these jobs to be engineers, but working in a technical company he does find the technical background helps in his work. Martin has three main responsibilities: meeting sales targets; managing his team of four people; and ensuring customer satisfaction.

In order to meet sales targets Martin is in touch with potential customers, discussing their requirements and proposing solutions. This may involve surveying buildings for LANs and specifying particular systems to meet the customer need. Because of his technical education Martin can do some of this work on his own, but in complex situations he will draft an initial specification and then discuss it with the technical department.

When the customer requires a system to be installed this is carried out by the customer support team. In these cases Martin will liaise with them and with the customer to ensure that all the equipment is available and that the requirements are fully understood.

Research Machines has a quality ethos, operating on the principle of Quality Improvement. As part of this, every group is required to have an on-going quality improvement project. For the sales department this may be something like reducing the time to answer customer enquiries. The projects are presented to the Sales and Marketing Director. At this time Martin has not set a project for his group, and this will be one of his management tasks when he has settled into the job.

Martin and his team provide customer feedback information to engineering, via marketing, as part of the product development programme. He feels that this way of distancing sales from engineering is important in the fast-moving environment of computer development. There is an inherent danger with direct links because of the need for sales to process the latest product while engineering are likely to be dealing with the next. However, Martin may have some direct contact when there is a need for engineering to talk directly to the customer and he is the link.

3.9.3 Case study: Simon Tu

Simon Tu Yeou Mou has been working as a quality engineer for Seagate Technology International in Singapore for one year. He returned to Singapore after graduating from a British polytechnic with a degree in mechanical engineering.

Seagate Technology employs 1200 people in its Tuas, Singapore, factory. This site is responsible for manufacture of high memory capacity computer disk drives which are made to Seagate designs from the United States. There are eight functional departments within the Singapore operation. Simon is employed in the Quality and Reliability Department. This department is responsible for defining the requirements and responsibilities within the manufacturing operations to assure conformance to specified requirements. This involves appraisal of all parts and finished goods, auditing of suppliers, in-process inspection, tools and equipment calibration control, and clean-room environmental auditing.

Simon's main responsibilities include implementation and monitoring of statistical process control techniques. To do this he has to work on the shopfloor occasionally, but he has a supervisor and two technicians reporting to him who do most of the shopfloor work. He is also responsible for carrying out quality system audits and establishing networks to disseminate quality system information. Obviously, these activities require Simon to liaise with all departments in the company.

Specific tasks include certification of all products from the United States to Singapore. This requires the inspection of all incoming parts and involves qualification of the plant's capability by process auditing. Simon carries out these tasks and then presents his audit results to a meeting in which all departments are represented. It is Simon's responsibility to identify any corrective action required and he will also explain what particular process requirements are. In addition, he may give guidance on how the requirement can be met. However, the responsibility for taking the action will rest with the department concerned.

A further aspect of Simon's job is running quality improvement meetings. These involve representatives from test, engineering, manufacturing, and quality. On all issues relating to product or process quality Simon has authority, including the ability to shut down the whole product line if he feels that is what is required.

As can be seen, Simon's job is very much concerned with people in other departments and then presenting information which requires action from them. He obviously has authority in the company but must exercise judgement before wielding that power if he is to maintain the respect of the managers to whom he is presenting information.

In business none of the functional areas described can operate in isolation and in the three case studies we have seen some examples of how the functions interact in three successful companies by looking at the role of one person in each company. As an example of these interactions we can focus on the Operations function and look at some of the issues that it must address working with other functional areas.

The operations function in any company will have to be able to communicate effectively and interact with every other function. The types of interactions required in a manufacturing company are as follows:

1. *Operations/Purchasing* These functional areas must interact in order to determine what parts, services, or equipment are required, and what are available, for manufacture.
2. *Operations/Sales and Marketing* The interaction here will be concerned with what products can be made, and when, and what are required for sale.
3. *Operations/Finance* These areas must work together to produce product costing and budgets. In addition, there will be payment of wages, determination of spending restrictions, stocktaking and so on.

4. *Operations/Product Development*   This is concerned with the process and skill constraints imposed on product development by manufacturing. It also involves development for manufacture and feedback on actual process and testing.

5. *Operations/Quality*   This is concerned with quality assurance procedures to ensure product quality meets the correct standards and with feedback on actual performance of product and processes.

6. *Operations/Personnel*   This interaction will involve many things including defining manpower requirements, recruitment, pay, and welfare.

The case studies show that there are also many other necessary interactions.

3.10 Summary

In the introduction to this chapter we identified the seven main functions that are required to operate a manufacturing business – purchasing and stores, operations, sales and marketing, finance, product development, quality, and personnel. We have examined each in turn and looked at some of the activities that form the function and at some of the factors that affect the way in which they are managed. We have also seen, by considering the role of three people working in engineering companies, that none of these functions can operate in isolation. In order for a company to transact business there need to be many interactions between the different business functions.

3.11 Revision questions

1. What are the main functions to be found in a manufacturing company? What information transfers are necessary between each function and operations?

2. What benefits can a company gain from operating a central purchasing and stores system, and what are the disadvantages of this method of operation?

3. Why is it necessary to recognize that there are different types of customers?

4. What are the categories into which customers can be defined?

5. How is the product development process affected by the category of customer that the company serves?

6. What is the role of marketing in the product development process?

7. In a project team charged with developing a new product what would be the role of a quality engineer?

8. Why is the ability to communicate well, both orally and written, so important for a development engineer like Matthew Hobbs?

9. Why is it necessary for an engineer, working in product development, to have an understanding of finance?

References and further reading

General
Batty, J. (1982), *Business Administration and Management*, Learnex (Publishers) Ltd.
Bennett, R. (1987), *CIMA Study & Revision Pack Stage 2 Management*, Pitman Polytech.
Lewis, D. (1988), *Basics of Business*, M&E Handbooks.
Needle, D. (1989), *Business in Context*, Van Nostrand Reinhold (International).
Purchasing
Baily, P. and Farmer, D. (1981), *Purchasing Principles and Management*, 4th edition, Pitman.
Operations
Harrison, M. (1990), *Advanced Manufacturing Technology Management*, Pitman.
Muhlemann, A., Oakland, J. and Lockyer, K. (1992), *Production and Operations Management*, 6th edition, Pitman.
Sales and marketing
Elvy, H. B. (1977), *Marketing Made Simple*, 2nd edition, Heinemann.
Kenny, B., with Dyson, K. (1989), *Marketing in Small Businesses*, Routledge.
Product development
Ray, M. S. (1985), *Elements of Engineering Design*, Prentice Hall International.
Quality
Dale, B. G. (ed.) (1994), *Managing Quality*, 2nd edition, Prentice Hall International.
ISO 8402: 1994 Quality vocabulary
ISO 9000 Series
  ISO 9000-1: 1994 Quality management and quality assurance standards
  ISO 9001: 1994 Quality systems
  Model for quality assurance in design, development, production, installation and servicing
  ISO 9002: 1994 Quality systems
  Model for quality assurance in production, installation and servicing
  ISO 9003: 1994 Quality systems
  Model for quality assurance in final inspection and test
  ISO 9004-1: 1994 Quality management and quality system elements. Part 1. Guidelines
Personnel
Farnham, D. (1987), *Personnel in Context*, revised edition, Institute of Personnel Management.

Part 2

THE MANAGEMENT OF ENGINEERING

Finance

Overview

It doesn't matter what you think about money, you need to be able to manage it. Corporate business and church fêtes both need to balance income and expenditure. No part of management goes without financial implications and professional engineers have needs for financial understanding just as other professionals do. In this chapter it is time to examine how the financial needs of an engineering organization may be met.

4.1 Introduction

To introduce this chapter we will start with a question. 'Would you like to make some money?' Not just earn a wage, most people can earn some sort of wage, but actually turn over money and keep some of it for yourself.

Those who apply themselves to the task of making money have one thing in common. They all realize that, among their many other skills, they will have to understand money. Most people know that to make money, goods have to be sold at a price that will cover materials, labour and overheads, and still leave some profit. However, very few people can look at a business and say whether a sensible asset distribution prevails or whether the accounting system is facilitating optimal decisions. Few understand that a healthy business can be rapidly bankrupt simply by not distributing its money correctly. Once we get into the realms of shares, debentures, mergers, acquisitions, taxation, investment policy or corporate financial planning all but a few are left behind.

Finance pervades nearly all activities in some way. A musician arranging a concert has to pay attention to its financial viability. A retail shop has more complex financial problems and needs to know, for example, how to price larger goods which take up more space, and therefore consume a greater proportion of the cost of this space, to the exclusion of other products. For a trading organization that takes in goods, processes them, and then sells them the financial complexities are immense and cannot be tackled by the novice. The most commonly quoted reason for the failure of new business is a lack of financial acumen.

This chapter will give you an introduction to finance. Engineers do not need to become skilled financiers in their own right but they must understand basic financial axioms in order to make good decisions. In the same way that a good farmer is knowledgeable about soil science and meteorology, a good engineer is knowledgeable about finance.

We will start by examining the need for monetary control. Before we embark upon an investigation of finance in detail we must examine what it is we expect to achieve through having a system at all. Section 4.2 examines this need. In Section 4.3 we introduce the subject of financial accounting and explain how this branch of finance serves one of the needs identified in the previous section, to provide financial information about the organization to those in the outside world. In Section 4.4 we examine the need for budgeting and explain how budgeting is performed and consider how financial management may be achieved through the use of budgeting.

In Section 4.5 we introduce the subject of management accounting and explain how this exists to provide financial information to those inside in the organization who have responsibility for its management. In Section 4.6 we introduce the business plan. This is the document one is required to present when applying for money. In this section we will see how the financial skills presented above may used to prepare a sensible business plan.

The learning objectives of this chapter are:

1. To explain why financial control is important
2. To describe financial accounting
3. To be able to understand the major financial documents
4. To explain management accounting
5. To develop an understanding of the budgeting process
6. To appreciate the overall structure of organizational budgeting
7. To understand the structure of a business plan.

4.2 The need for monetary control

Money is required in the administration of all organizations. Some organizations seem to have making money as their sole aim while others, such as charities, seek to encourage plentiful contributions and then distribute them in accordance with a perceived need. Even though charities are not trading, they still have to be skilful handlers of money. It doesn't matter what an organization is trying to do, it must be able to control its finances. The more complex the situation, the greater the need for good financial control.

We shall now consider a simple business to see how the need for a clear financial management system comes about and requires more than a simple personal finance system to be effective.

4.2.1 Inadequate financial systems – a case study

Jake has set up his own business operating out of his backyard. He is renting his house and has finished studying for a higher degree in music and so has some spare time. He is convinced that with his hobby-based knowledge of digital electronics he can develop a home hi-fi business based on combining proprietary supplies of hi-fi equipment with his own creations to produce individual systems for each customer. He calls the enterprise 'Jake Sounds'.

He soon has an order from a friend. He discusses it, and produces a design. Jake buys the components from various suppliers and assembles the equipment. Some of the components have to be sent off for machining, painting, silk-screen printing and other processes with local companies, others arrive ready for use.

To keep track of finances Jake has a filing system consisting of three files. One file contains a record of all purchases, another a record of all sales. The third contains a summary of the other two; this is the most important one to Jake, since it contains the total amount received and the total amount paid out. He is looking forward to seeing the first number minus the second become very big! The first job is finished in only six weeks and the total amount spent was £1200. Jake notices that he has a few components left over but they cannot be sold back to the supplier and so he decides not to give the customer any discount. He reads a textbook on pricing policy and discovers that it is normal to charge 2.5 times the cost of materials and labour to the customer in his industry. He costs his own time at £8 per hour to give £960. Adding this to £1200 and multiplying by 2.5 gives him £5400, which terrifies the customer so he charges only £4500. Filling in his books he calculates in the third file that he has made £4500 minus £1200 minus £960 = £2340 clear profit. As this is after paying his own wages of £960 Jake is delighted. The rent and bills for the premises can easily be covered for the whole year with just a few such jobs, and after that it's all profit in the bank for the business.

After operating on a similar basis for some time Jake makes the following summary of his business. The order book is growing and he can no longer do all the work himself. Nearly all customers want a mixture of standard equipment and his own custom designs but these designs are practically never the same as something built previously. The accounting system is adequate.

Because of these factors he hires three people and divides the company into two sections, though they are all physically still in one room. The first section handles all the custom equipment and the second purchases all the standard equipment and combines it with the custom equipment to form the complete system. This is then tested, and corrected if necessary, prior to shipment. Each department orders the parts it needs which, when delivered, are placed on shelves covering one wall of the room. All purchases are recorded in the file, as are all the sales, just as before.

Two orders are processed sequentially in this way to allow everyone the opportunity to follow the system. There is some confusion sometimes over what belongs to whom on the shelves, and it has become necessary for the three of them to look up the cost of goods used on each system in order to set the selling price accurately. These points are not considered to be serious and they continue operations, having had a profitable three months.

They now have many orders and some have to be worked on simultaneously. Jake hires a buyer to take the load off the engineers, leaving them free to concentrate on the equipment. This works very well and whenever the engineers need something they ask and it soon appears on the shelves. The buyer is always helpful and always seems to know what's going on.

Before long, the next quarter is over and it is time for Jake to assess how things are going. The company has completed seven systems this quarter, which makes the productivity less than the 'one system per six person-weeks' that he used to average on his own. Nonetheless, he calculates that at about £2300 profit per system that should make £16 100 clear profit to invest back in the business after wages have been paid. By the time he starts making up the books, Jake is already planning how to make his business grow faster with a new marketing initiative. He plans a profit-related bonus scheme to reward the industry of his employees, and a bottle a Château de la Tour 1961 for himself. Imagine then, his disappointment when he finds that the company has made only £1225. He cannot believe his eyes and promptly calls a meeting to discuss the matter.

Several points arise in the discussion, starting with the pile of components on the shelves which seems to be bigger than before. Two particular jobs were especially difficult to get right, the components used were not extraordinarily expensive but several sets were blown during manufacture. No one can put their hand on their heart and say accurately how much each system cost. Jake can certainly find out how much all the combined systems cost by totalling the 'purchases ledger', but he cannot now find out how much was spent on materials for each. He looks up total expenditure and finds that much more has been spent than he expected.

One of the employees comments that it is no longer possible to know exactly how much time is spent on each job. Previously, when there were only one or two systems, it was easy to keep track, but now he freely admits that he divides his time between all the jobs in progress and hopes it all balances out. The others comment that even though the two difficult jobs consumed most of their time, they still allocated their time evenly among the five jobs in progress at the time. Jake is very concerned at this and the unknown material costs since it means that the selling prices were fixed using the wrong information. Therefore the profit margin for each job will be quite different from his original estimates.

The team turn to the stores and purchasing system. The buyer explains that, where possible, he has been buying in bulk for economy. This is obviously good, but it has caused some difficulty. Kate mentions that she was about to ask for some components to be ordered when she saw some in stock, and so used those. The components were much the most expensive in the system, and because of the discount the system must have been much cheaper to make but the selling price was based on the old price. On hearing this, another engineer is enlightened. He ordered a matched set of output Stegatrons for one difficult system and later found some of them missing and had to order even more – they were really expensive. It transpires that any profit that might have been made on Kate's system will have been lost by filling it with expensive and unnecessarily matched Stegatrons.

The engineers also noticed that there is room for improvement in the accounting and stock control system. They even found an example of the same delivery of ten components (worth a considerable part of the material cost) being included in the costing of two different systems even though each system used only one of the components. Another engineer had looked at this information and thought that all the components must have been used up, and so ordered some more. There were then eighteen units in the store. This was a really poignant example, since the company had only orders needing six more of the particular component and the remainder would probably become scrap, since the manufacturer would very soon be replacing the unit with a new version that the customers would want in their systems.

Some chartered accountants who have been hired to audit the company's books sent a letter to Jake asking the following questions:

- What is the value of your stock?
- How much profit did you make on each of the jobs?
- What is the component wastage rate?
- How much money is outstanding in unpaid bills?
- How much money is outstanding in uncollected debts?
- How much of the purchases were for components and how much for tax-deductible assets or equipment?
- Why is one particular payment to a supplier much bigger than any other?
- How much profit have you made overall?

Jake realizes that he cannot answer any of the questions, not even the last. He lowered his face into his hands and made manifest his heightened state of aggravation and woe. 'What are we going to do?' he asks. He called a consultant to help him understand what had happened. The consultant confirmed that the business was suffering from an unacceptable accounting system, 'If there had been a half-decent one in place the problems you now face would have been obvious to you as they occurred rather than all this time later. Think about this idea,' the consultant explains.

The buyer orders the items and pays for them out of the company bank account. When the goods are delivered they are 'booked' into stores, meaning that a record is made of the items in a file. When the engineers need a component they go to the store, take the part required and then record in the file what has been taken, its value and the job on which it is to be used. They also decrease the stores count so that the next person knows accurately what there is and what it cost. The stores becomes a control system, not just some alphabetical shelf space. If an engineer knows of a future requirement, existing stock may be earmarked in the book for use and so not used by someone else. If the engineers all use a time sheet to record how much time is spent on each individual job you will then be able to easily work out how much each job is costing.

The consultant said that he found £6000 of material in the store. Jake is delighted until he hears that £6000 is what was paid for it and not what it is worth now. The consultant said:

> The value of the stock in terms of resale value is only about £1000. Great savings could also be made by organizing the timing of your purchases, as you have already noted. You need to schedule your production to ensure that you take advantage of the discounts available. Make a list of all the components in each of the systems and get the buyer to order them sufficiently early to be sure they arrive in time without paying any premium for rapid delivery if the order is placed a little late. You could even note down the typical delivery times of your suppliers and order parts in a scheduled way so that they arrive only shortly before they are needed. In this way you could minimize your stock. The buyer would then know the best time to order a given component and would only buy early if any discounts or special terms that the suppliers may be offering make it worthwhile.

The consultant's face darkened as he turned to Jake and said, 'Worse than this, however, and an unforgivable error in a manufacturing company, is the way the selling price has been based on the wrong data.' The consultant estimates that the two really difficult jobs were actually sold at a loss of about £500 each. He guesses that the profit was nearly all made on three other jobs whose selling price was set too high. This was because the selling price was based on estimates which turned out to be wrong. Jake remembers that one of these customers was rather concerned at the price and will probably not now do repeat business.

Jake is worried about how he can answer the questions of the auditors but is also resolved to sort this out so that his company may become highly profitable once again. He asks one last question of the consultant. 'But how come I was alright before, the accounting system was fine and I always knew where I was?' The consultant replies:

> Yes, you were alright, but that was because you had all the information you needed. The company was simple so the accounting system could be too. Your company has changed and just because something served you well under one situation it doesn't mean that it will continue to serve you well under a new one. You used to know everything about each job and that is still what you need to know. All you have to do is have a stores system, have a file for each job into which you book all stores requisitions for that job and into which you book all the time spent on each project. Then you will have all the information your company needs to satisfy the tax inspector and make the best business decisions for the future. No one likes paperwork so have only enough to meet your needs, but do make sure you have sufficient for the needs of the business.

4.2.2 The ideal financial system

We know that money is the lifeblood of all organizations. It does not matter what the organization's purpose is, it will still have to manipulate money in order to meet it. There are good and bad ways to handle money and the ideal financial system will clearly be the one that enables the best manipulation of the organization's finances to be achieved.

For instance, imagine a manufacturing company that purchases all its raw materials with 60 days' credit, the assembly and shipment takes 20 days and the product is sold within 5 days, on 30 days' credit. This leaves 5 days clear after receiving the customer's payment before the bill from the suppliers arrives. Such a business would not have to borrow money to start production, could cope with any number of orders and would have no financial limitations on its rate of expansion. If the same company had to pay the suppliers well in advance of receiving payment from its customers the situation would be very different. In order to produce each product the company would have to borrow the materials cost of the product for the period of manufacture, and the number of products it could make at once would be limited by how much it could borrow. The rate of growth would be limited by having to borrow against tomorrow's larger profits in order to purchase today's larger material bill. These differences do not necessarily have anything to do with the design of the product, its success in meeting the customers' needs or the production arrangements. They do, however, depend directly on the way in which the organization negotiates and operates its finances. It is clear which sort of arrangement is preferable.

An ideal financial system is defined as 'the most cost-effective method of ensuring that the organization can operate its financial affairs in such a way that the organization's overall goals may be best served'. Such a system must consume an acceptably small portion of the organization's resources while providing for its financial needs. These needs may be external, as in the requirement to produce annual audited accounts for the state, or they may be internal. A manufacturing business needs to know, with sufficient accuracy, how each area of the business is performing in order that good areas may be left alone and bad ones rectified. These two needs are often separated into two separate subjects. The first, provision of information for people outside the organization, is called financial accounting; the second, the provision of information for those within the organization, is management accounting. Both are a balance between providing too much information or too little. Too much will cause unnecessary expense, be confusing and probably be wasted. Too little or the wrong information will compromise the quality of decision, lead to irrational actions and might even risk the organization's survival.

4.3 Financial accounting

Financial accounting is the name given to the preparation of financial reports predominantly for external needs. There are two main bodies interested in this information. The first are people who invest in the organization in some way. Such people are clearly interested to know how their investments are performing. In the case of new investments, an assessment of the likely profitability will be prepared and the financial performance of the organization will be of key interest. The second body is the state, and its interest is twofold, first, because it requires tax to be paid on the profits of trading, and second, because the state regulates trading and seeks to prevent unscrupulous organizations exploiting investors and employees. It is for these reasons that there are legal requirements to produce audited accounts.

To meet these needs various financial documents have been developed and in this section we shall consider some of them. Although there are many financial documents we shall restrict ourselves to three of the most important: the balance sheet, the profit and loss account and the cash flow projection.

4.3.1 The balance sheet

The balance sheet is a very important financial document. It not only lists all the different areas in which money belonging to the organization currently resides but also specifies the amount in each. Since money is constantly flowing around the organization the balance sheet can only apply to one particular instant. The best conceptual understanding of the balance sheet is gained by thinking of it as a financial snapshot of the organization or, perhaps, as a single frame from a movie film of the organization's finances. Because of this the balance sheet is prepared for a particular day, usually for the day that ends the organization's financial year.

This snapshot nature sets limits on what may be inferred from a balance sheet. For instance, if you walked into a business one day and saw masses of money being made using very little stock, this snapshot would make the business look profitable. Watching on another day, however, you may observe that things are quite different and that now that the company actually has a very poor rate of converting stock into profit. You would then have to revise your optimistic appraisal of the organization. For this reason, the balance sheet is not used on its own when assessing an organization but is considered together with other documents such as the profit and loss account.

The balance sheet takes its name from the fact that it has to balance two amounts of money. It operates on the principle that if you have something it must have been paid for somehow. For instance, if you own a building you must have once had some money with which you bought it. To represent this on a balance sheet one side would show where the money came from in the first place, perhaps a loan, while the other side of the sheet shows what has been done with it, in this case converted into an asset. The basic axiom of the balance sheet is therefore: 'Assets equal the sources of the funds.'

The balance sheet: preparation and understanding

To understand how a balance sheet is actually prepared in practice we shall now follow Jake Sounds from Section 4.2.1, as Jake prepares a balance sheet to sort out the mess he got into. To do this he went back to the beginning when he started the company.

Jake invested £1000 of his own money in the company. His company therefore had a source of funding and an asset. Jake makes up the balance sheet as follows:

Balance sheet for Jake Sounds on day 1

Assets	*£*	*Source of funds*	*£*
Cash in account	1000	Own capital	1000
	1000		1000

At the time Jake felt that this was not enough and persuaded his bank to invest a further £800. This happened on day 5 and the balance sheet was updated:

Balance sheet for Jake Sounds on day 5

Assets	*£*	*Source of funds*	*£*
Cash in account	1800	Own capital	1000
		Loan from bank	800
	1800		1800

Although there is £1800 in the account the balance sheet makes it clear that the asset is made up of two parts, Jake's own investment of capital which need not be repaid, and the bank loan, a liability, which will have to be repaid.

On day 9, Jake bought some stock: £450 was spent with a local shop. The effect was to reduce the company's cash while increasing the value of the stock. There was no effect on the sources of finance:

Balance sheet for Jake Sounds on day 9

Assets	*£*	*Source of funds*	*£*
Cash in account	1350	Own capital	1000
Stock	450	Loan from bank	800
	1800		1800

Jake then bought a further £850 worth of materials with his credit card. This meant receiving the stock, and paying the bill later. Of course the 'creditor' was owed money and the balance sheet must reflect this. Jake decided that the creditor must be a 'negative asset' because a consequence of acquiring a creditor is that money from the bank account will be required to pay off the bill and so it should be 'earmarked' for this use immediately.

Jake then noticed something interesting. A balance sheet will still mathematically balance, regardless of whether an existing item is included as a positive value on one side or as a negative one on the other. One could, for instance, show a creditor as a positive source of finance and still have a balance. In fact this would be an equally fair way of looking at the creditor since, like the bank who gave the loan, the creditor supplied an asset on condition that its value would be repaid later.

Jake realized that once a convention is adopted one must stick to it throughout the balance sheet if it is to remain consistent and he decided to regard creditors as negative assets. On day 15, stock increased by £850 but this was offset by £850 worth of creditors and the total assets of the business remained unchanged:

Balance sheet for Jake Sounds on day 15

Assets	£	Source of funds	£
Cash in account	1350	Own capital	1000
Stock	1300	Loan from bank	800
Trade creditors	(850)		
	1800		1800

(Note: It is usual to show negative amounts in brackets.)

The first customer took delivery on day 31 and the business acquired its first trade debtor. The selling price was £4500 and this amount must be entered on the left-hand side of the balance sheet since a debtor, being someone who owes you money, is clearly an asset. Jake almost forgot that the transaction also resulted in a loss of stock equal to the value of the components in the system.

Materials worth £1200 were used in the system. The balance sheet is now made up to take account of these transactions. Assets have risen by £4500 and stock has fallen by £1200. However, both these items are on the left-hand side, and since they are not equal, the balance sheet will no longer balance. Jake cannot see what has gone wrong but is convinced that a source of finance must be found to balance the change in assets. Finally, he realizes that the product has been sold at a profit. Profit is a source of finance, the very source of finance he is trying to bring about. The amount of profit is easily calculated by making up the left-hand side of the balance sheet and then entering the required amount to make the sheet balance. In this way Jake is sure that all the deductions that should be made have been included and the profit figure is accurate. The result of these actions is to increase assets by £4500, reduce stock by £1200 and create a profit to balance the sheet. A new and important category must therefore be entered on the balance sheet to display this profit and is shown below as 'Profit and Loss Account'. Clearly, it is possible for a loss to be made and in such a case the figure entered would be negative. This entry in the balance sheet should not be confused with the separate financial document of a similar name described below in Section 4.3.2. Although the two things are related they are, as we will see, different:

Balance sheet for Jake Sounds on day 31

Assets	£	*Source of funds*	£
Cash in account	1350	Own capital	1000
Stock	100	Loan from bank	800
Debtors	4500	Profit and loss account	3300
Trade creditors	(850)		
	5100		5100

Jake's wages were £960 for the first six weeks. The above balance sheet was made up for day 31 and the wages were not yet due and so the profit looks larger than it truly is, illustrating a management opportunity for 'dressing' the books. This can involve more than simply choosing a particular day on which to prepare the balance sheet. It can extend to running down stock near the year-end or squeezing credit arrangements together with an aggressive round of debt collection in order to make the balance sheet unusually favourable on the day of preparation. Because of such possible ambiguity in meaning, anyone assessing a company never depends solely upon the balance sheet but uses other information to support the assessment.

On day 40 the business receives payment from the customer (a debtor) after acceptance testing is complete. The supplier (a creditor) is also paid. At the same time, an appraisal is made of the remaining stock, which proves to be useless leftovers, changing the value of the company. An asset previously valued at £100 turned out to be worth nothing. If the value of stock has changed, something else must change too in order for the balance sheet to remain in balance. The simplest way to visualize this stock revaluation is to imagine that a product with a material cost of £100 has been sold for £0. The profit and loss account must therefore be reduced to pay for the loss.

The effect of all this is to increase the cash in hand by £4500, reduce the debtors by an equal amount, reduce the creditors by £850 and reduce the cash by the same amount. Stock is reduced by £100 and so is the profit and loss account:

Balance sheet for Jake Sounds on day 40

Assets	£	*Source of funds*	£
Cash in account	5000	Own capital	1000
		Profit and loss account	3200
Stock	0	Loan from bank	800
Debtors	0		
Trade creditors	0		
	5000		5000

Finally on day 43 the wages are paid. The effect of this is to reduce the amount in the bank by £960, which must clearly come from a source of finance, the profit and loss account:

Balance sheet for Jake Sounds on day 43

Assets	£	Source of funds	£
Cash in account	4040	Own capital	1000
		Profit and loss account	2240
Stock	0	Loan from bank	800
Debtors	0		
Trade creditors	0		
	4040		4040

We have now followed a balance sheet for a particular company through a process of money creation. Initially the company purchased raw materials, work was performed with the expectation of payment, a sale was made and finally, cash was received. The cash was used to pay off the creditors and the remainder was kept as profit. In reality, organizations do not make up a new balance sheet every time a transaction is conducted and the balance sheet contains much more information than in the simple example above.

Five years later Jake Sounds is doing very well and Jake has prepared the balance sheet for the annual accounts shown below. Like all balance sheets, assets equal the sources of their finance. This balance sheet also shows last year's figures for an easy assessment of performance. There are many legal requirements imposed upon the preparation of an audited balance sheet and the books listed in the further reading section at the end of the chapter include an explanation of these.

Jake explains what each section means in the following paragraphs:

(a) *Tangible assets*: These are under the general heading of 'Fixed assets' and so list the physical assets such as plant, machinery or buildings. Depreciation is allowed for and the value of a given asset falls with time, the deprecation being paid for from the profit and loss account. Occasionally, particularly when a company is being sold, a financial term called 'Goodwill' is listed is a fixed asset. Goodwill reflects the increased worth of a business, over and above its asset value, by virtue of some particular benefit. This might take the form of an especially loyal customer base who are prepared to pay over the market rates for the organization's products. The consequence of such an advantage is that the business is really worth more than the basic asset value since it has the capacity to generate more profit than would normally be expected for such a business. Goodwill is therefore defined as the difference between the sale price and net worth (net assets) of a company. In non-accounting terms this is the difference between what a buyer is prepared to pay and what the books indicate the company is worth. This difference is quantified at the time of sale and is recorded as 'Goodwill'.

Balance sheet for Jake Sounds at 30 April year 5

		Year 5 £k	Year 4 £k
	Fixed assets		
(a)	Tangible assets	35.6	28.6
(b)	Investments	4.1	2.6
(c)		39.7	31.2
	Current assets		
(d)	Stocks	45.6	41.3
(e)	Debtors	15.6	14.3
(f)	Cash at bank and in hand	8.9	6.9
		70.1	62.5
	Current liabilities: amounts due within 1 year		
(g)	Debenture loans	0.5	1.2
(h)	Creditors	20.5	17.4
(i)	Net current assets	49.1	43.9
	Total assets less current		
(j)	liabilities	88.8	75.1
	Liabilities: due after more than 1 year		
(k)	Finance loans	1.6	1.8
(l)	Creditors	0.8	0.6
(m)	Provisions for liabilities	7.5	6.8
(n)		78.9	65.9
	Capital and reserves		
(o)	Called-up share capital	17.3	16.1
(p)	Reserves	0.5	0.3
(q)	Share premium account	7.6	7.1
(r)	Profit and loss account	53.5	42.4
(s)		78.9	65.9

(b) *Investments*: Listed under 'Fixed assets' these relate to long-term investments, usually meaning longer than one financial period, such as loans made to other companies. Short-term investments, also known as 'money market investments', are shown as current assets.

(c) *Total fixed assets*: The sum of (a) and (b) above.

(d) *Stocks*: The stock of a company is all the useful components and raw materials used in manufacturing the product and owned by the company. Some stock items devalue with time and, where applicable, correction must be made for this. Work In Progress (WIP), means materials and equipment that are not

contained within the stores area because they are actually being worked on at that moment. Material in this situation must be accounted for and it is included in the value for stocks.

(e) Debtors: Someone who is shortly to pay you money is clearly a current asset. It is common to find a small percentage rate of non-payment, so-called 'bad debts' and a 'provision' for this may be made, (see (m) below).

(f) Cash in hand: No possession can more clearly be a current asset than cash in hand or at the bank!

(g) Debenture loans: A debenture loan, like a share issue, is a way for the company to raise money. The company enters into an agreement with the lender to repay money at a fixed rate over a period of time in return for a lump sum. Where an asset of the company is offered as collateral to the lender the debenture has become a mortgage. In this case the lender takes possession of the asset in the event of a default on the loan repayments. A debenture is issued for a specified period of time. The length of time will decide whether the debenture should be listed under current liabilities (as here) or below (see (l) and (m)), although debenture loans are normally long term. They can also be bought and sold on financial markets, just as shares can.

(h) Creditors: Creditors are people to whom the organization owes money. The normal period of credit is 30 days. Creditors, together with debenture loans, are normally regarded as 'negative assets' in the sense that they must soon be paid for and will consume cash. Such 'negative assets' are called 'liabilities'.

(i) Net current assets: The net current assets are calculated by adding or subtracting from the current assets of the business the amount currently outstanding. This figure is of interest since it represents the pool of finance from which creditors can be paid. A company can have great wealth tied up in its long-term assets but still find difficulty meeting its operating expenses if its net current assets, sometimes called 'working capital', are too small.

(j) Total assets less current liabilities: This number is the total assets of the company less the liabilities falling due this financial period. It is the sum of (i) and (c). It is of interest to potential purchasers of the company since it gives an indication of the value of the company today, without including money owed in the long term which might be rescheduled or renegotiated.

(k) Finance loans: One common source of funds is a long-term finance loan. Such a loan is classed a negative asset, rather than a source of finance, since it has associated repayment. Often such a loan will have scheduled repayments lasting many years.

(l) Creditors: It is quite possible to acquire long-term creditors in the same way as short-term ones. The purchase of major machinery, for example, frequently involves paying the supplier over some years.

(m) Provision for liabilities: Prudent management will make an allowance this year for impending liabilities next year. This entry on the balance sheet is used to show money set aside for future liabilities that have arisen out of past transactions and are already known. Provisions for liabilities are also called 'contingent liabilities'.

(n) Net assets (sometimes called 'total assets'): The title of this entry is self-evident given the nature of the balance sheet and is not usually given a heading.

(o) Called-up share capital: In this section of the balance sheet the source of the finance used to purchase the assets above is made clear. A common way for an organization to raise money is to issue shares. The money so raised is shown as called-up share capital. This account contains capital raised from the sale of shares at their nominal value.

(p) Reserves: The total reserves of a company consist of the share premium account and the reserves transferred out of profit at the discretion of the directors. Together they comprise the shareholders' funds. This, along with the share capital, is the value of the shareholders' investment in the company. Reserves also include balancing amounts of money caused by, for example, revaluation of property.

(q) Share premium account: This shows the amount of money raised by issuing shares at a price above their nominal value. Such shares are 'issued at a premium', hence the name. Double-counting must be avoided and so the money raised by share sales at the nominal value is included elsewhere.

(r) Profit and loss account: Once the shareholders have ratified the decision of the board of directors concerning the amount of any share dividends which are paid for in cash there will be an amount left to balance the balance sheet. This will be the profit and loss figure for that particular day. This is calculated as (n) minus the sum of (o), (p) and (q). This number is distinct from the separate financial document called 'profit and loss account', sometimes called the 'profit and loss, and appropriation account'. In the first financial year the value on the balance sheet will be the same as the retained profit figure from the profit and loss account. In subsequent years the figure in the balance sheet will show the cumulative amount of money raised by the organization over its trading life, while the profit and loss account will indicate the profit generated and retained during only the previous year. Just as the share entries show the total sources of finance raised by share issue, the profit and loss entry of the balance sheet shows the total sources of finance raised by profitable trading.

(s) The total sources of finance: Like the net assets the title of this entry is not usually printed out. It is, by definition, equal to (n).

4.3.2 Profit and loss, and appropriation account

It is the objective of most traders to make a profit. They aim to do this by buying things, doing something to them to make them more valuable and then selling them at a sufficiently high price to cover not only the materials and labour costs of manufacture but also the costs of running the business. The money left over after these have been paid is profit. Profit may be disposed of however the organization desires, perhaps as payments to shareholders, purchase of new assets for the company or bonuses to employees. It is usual for the owners to employ the profit in a way that benefits their organization. The profit and loss account is the financial document which shows how much profit was made during the last financial period.

Jake has also prepared a profit and loss account for year five of Jake Sounds, and this is shown below. Unlike the balance sheet, the profit and loss account applies to a period of time and shows the cumulative effect of all the transactions over a period, usually a year. It records how much has been spent on, and how much has been received from, sales. Jake's account starts with turnover (all incomings) and subtracts net operating costs (all outgoings) to leave the profit. The rest of the account shows what happened to the profit ending with the amount retained by the organization for reinvestment. Dividends are payments made to shareholders. The profit retained by Jake Sounds in year 5 can be seen to agree with the balance sheet figures and is equal to the difference between the profit figures for year 5 and year 4 (53.5 − 42.4 = 11.1).

Profit and loss account for Jake Sounds at end-year 5

	Year 5 $£k$	Year 4 $£k$
Turnover	90.7	83.8
Net operating costs	71.5	66.6
Profit before taxation	19.2	17.2
Taxation	6.2	5.8
Profit for the financial period	13.0	11.4
Dividends	1.9	1.7
Profit retained	11.1	9.7

The balance sheet and profit and loss account together form a meaningful summary of the organization's financial activities. English law requires that both documents are produced and contained in the annual directors' report.

4.3.3 Cash flow projection

During his fifth year of trading Jake plans to buy new premises for the business. He has been operating out of his own house since the company began and now finds that the business is just too big to stay there.

Jake decides to borrow the money for this from the bank. They have asked him for some sort of justification for the loan and want to be convinced that Jake will be able to repay the loan together with the interest. To do this Jake provides copies of the balance sheet, profit and loss account, a report on how the new building will be used, and a cash flow projection. This last document is of greatest interest to the bank.

Cash flow projection for Jake Sounds

Year	Sales	Cost of sales	Profit after tax	Loan repayment	Loan remaining
5	–	–	–	–	30
6	105	80	19	2	28
7	125	95	22	10	18
8	130	89	33	12	6
9	130	87	35	6	–
10	130	87	35	–	–

In the above cash flow projection one can quickly see how many sales have to made to allow Jake to repay the loan. The amount of profit consumed by the loan is clear and can be seen to be acceptably small. From a table such as this it is possible for the bank to cross-examine Jake and find out the likelihood of his being able to repay the loan through their thorough understanding of business.

In general, cash flow projections are used to cover any situation where the distribution of cash is important. It is important to realize that it is perfectly possible for a healthy business to become insolvent simply by not making sure that sufficient cash is available where it is needed. In the same way that an army must not only have sufficient food but must have it distributed appropriately among its soldiers, an organization must not only have enough cash, but must have it distributed correctly to meet the expenditures for which it is required. If, for instance, there is not enough cash to meet the creditors who require imminent payment, insolvency may result even though the value of the whole company far exceeds the amount it owes.

There will often be intense debate about the numbers in the cash flow projection, particularly the estimated number of sales. A bank making a loan will be keen to understand what evidence the organization has for believing the predictions. A statement based on market surveys, previous experience, an analysis of competitors and customer research is much more likely to convince a lender to part with money than numbers made up by irrationally optimistic owners of new businesses.

4.3.4 Financial accounting ratios

We have seen how a balance sheet and profit and loss account can be prepared. We now have the ability to understand something of an organization simply by looking at its published accounts. There are two parties who, in general, are interested in using the financial documents of a company to make decisions. The first is the organization itself. The senior management of an organization may obtain copies of the financial documents of other organizations for comparison. They are concerned that their own organization is going to fare well and seek to ensure that good decisions are taken by understanding the behaviour of their rivals. The second group of people are those interested in collaborating with the organization in some way. For instance, these might be investors anxious that their investment will be profitable or be customers wanting to confirm the financial stability of a supplier.

The published financial documents of an organization contain much information, and the interpretation of them is a skilful process. To assist in analyzing financial documents various ratios of the numbers contained within them have become commonly accepted as measures of important characteristics. The ratios are important in two ways, first, in absolute magnitude which betrays current performance, and second, their trend with time, which indicates likely future performance. Every section of industry has its own 'normal' values for these ratios and the relevance of comparing one organization with another falls as the commercial differences between them increase.

There are a great many such ratios, too many to consider in this text. However, we will look at some of the most important, both as an introduction and to illustrate how ratio analysis in general operates.

- *Return on capital employed (ROCE)*: This ratio is defined as:

$$\frac{\text{Pre-tax profit}}{\text{Capital employed}}$$

Pre-tax profit can be read directly from the profit and loss account. It is common to use a profit figure before taxation and before interest in this ratio since this removes the investment performance of the organization and assesses more directly the ROCE. The capital employed means the capital used in the creation of that profit, which is defined as total assets less current liabilities. The ratio is clearly of interest both to the organization and to those who invest in it in any way. It indicates how much money was used in the generation of the profit. If the ratio is very large, the organization makes a lot of money without using much in the process. If it is smaller than the prevailing rate of interest, investors would be better advised to place their money in a building society. If it is below the rate of inflation, the value of an investment in the organization is actually decreasing.

- *Profit margin*: This ratio is a basic performance indicator of the potential for the ROCE. It is defined as:

$$\frac{\text{Trading profit}}{\text{Value of sales}}$$

If the profit margin is high then each sale is generating much profit and if the rest of the company is efficient then the ROCE should be good. If it is low, then no matter how efficient the administration of the organization, a high ROCE is unachievable. The ratio can be expressed in many ways and is improved by making it relate as closely as possible to the profit from sales without including any other activities of the business. Because of this, the 'gross profit margin' is often used, which is defined as:

$$\frac{\text{Value of sales} - \text{cost of sales}}{\text{Sales value}}$$

- *Stock turnover*: This is defined as:

$$\frac{\text{Sales}}{\text{Average inventory}}$$

This ratio indicates how much time stock is typically spending within the organization. Sometimes this is described as how fast the stock is 'being turned'. This factor has much importance in production control, financial efficiency and the market responsiveness of the organization. In some industries it may be very fast, with the entire stock being purged every few weeks, in others it is much slower. In general, the ratio is very dependent on the nature of the business, but all organizations seek to minimize it since a large value implies that a lot of stock is being held which is financially less efficient.

- *Gearing*: This ratio is defined as:

$$\frac{\text{Fixed-term loans}}{\text{Net assets}}$$

This ratio is used to express how heavily an organization is committed to repaying borrowed money. When the ratio is high, most of the profit is being diverted into repaying loans. If the ratio equals zero, the organization has no repayment commitments and is entirely self-funded.

- *Liquidity ratio*: Sometimes this is called the 'acid test' and is defined as:

$$\frac{\text{Current assets} - \text{stocks}}{\text{Current liabilities}}$$

This ratio measures the organization's ability to meet incoming debts. The top line indicates ready cash and the bottom, incoming bills. In accounting, 'liquid' is used to mean 'readily converted into cash'. The current assets (also called liquid assets) may be defined as either cash plus debtors plus short-term investments or current assets less stocks less prepayments. If the ratio is less than one, it is almost certain that the organization will receive more incoming bills than the ready cash can pay. This can lead to instant insolvency. A value of about two is considered safe. Much higher than this and the organization is unnecessarily well covered and probably has capital tied up in its bank account not doing any useful work.

- *Output per employee*: Many versions of this ratio are used, and a common one is:

$$\frac{\text{Value of sales}}{\text{Number of employees}}$$

In general, all the ratios provide a crude overall indicator of the organization's efficiency which allows comparison with competitors. The ratio values are very dependent on the nature of the business and all markets have their own minimum safe levels that organizations should exceed in order to be financially viable in the long term.

4.4 Budgeting

4.4.1 Master budgets

We have now seen how important the financial condition of an organization is. In order to achieve its mission statement it is crucial that the financial arrangements of the organization serve its needs. Such arrangements will not just happen alone and they must be planned for. The organization must define the financial position it intends to achieve and then head towards it. It is not sufficient to simply aim to make each sale at a certain profit or perhaps to have a target output per employee. These are certainly sensible things to control but in themselves are not enough. To be successful, the organization must start with comprehensive plans that make clear the route to the desired value of all the important financial parameters. Of course, many things may go wrong along the way and simply having the plans by no means guarantees success. However, the organization is much more likely to achieve its goals if good plans are prepared first. Financial plans of this kind are called budgets.

A master budget is the starting point and is the name given to a balance sheet, profit and loss account and a cash budget which have been prepared, as a group, to describe a desired future financial position. A cash budget is a projection of how much money will be in the bank account over time as a result of all the activities that the organization is planning to undertake. From this budget, important issues such as the desired asset value of the organization, the desired sales' value, the desired ROCE, and all the other important values can be seen.

Once prepared, the master budget enables the objectives of each department to be quantified. A self-consistent group of objectives can be prepared which, when complete, will result in the desired financial situation. In this way the organization's finances are directed at achieving the mission statement. This is clearly a superior approach to simply undertaking a lot of financially profitable ventures and seeing how much money can be generated.

4.4.2 Re-employing the profit

One of the figures within the master budget needs particularly careful consideration – the profit. Profit is usually re-employed in the organization to assist in achieving the company's goals. The organization will normally have many exciting possible enterprises into which resources could profitably be directed. It is a question not of finding something profitable to do with the profit but of choosing wisely a portfolio of the activities that will provide the greatest likelihood of achieving the mission statement. Once the master budget has been prepared and the budgeted profit is known, the organization must choose exactly how this money will be re-employed as it is generated.

Case study

Every year the Tachatronics Corporation have an intensive month of budgeting in which plans for the next year are agreed. Last year was good and prior to the 'budget month' the management met and expanded the mission statement to include the Japanese market from which they had previously excluded themselves. Next year's budgeted profit is £400 000.

The company's five divisional managers put in preliminary budgets, each manoeuvring for the most cash, totalling £615 000. The budgets are for money to spend on activities that meet the departmental objectives and do not include salaries. The senior management decide that to penetrate the new Japanese market a new 'Oriental Sales Division' employing four people should be created. The five existing divisions are Sales, Production, Research, Service and Administration. Money cannot be spared from the modest Service or Administration budgets and so priorities must be set between the remaining departments, first, to allow for the new Oriental Sales Division and second, to bring the total budgeted expenditure of profit within the £400 000 available.

The senior management consider the provisional plans of each department. The production division has several projects in progress. First on the list is the capital-intensive acquisition of a new blanking machine tool while lower down is a value engineering programme. Sales forecasts are not as large as when the new machine was first considered. The Research division have had four major programmes in progress and plan to continue all of them with increased expenditure, arguing that they will all produce products. Many such details are considered and tough decisions about which projects will stay and which will go are agreed. The decisions are not unanimous and the board have to face the fact that they cannot afford to follow all the interesting avenues that have been presented. After several days, a package of objectives is chosen for the next twelve months, and the following draft divisional budgets then result:

- *Sales £54000:* Objective: close contracts worth 10% more than last year's contracts while using the same number of staff. This is to be achieved through the introduction of new technology to the division.
- *Oriental Sales £57 000:* Objective: close four contracts in Japan for existing products. Do this by identifying an agent already trading and enter the market by collaborating.
- *Production £115 000:* Objective: complete all deliveries within two months of receiving sales order and cut direct production costs by 7% at the half-year by introduction of value engineering programme.
- *Research £105 000:* Objective: produce two complete products, including specifications, drawings, test data and any other documentation required to specify the products completely. The two are to be chosen from the five currently in progress. This selection is to be done in conjunction with Sales to identify best choices and complete reports should be written on the other three. The first product is to be completed by the end of month 4, the second by the end of month 9.
- *Service £59 000:* Objective: answer all calls within two hours. Have an engineer on-site within 24 hours for Europe and the USA. All breakdowns to be repaired in two days of service engineer's time.
- *Admin £10 000:* Objective: complete all administrative tasks within one day of each request. Ensure payroll activities run error-free for the whole year.
- TOTAL: £400 000

The divisional managers discuss the draft corporate plans and after two weeks of intensive planning and organizing with their teams they put forward some modifications, particularly to time scales of introduction. The managers have each divided the objectives into separate projects and have plans for each. The plans are presented to the board in turn at which some further modifications are made, but within the month plans for every division have been agreed in detail. Senior management are confident that the objectives serve their mission statement and the divisional managers are clear both about what has to be done and what financial resources are at their disposal.

4.4.3 Individual budgets

Once divisional budgets have been prepared in the way described above, all the departmental managers have their own resources allocated in an indisputable way. Every manager knows what is expected of their own department and every manager knows how much money is available for use in achieving their objectives.

Chapter 8 deals with the planning of projects and describes some techniques for managing them. By using these project planning techniques and estimating the cost of each undertaking a departmental manager is able to produce an estimate of the total financial requirements of the projects. Such estimates are the basis of the budget allocated to the department. As the financial year progresses, each department directs its resources at achieving its objectives. If all departments are successful the objectives of the organization are met. Of course, things never go exactly as predicted and it may become necessary to readjust budgets during the year. This does not mean that the budgeting process was in some way wrong or is pointless, rather it shows that it is difficult to be accurate when planning the future and it is certainly true that it is better to know exactly what it is that one is changing and why.

4.5 Management accounting

In this section we will look at some of the financial data and information generated and used by engineers in the areas of costing and investment appraisal. We will also consider what is meant by depreciation and how it is calculated. These are the most important topics in management accounting. The information that is generated through management accounting techniques is not normally divulged to people outside the company since it is usually sensitive. Knowledge of issues such as pricing structures or cost of manufacture could be used effectively by the organization's competitors, or customers, to undermine profitability.

4.5.1 Costing

Costing is the process of calculating how much something costs to make. The most important use of this information is in determining a realistic selling price, usually with the aim of making a profit. In addition, it allows you to determine which product areas are worth investing in and provides information on which to base management decisions relating to making or buying-in parts. It also allows you to decide which products need further development to give them the desired financial profile. The term 'costing' generally applies to existing products where manufacturing techniques are known and real data may be used. Before this time, costs are prepared by 'estimating'.

The role of engineers in costing and estimating can be very significant, depending upon the type of company and products that are involved. Invariably, if you design a product you will have to calculate how much it is going to cost to make so that the company can decide whether or not to proceed with it. In a company that manufactures customized products an engineer may also be involved with estimating costs for customer quotations.

Costing is not as straightforward as it might at first appear. Most people instantly think of the sum of the components of cost, materials and labour, for example, but these by no means explain how much something costs to make. What about necessary services that do not directly contribute to the product (lighting and heating, for example)? One might suggest that these should be totalled up and then distributed evenly among the products of the organization but what if the products consume them unevenly or some not at all? Costing requires that you consider all the factors affecting the cost of the product, i.e. the amount of labour involved in its manufacture, the parts that go into it, any services that have to be bought in so that manufacture may proceed, and how much it is costing to keep the factory operating in order to make that product. We will now explain two methods that are used for calculating costings. The first is based on direct and indirect costs, the second on variable and fixed costs.

Direct and indirect costs

In this method the cost of a product is divided into two components: direct costs and the indirect. The total price for the product is calculated by first producing a cost for each element. In the first case this is relatively straightforward but in the second it is more complex since it is necessary to find a fair way to distribute the indirect costs (e.g. lighting and heating bills) to the different products. The total cost of the product is produced by adding the two components together.

● *Direct costs*: Direct costs are those that can be directly attributed to the product for which a cost is being prepared, and can be considered under the headings of labour, materials and expense. Direct labour is that usually done by operatives or technicians who book their time to the job that they are working on. Costing of direct labour will involve determining the amount of time spent on that product and then deciding what labour rate applies. The time information can be gathered by reviewing time sheets or route cards. The labour rate is often more difficult because of the variability in the wages that different people will be paid. One way of dealing with this problem is to calculate a standard labour rate, based on the average wage bill, which will then be applicable for different processes or workshops. Some companies simplify this further by having one labour rate for all direct labour in the factory.

Direct materials are those which go only into that product. For example, if you imagine a china cup, the direct material will be so many grams of china clay. The direct materials will be detailed in the bill of materials prepared for the product or part. Direct material cost can be determined by considering the cost of making or buying the actual parts that were used in that product, although this may entail some rather onerous tracking of parts. Alternatively, when you are using parts from stock which may have been purchased at different times and at different prices you can use standardized costs. Cost standardization can be on the basis of the cost of the last delivery of parts, although this may lead to discrepancies where there are large fluctuations in price, or it could be on the average cost of stock held.

Direct expense relates to costs incurred by buying in services for that particular product and so would include subcontract machining costs, printing costs, finishing costs and so on.

EXAMPLE

Calculate the direct costs involved in the manufacture of product A from the information below:

Material costs	£5.50
Standard labour rate	£7.50/hour
Finishing costs	£2.00
Total processing times	Turning centre 14 minutes
	Assembly 3 minutes

Answer:

$$\text{Direct labour} = \frac{17}{60} \times 7.50 = £2.13$$

Total direct costs $= \text{materials} + \text{labour} + \text{expense}$
$= £5.50 + £2.13 + £2.00$
$= £9.63$

● *Indirect costs*: The indirect costs, also called 'overheads', are the costs that are associated with operating the business but that cannot be directly assigned to product or services being provided. As with direct costs there are three categories of overhead: materials, labour, and expense. For example, indirect materials might include things like solder flux, rags and cutting fluid. Indirect labour would include design engineers, supervisors and secretaries. Indirect expense would include heating, lighting and rent for premises. These are all costs to the business and the way in which funds will then be acquired to cover them is called overhead recovery and requires a two-stage process which ends up with a charge for overheads being made on products. The two stages involved are apportionment to cost centres and absorption by product.

● *Apportionment of overhead to cost centres*: A cost centre is a part of the business that can be identified for the purpose of determining costs. A cost centre may be a whole factory, a particular department, or a specific machine. Each cost centre should bear a fair share of the indirect costs with the costs being apportioned on the most realistic and convenient basis that can be devised. Typical bases for apportionment might be floor area for apportioning rent, the number of employees for indirect labour and personnel costs, and the book value of assets for depreciation and insurance.

EXAMPLE

Apportion the overheads for a company having the following departments:

	Department overhead	People	Floor area (m²)	Direct materials
Personnel	£50 000	4		
Purchasing & Stores	£75 000	6		
Building Services	£20 000	3		
Light Machine Shop	£85 000	12	500	£200 000
Assembly Shop	£65 000	20	1 000	£130 000
Press Shop	£90 000	12	1 000	£150 000
Totals	£385 000	57	2 500	£480 000

The only areas that would be directly involved with product are the three workshops and therefore the overheads for the other areas must be apportioned to these.

Considering personnel:
This has an overhead cost of £50 000 and because it is dealing with people in the factory a fair way to apportion this would be on the basis of the number of staff in each area. There are 53 people in the company, excluding Personnel, and the overhead is therefore distributed on the following basis:

Purchasing & Stores: $\dfrac{6}{53} \times 50\ 000 =$ £5 660

Building Services: $\dfrac{3}{53} \times 50\ 000 =$ £2 830

Light Machine Shop: $\dfrac{12}{53} \times 50\ 000 =$ £11 321

Assembly Shop: $\dfrac{20}{53} \times 50\ 000 =$ £18 868

Press Shop: $\dfrac{12}{53} \times 50\ 000 =$ £11 321

This apportionment now gives the total overheads for the Purchasing & Stores as £80 660, and Building Services as £22 830. These will be apportioned to the three workshops.
     The Purchasing & Stores overhead will be apportioned on the basis of

direct material cost and the Building Services overhead on the basis of floor area. This gives:

	Department overhead	*Purchasing overhead*	*Building overhead*	*Total overhead*
Light Machine Shop	£96 321	£33 608	£4 566	£134 495
Assembly Shop	£83 868	£21 846	£9 132	£114 846
Press Shop	£101 321	£25 206	£9 132	£135 659
Total				£385 000

All the overheads are now apportioned to production departments from which they can be included as a cost of the product.

● *Absorption of overhead by product*: In order to include the indirect costs in the product costing they are charged to individual products passing through the cost centre to which they have been apportioned. This is called overhead absorption because the costs are absorbed by the product cost. To calculate overhead absorption an absorption rate, also called overhead recovery rate, is calculated which is based on a standard or expected rate, or volume, of production. This recovery rate is then applied to the actual rate of production. Common bases for overhead absorption are direct labour (hours or cost), units of output, and machine hours. These would be used by calculating appropriate recovery rates as described below.

● *Material rate*: In this method the overhead is divided by the actual or expected cost of the direct materials.

EXAMPLE

For a light machine shop the total annual cost of direct materials is expected to be £200 000 and the annual overhead to be absorbed is £134 495. Thus we can calculate a recovery rate as:

$$\frac{£134\ 495}{£200\ 000} = 0.672$$

Therefore for every job done in this shop we add an overhead charge of £0.672 for every £1.00 of direct materials. Thus, for job ABC:

Direct labour	= £4.25
Direct materials	= £7.25
Overhead (£7.25 × 0.672)	= £4.87
Total cost	= £16.37

- *Labour hour rate*: In this method the overhead is divided by the total labour hours estimated.

EXAMPLE

The annual overhead for an assembly shop is £114 846 and the estimated number of person-hours to be worked in a year is 35 520. Therefore the labour-hour overhead recovery rate will be

$$\frac{£114\ 846}{35\ 520} = £3.23 \text{ per labour hour}$$

Thus for job ABC:

Direct materials	= £7.25
Direct labour	= £4.25
(30 minutes)	
Overhead $\dfrac{30}{60} \times 3.23$	= £1.62
Total	= £13.12

- *Machine-hour rate*: In this method the costs of owning a machine are to be absorbed by charging them per hour of operation expected.

EXAMPLE

The running costs of a turning centre are as follows:

Power	= £0.25 per hour
Consumables	= £400 per year
Maintenance	= £850 per year
Tooling	= £3500 per year

In addition, the machine is depreciated at £4000 per year. The expected run time of the machine is 242 days per year with one eight-hour shift each day:

Machine costs per year	= 400 + 850 + 3500 + 4000
	= £8750
Total hours per year	= 242 × 8 = 1936
Machine overhead recovery rate	= $\dfrac{8750}{1936} + 0.25$
	= £4.77 per hour

Thus for job ABC:

Direct materials	= £7.25
Direct labour	= £4.25
Machining time 30 minutes	
Overhead $\dfrac{30}{60} \times 4.77$	= £2.39
Total cost	= £13.89

Full costing

A full costing is the cost of a product based on both direct costs and overhead costs, as shown in the previous examples. It is an onerous and time-consuming task, particularly because of the overhead apportionment and absorption calculations and the subjective way in which bases for these calculations are chosen. However, it is only by following such a procedure that a rational answer can be given to the question 'how much does the product cost to make?'

4.5.2 Marginal costing

We have looked above at how the real cost of a product may be calculated, a cost that includes not only the direct elements but also the indirect ones. Sometimes the cost of a product is divided in a different way, into 'fixed' and 'variable' costs. The variable costs are those that vary in proportion to the amount of output and so would include direct materials, use of temporary labour, and overtime charges. The fixed costs are those which have to be paid irrespective of output and would include the cost of employing people and having the premises and machinery available.

By considering the fixed and variable costs associated with manufacture you can determine the break-even point for the manufacture of any product. This is the point at which the income from sales is equal to the cost of manufacture, where no profit or loss is made, and the company breaks even. A break-even chart is shown in Figure 4.1. Such a chart is used to display the profitability of a particular product and decide whether or not its manufacture should continue. When the costs of a product are analyzed in terms of fixed and variable costs the process is described as a 'marginal costing'. As we shall see below, this different way of looking at costs can be very valuable.

The break-even point can be calculated using the formula:

$$N = \frac{F}{S - V}$$

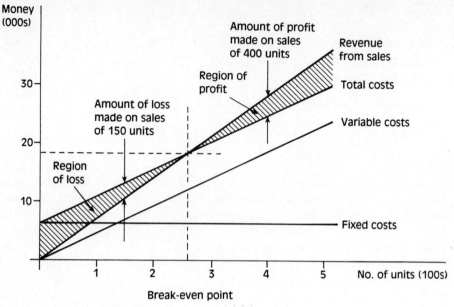

Figure 4.1   Break-even chart

where

> N = number of sales to break even
> F = total fixed costs
> S = selling price per unit
> V = variable cost per unit

To understand the advantage that marginal costing offers over full costing we shall consider an example. Imagine a manufacturing company who makes two products, A and B. The costs for these two products are shown in Table 4.1.

The managers have the option to increase production since sales of both products are going well and a factory reorganization has liberated a little more space. The question is then which product should be put forward for an increase in production. Naturally, the managers will wish to increase production of the

Table 4.1

Product	A	B
Fixed costs (£/unit/period)	5000	7500
Variable costs (£/unit)	6.5	6
Current volume (units per period)	2000	1900
Sales price	13	14

most profitable product. This is calculated by first taking the total money received from sales and then subtracting the cost of making those sales. The cost of sales comes in two parts, the fixed and variable costs (see Table 4.2). By this means the choice is simple. Product A makes a more sensible candidate for a production volume increase since it is the most profitable.

In fact, however, for our example this answer must be wrong. Look again at the data in Table 4.1. Product B has lower variable costs than product A and, what's more, it sells for a higher price. These two facts mean that each unit of B sold will produce more profit than each of A. Yet the result in Table 4.2 shows that A is more profitable. How can this be? The answer is simply that in the above analysis the fixed costs are considered over the whole batch. Each unit of production is made equally responsible for paying the fixed costs. Accountants will argue that this is the wrong way to look at what is happening. In fact what happens is that the product produces an amount of money equal to the difference between the revenue received from sales and the variable costs. Initially this is used to pay off the fixed costs; but, once these are paid off, it is all profit. Accountants call this amount of money 'contribution'. When looked at in this way, it is clear that the managers in our example need to choose the product whose contribution is the greater. For our example these are shown in Table 4.3.

The analysis in Table 4.3 clearly shows that in fact product B makes a much better choice for an increase in production volume. However, one must be certain that the products in question are in fact beyond the break-even point. It would clearly be foolish to suggest expanding the production of a product that was operating at a loss. (Doing so would certainly reduce the loss being made and might even be enough to bring the product into profit. However, products that are as unprofitable as this are usually selected for more radical treatment than a change in production volume.) This can be checked by referring to the analysis in

Table 4.2

For product A:

$$\text{Profit} = (2000 * 13) - (2000 * 6.5) - 5000$$
$$= 8000$$

For product B:

$$\text{Profit} = (1900 * 14) - (1900 * 6) - 7500$$
$$= 7700$$

Table 4.3

For product A: Contribution = £13 − 6.5
= £6.5

For product B: Contribution = £14 − 6
= £8

Table 4.2 from which it is clear that, indeed, both products are making a profit overall.

Marginal costing is therefore an alternative to full costing and is in some ways a better one. Unless the costs of production are prepared as fixed and variable costs (instead of direct and indirect costs as in Section 4.5.1) it is not possible to calculate contribution easily. Its big advantage is that it accounts for the fact that when one changes a production batch size it is the profit or loss associated with the last unit of manufacture that should be used to judge the merit of the change and not the resulting average of the whole batch. On the other hand, one cannot get away from the fact that the fixed costs of production have to be met and full costing presents this aspect more clearly. Any system of accounting that looks only at marginal costing can place so much emphasis on contribution that simple determination of whether a product is breaking even or not can be forgotten.

The important issue is being clear about how the accounting data was prepared and what may therefore be inferred from it. Engineers would not dream of putting forward a product for manufacture without having measurements that confirm the device operates as designed. Similarly, accountants will not present a financial view of performance without first having taken the correct measurements and treated them in the correct way.

4.5.3 Case study – Jordan Greenmore

Jordan is a graduate in engineering working for a company called SAE Instrumentation. She did a degree in electronic engineering at Leeds University and is now employed as a member of the engineering design team at SAE, which makes a range of analytical instruments. The company has two major product ranges, one for gas analysis and the other for flame spectroscopy of organic solids. The customers range from research laboratories and departments buying one-offs to OEM customers who purchase many tens of units a year and then incorporate the product into their own equipment.

Jordan was employed when her predecessor left to go to America and she inherited a product that was in considerable disarray. The in-line photo-metric analysis (IPMA) system is SAE's rather weak attempt to attack the in-line market. This market area is heavily covered by SAE's competition and SAE really only entered the market ten years ago because it felt that as a major supplier of analytical equipment it should have a presence in this area. The difficulty was that the in-line techniques used in the product are not like SAE's other products, and with little experience of them and a strong competition it was not surprising that results were not astounding.

Jordan's boss has seen sales creep along and has felt for some time that the product might do much better if given some real support. He finally won his battle last year and the board approved a year of engineering time to spend developing the product as he thought best, but with particular emphasis on cost reduction since the unit was expensive when compared with the competition. The result was a year's work for Jordan on the IPMA.

In the past the machine has been made in batches of two or three. The heart of the system, the photo-multiplier and transport assembly, is reasonably well documented and can be built reliably. The rest is virtually done by the technicians who have a file of the components used such as the enclosure, the support, the control electronics and test procedures, etc.

Jordan and her boss agreed at the start that the main problem with the product was cost. The performance of the unit was as good as that of the opposition but it was far more expensive. In fact, the unit's performance was significantly better from a resolution point of view but in the application for which it is used the extra resolution is not really necessary. Consequently, only those who had some special reason for buying SAE equipment became customers. The reason was that the analysis electronics that gave the resolution was taken straight from another product and it was thought cheaper to do this than have a separate, lower-resolution, cheaper unit for the IPMA.

Jordan's first job was then to look at the product costs. The company used a 'prime cost model' for costings. Basically this means that you add up all the direct costs of the product, material and labour and call this the 'prime cost'. This prime cost is then expressed as a percentage of selling price. The company model says that products should have a prime cost of about 40%. Overheads are assumed to be equal to prime cost and these are therefore typically 40% as well. This leaves a model of 20% for profit. Naturally, these ideals vary a little from product to product. When this model is applied to the IPMA a prime cost of 47% is obtained and this means that profit is only 6%, clearly too low. Everyone knew this and many people thought the IPMA should be dropped from the product range and the spare effort employed on more profitable projects.

Jordan first prepared a bill of materials which took nearly three months to complete. She used the existing data for the analysis electronics, and the photo-multiplier and transport system. The technician's files were useful for much of the rest but a lot had to be obtained simply by talking to the people involved and finding out what they actually did. After three months she had a bill of materials broken into subassemblies and had worked out the cost for each item. The results made interesting reading and she was asked to make a presentation with her boss to the technical director and a senior accountant.

In the presentation they explained that since much of the product was bought in the existing company model of prime cost was not appropriate. The model worked well for other products where the vast majority of component manufacture was done in-house. The IPMA was not like this, nearly 70% of the value of the components was from bought-in assemblies from suppliers. These parts were simply assembled into the product with an average labour time of less than one hour yet they were being costed with an overhead equal to their prime cost. The parts were being charged a lot of money for an overhead with which they were simply not connected.

The bill of material showed that the assemblies at the heart of the product accounted for only 20% of the prime cost. These were the analysis electronics and the photo-multiplier and transport assembly. In the past everyone had concentrated on these units in discussions of cost reduction of the IPMA because they were thought of as the important part of the product. Since they were standard parts that were profitable in other SAE products and therefore could not be changed it was concluded that nothing could be done to reduce the cost of the IPMA. However, as Jordan had shown from the detailed product costing, the proper way to view the IPMA was as a very expensive packaging of some really quite cheap central parts.

The meeting moved on to examine how best to reduce the prime cost of the IPMA and so be able to decrease the selling price and stimulate more sales. At the meeting several points were agreed. First, a separate costing structure would be established by the accounts department to apply specifically to the IPMA and provide a true cost of it. Second, the decision to use existing in-house assemblies for the analysis electronics and the photo-multiplier assemblies was sensible. Third, Jordan would now concentrate on reducing the cost of the bought-in components. A threshold reduction of 15% was set since this would permit the sales price to fall below that of the competition. However, Jordan was certain this could be achieved. With her knowledge of the suppliers a target reduction of 22% was agreed as being a likely amount to be achieved within three months. Fourth, when the cost reduction from supplies was achieved Jordan would bring all the product documentation up to date and production of the new version would begin in six months' time, the batch after next.

The technical director closed the meeting by saying that he was very impressed with the work and would be discussing ways to promote the product with marketing. He said that at last someone understood the IPMA, and now that all the mystery about its costing had been cleared away one could clearly see a good product. Inwardly, however, he was quite concerned. After all, if a proper investigation of costs on a relatively minor product such as the IPMA revealed all this, how much more might be wrong with the costings of the other products?

The following months saw an examination of the likely errors in the prime cost model across the entire product range. The review showed that the IPMA was, in fact, the product that stood to have the least accurate product cost if the prime cost model was used. For the main products the error was less than 3%, and further examination was not worth the effort. One other product was identified that might have an error of 15% and it was agreed that Jordan would investigate this next.

4.5.4 Investment appraisal

Investment appraisal, which includes any capital expenditure, is a way of analyzing the financial value or cost of investment decisions. It is a technique which looks at the financial return on an investment. This type of technique is not to be used on its own, as there will always be many factors that affect a decision. However, it does have value in that it is an objective assessment. Investment appraisal will help, particularly in comparing investments and in forcing you to think about the cost of a particular decision, although appraisal is not a costing exercise in itself.

As an engineer you may be involved with considering such things as the cost and returns associated with the purchase of a new piece of equipment or with a project for introducing a new system or product. We will look at two techniques that can be used for investment appraisal, the payback period and discounted cash flow.

Payback period

The payback period is the amount of time that a project must run for in order that the cash generated will repay the initial investment. Payback is calculated on the absolute values of the expected income after tax.

EXAMPLE

Investment = £1000
Annual return = £200

$$\text{Payback} \quad \frac{1000}{200} = 5 \text{ years}$$

This is a very simple technique and is therefore widely used. It shows that early returns are generally preferable to longer-term ones. However, it has major drawbacks in that it does not recognize that money devalues with time, it does not take into account any earnings after the payback date, and it ignores the timing of earnings prior to the payback date.

EXAMPLE

		Project A	Project B
Capital investment		£7 000	£7 000
Returns:	Year 1	£5 000	£2 000
	Year 2	£2 000	£5 000
	Year 3	£2 000	£6 000
	Year 4	£8 000	£3 000
Total		£17 000	£16 000
Payback period:		2 years	2 years

In this example the payback periods are the same since after two years both projects have returned the original £7000. Yet the total returns are different in both amount and profile. We need to know which one is best and by how much.

Discounted cash flow

Discounted cash flow looks at the earnings over the life of the project and the time at which the returns are received, and it allows for the fall in the value of money over a period of time. The fact is that £1.00 today is worth more than £1.00 would be in a few years' time. This comes about not because of inflation, which evens out over a period of time and also affects both costs and profit equally, but because £1.00 invested today will attract interest at some rate and therefore the investment will increase in value.

If the interest rate is 10% we can invest £5.00 today and in five years' time the investment will be worth £8.05, assuming that the interest is immediately reinvested at the same rate and that the capital is left untouched. In a similar way you can work backwards and say that if the interest rate is 10% then £5.00 payable in five years' time will be worth £3.10 today. This is called the present value. Tables are available which allow you to look up 'present values'. However, they can be calculated using the formula

$$P = \frac{S}{(1 + i)^n}$$

where

$P =$ present value
$S =$ sum in the future
$i =$ interest rate (also called discount rate)
$n =$ number of years

EXAMPLE

An investor can get 8% interest per annum by investing money in a building society. The investor is offered an alternative investment by another company which will provide £388 at the end of each of the three following years. In order to calculate the value of this second investment you need to assess what it is worth today and then compare this with the cost of the investment as follows:

$$P = \frac{388}{1.08^1} + \frac{388}{1.08^2} + \frac{388}{1.08^3}$$
$$= £1000$$

The present value of the investment is £1000 therefore if the capital cost of the investment is less than £1000 then this will make more money than if the capital was invested in a building society. The difference between the present value of the returns and the capital cost is called the net present value (NPV).

EXAMPLE

An investment of £13 000 yields returns of £4000 per year for the first three years and £2000 per year for the next two years. By calculating net present values determine whether this is a worthwhile investment, if the money could otherwise be invested at 5% per annum, or at 10%.

At 5%:

$$P = \frac{4000}{1.05^1} + \frac{4000}{1.05^2} + \frac{4000}{1.05^3} + \frac{2000}{1.05^4} + \frac{2000}{1.05^5}$$

$$= £14\ 105$$

Therefore, at 5% net present value:

$$= £14\ 105 - £13\ 000$$

$$= +£1105$$

If the money could only be invested elsewhere at 5% then this is a worthwhile investment as it would bring a positive net present value, i.e. a profit.
At 10%:

$$P = \frac{4000}{1.1^1} + \frac{4000}{1.1^2} + \frac{4000}{1.1^3} + \frac{2000}{1.1^4} + \frac{2000}{1.1^5}$$

$$= £12\ 555$$

Therefore, at 10% net present value:

$$= £12\ 555 - £13\ 000$$

$$= -£445$$

If the money could be invested elsewhere at 10% then this investment is not so profitable and the other investment would make more money.

Internal rate of return
This is the discount rate, or rate or interest, at which the net present value is zero. It is the point at which you break even. It can be calculated simply as shown below, which uses the previous example.

At 5% NPV $= +£1105$
At 10% NPV $= -£445$

Using this information we can draw a graph as shown in Figure 4.2. IRR = interest rate at which NPV = 0. By simple extrapolation:

$$\text{IRR} = 5 + 5 \times \frac{1105}{1105 + 445} = 8.56\%$$

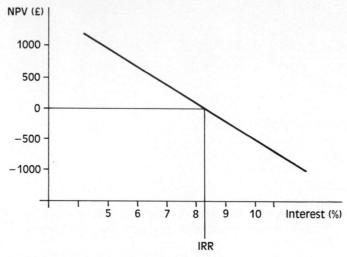

Figure 4.2 Determination of internal rate of return

The internal rate of return is 8.56%, therefore if your money could earn more than 8.56% elsewhere over the period of the investment, you would be better advised to take that alternative rather than the investment proposal described above. However, if you believe that the money would earn less than 8.56% then the proposal is a good investment since you will earn less elsewhere.

4.5.5 Depreciation

When considering investments they are looked at in terms of the cost to buy and then in successive years they are considered as assets to the company which have a decreasing value. The term used to describe this reduction in value is depreciation, and there are accounting techniques which are used to calculate it. Some factors that give rise to depreciation are wear and tear, shelf-life, and damage due to misuse. Depreciation must be incorporated into the balance sheet to show the reduction in asset value. Stock is not depreciated because it is valued annually using a stocktake.

There are many ways of calculating a charge for depreciation, and companies can choose which one they like but they should be consistent across the company. We will look at three methods that are used.

Straight-line method
In this method the asset is depreciated by equal annual charges spread over the asset's estimated life, as shown in Figure 4.3.

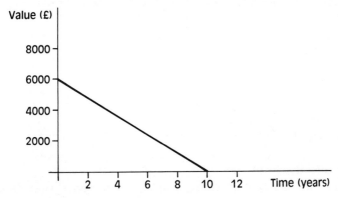

Figure 4.3 Straight-line depreciation

EXAMPLE

Cost $= £6000$
Scrap value $= £400$
Estimated life $= 10$ years
Annual depreciation:

$$\frac{6000 - 400}{10} = £560$$

Reducing-balance method

In this method the depreciation charge is a constant proportion of the balance remaining at the end of each accounting period, as shown in Figure 4.4.

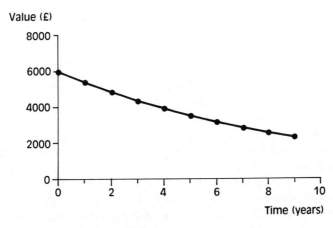

Figure 4.4 Reducing-balance depreciation

EXAMPLE

Initial cost = £6000
Depreciation rate = 10%

Value	Year
£6000	0
6000 − 10% = 5400	1
5400 − 10% = 4860	2

Production unit method

In this method the depreciation charge is based on the number of units of work produced, multiplied by a constant rate for each unit.

EXAMPLE

Cost $\quad\quad$ = £13 000
Scrap value $\quad$ = £500
Estimated production = 250 000 units
Depreciation charge

$$\frac{13\,000 - 500}{250\,000} = 5\text{p/unit}$$

4.6 Business plans

One of the reasons engineers sometimes become interested in finance is because they want to start a business of their own. In the opening lines of this chapter the question of starting one's own business was raised. We have now studied finance and have an understanding of its important concepts. We are therefore in a position to understand business plans. In this section we shall introduce the concept of a business plan and describe the information that it must contain.

4.6.1 The purpose of the plan

The fundamental purpose of the plan is to convince other people who own capital to invest in your business. The secondary purpose is to be the plan which the business owner uses to ensure that the business achieves its aims.

Naturally, if the plan is to convince anyone to lend you money they will need to be satisfied of various concerns. First, they must be sure that the business you propose will make sufficient money not simply to sustain itself but also to repay any loans. If the investors are being offered shares in the business rather than simple repayment they will be concerned that the business has the potential to grow in the medium to long term. These worries in turn depend on the customers of the business. The investors will need to know who would want to buy the product and how many customers there are, and why they should buy the products of this company rather than those of a competitor. These are hard questions but if they cannot be answered satisfactorily on paper in a report, how much less hope is there of the business actually working in practice?

4.6.2 What should be in the plan

The precise details of what will go into the plan depend on exactly what the product in question is and how the business will be established. However, the following points are certain to need addressing in some form or other:

1. *A summary of the business*: This should describe the business in overview so that someone who has never been involved with the business area can understand the proposal. The customers should be described and their reasons for wanting the product or service. You should include overview figures such as the total market for the product and the total investment sum required. This section is normally quite short, only two or three pages.
2. *A description of the product*: A description of the product is required. The report should not include technical calculations, etc. rather a description of the products' usage, its life and the ways in which it might change. Most importantly, what is different about this product from all the others to which the customer has access?
3. *The market*: The market for your product must be described. Your customers are the most important part of your business and if you are to persuade anyone to lend you money you will first have to convince them that you know your customers. The things you must describe are the market size, how it breaks down, the spending power of each section, the preferences of each section, your competitors and external influences such as new legislation.
4. *Management*: You must describe how your business will operate. How will sales be obtained? How will manufacture occur? What are the processes that you will undertake? Who will be your suppliers? Where will you be based? All the operational details associated with running your business must be described.
5. *Financial details*: This section must include the financial documents that you offer to convince the reader of the profitability of your business. Naturally, this must include a discounted cash flow forecast which shows the net present value of the total incomings, total outgoings, loan repayments, and profit. Also included must be a monthly cash flow forecast, the balance sheet and the profit and loss account. These should be presented for the next three years at a minimum and, more usually, five.

4.6.3 Preparation of the plan

The preparation of a good business plan is a major undertaking often lasting a year or so. Naturally, the person who should prepare it will play the major role in running the business since it is they who must ultimately operate it. However, it is advisable to get help on the areas where knowledge is weak rather than trying to research everything oneself – for example, taxation and VAT, legal implications of trading, legal requirements of accounting, product liability and insurance. For many people who want to prepare a business these subjects are a mystery and much the best thing is to enlist the help of a professional lawyer or accountant, for example. There are many sources of help for individuals wanting to start a business (the Department of Trade and Industry (DTI), the banks, local enterprise support initiatives, books, etc.). Some accountants and banks even run training and offer consultancy.

Naturally, the business plan cannot foresee all the problems and changes that the new business will face, and one cannot know the future. For this reason, the business plan will need updating as times change. In practice, this is difficult to achieve. When people start out in their business the plan consumes all their time and a good job is made of it. After a year or two of successful trading the busy commercial life makes finding time for updating the plan difficult. The importance of the business plan does not reduce, however, and companies that do well are characterized by maintaining the initial high standard of business plan for each subsequent year of trading.

4.7 Summary

In this chapter we began by examining where the need for financial control comes from in the first place. It is clear that, even as individuals, we need some sort of financial control simply to ensure that we avoid living beyond our means. In the case study we found that a business like that of Jake Sounds has need for a far greater degree of control than we as individuals. We finished Section 4.2 by defining what an ideal financial system must provide.

In Section 4.3 we examined the area of the financial profession which provides information about the organization to those in the outside world. In the section we explained the use and preparation of the three most important financial accounting documents. These were the balance sheet, the profit and loss account, and the cash flow projection. By the end of the section we were able to prepare these documents for ourselves. The interpretation of such documents is a skill in its own right and in the last part of this section we met the concept of accounting ratios, which provide an indication of how particular areas of the business are performing.

In Section 4.4 we saw how an organization can use the budgeting process to control its activities in a logical way. We found that there was common ground between the use of budgeting and management by objectives. In the case study we saw how a particular organization set objectives for each department and budgets at the same time. At the end of the budgeting process each department had a very clear and unambiguous statement of what was to be achieved over the next year and what financial resources could be employed. This process was relatively rapid and by using budgeting in this way planning the next year consumed an insignificant portion of it.

In Section 4.5 we explained how financial data required for internal use may be prepared. The managers of the organization must have a rational basis upon which to judge the value of future investments and with which to prepare costings of present products. This section explained the use of the management accounting techniques of costing, investment appraisal, and depreciation.

In Section 4.6 we briefly reviewed the preparation and importance of business plans.

4.8 Revision questions

1. Why is profit not an asset?

2. In the example of Jake Sounds a balance sheet was prepared for a very few financial transactions. How might preparing a balance sheet for a company that has conducted hundreds or even thousands of transactions differ from the way the one in the example was prepared? How might this process change further if the size of the organization becomes larger still?

3. When preparing a summary of an organization's financial activities, what advantages and disadvantages of using a balance sheet are there?

4. Arrange the following data into a balance sheet:

Finance debt	325
Long-term mortgage	150
Permanent equipment	600
Stocks	1200
Profit and loss account	1617
Cash	850
Creditors	1500
Three-month bank loan	75
Called-up share capital	460
Depreciation on equipment	23
Buildings	500
Reserves	50
Debtors	1050

5. In an essay of less than 2000 words (i) list the sorts of records an engineering company must keep in order to make the preparation of a balance sheet possible and (ii) from your knowledge of the various departments within a company, discuss the ways in which the different departments might provide these data.

6. A balance sheet is often used by analysts to examine a company from a financial point of view. Describe the strengths and weaknesses of the document for this purpose.

7. What is the basic equation of a balance sheet? How does a balance sheet work? Why is it needed? What advantages does it bring? What disadvantages does it bring?

8. How is a profit and loss account arranged? What does it show? Why is it useful? How does it differ from a balance sheet? What is the difference between a profit and loss account as a financial document and the entry on a balance sheet called 'profit and loss'?

9. When might a cash flow projection be useful to an engineer? How is it prepared? What limits the accuracy with which it might be prepared?

10. Why is the cost of a product not as straightforward to calculate as it appears at first? Describe two methods of preparing product costs.

11. What do the following terms mean?

    Direct costs
    Break-even point
    Contribution
    Direct labour
    Discount rate
    Indirect costs
    Internal rate of return
    Machine hour rate
    Marginal costing
    Overhead recovery rate
    Production unit method

12. Why are budgets prepared? Describe a process by which they may be prepared. What advantages do they bring? How do budgets affect senior management? How do they affect engineers in the organization? How do they affect junior staff?

13. Eruco Ltd manufacture luxury cat toys. Their electric mouse department is particularly profitable, producing 150 000 mice per year at 30% profit. The bill of materials for an electric mouse gives the following information:

Part	Quantity	Cost/unit
Fur fabric	0.0225 m²	6.50
Motor	1	0.20
Revolving eye	2	0.03
On/off key	1	0.03
Plastic tail	1	0.01
Frame	1	0.02

The mice are made in batches of 500, each taking 43 hours. Labour is charged at £5 per hour. Overheads for the mouse department are £73 000 per annum and are absorbed on a direct materials basis. Calculate the manufacture cost and the selling price of an electric mouse.

14. The overheads for a subcontract machinist are:

Cost centre	No. of people	Direct material	Total overhead
Admin & finance	3	0	75 000
Sales	1	0	25 000
Capstan shop	5	80 000	90 000
Press shop	5	60 000	120 000

Apportion the overheads to the cost centres and calculate the cost of job XN01 on the basis of the following information.

Job XN01: Direct materials £15.13
Direct labour   Capstan shop   30 min @ £6/hour.
                Press shop   15 min @ £7/hour

Overheads are to be absorbed on the basis of direct labour. The factory works 47 weeks a year and operates an 8-hour day.

15. A company is offered two investments. The first requires a capital outlay of £40 000 and will bring a return of £12 000 per annum for the following five years. The second investment requires a capital outlay of £60 000 and will bring returns of £15 000 per annum for the first two years but this will fall to £10 000 for the next six years. Assuming the rate of interest is 10%, calculate the net present value of each of these investments. On the basis of net present value, which would you suggest is the best investment for the company? Would your advice change if it was likely that the interest rate would increase to 15% during the life of the investments?

Further reading

Droms, W. G., *Finance and Accounting for Non-financial Managers* (1990) 3rd edition, Addison-Wesley. A readable 250-page textbook introducing finance and accounting and requiring no specialist knowledge. The book is in six sections. The first introduces financial management and tax. The second covers the fundamentals of financial accounting, the third financial analysis and control. The fourth section covers decisions regarding working capital, the fifth long-term investment decisions and the last long-term financing decisions.

Chadwick, L., *The Essence of Management Accounting* (1991), Prentice Hall International (UK) Ltd. A 170-page distillation of management accountancy that is ideal as preparatory reading for a short course or for reference. It is very compact and synoptic in its approach.

5
Product development

Overview

In this chapter we shall examine the way in which the process of product development is performed and managed. However, since this is a study of management we shall not be looking at the technical aspects of design, where engineers put their technical skills to work designing products.

Producing a new product is a risky venture. There are always uncertainties, and the process must be managed to prevent wasted resources. Product development can easily degenerate into a bottomless pit of expenditure – there is no upper limit on the amount of resources engineers can consume in developing a product. Careful control is necessary to ensure that the engineers are confident of success while the venture as a whole remains profitable.

5.1 Introduction

Imagine the appearance of a state-of-the-art electronics company forty years ago. The new technologies of this period would have been high-current valves. These small glass tubes containing intricate metal foils, heaters and grids were the staple diet of all electronics students of that time. The companies that manufactured them were specialists in precision metal forming, in glass blowing and in achieving vacuum-tight seals between the glass and the metal pins that must go through it. Today, these same companies manufacture transistors, silicon chips and microprocessors. Many of the specialist skills they had forty years ago are completely useless in today's electronics markets. Indeed, if an employee travelled in time from the past of the company into today's office they would find it hard even to understand what was taking place, let alone contribute to it.

In the example above it is clear that a lot of product change has taken place. In this particular case there have clearly been great changes in technology that have allowed these developments in products to occur and it is common to find people thinking that it is only technological advance that leads to product development. This is not true. An employee in the wine business who travelled in time from forty years past to the present would find the contemporary scene just as alien as would the electronics employee, yet wine is made in an almost identical way now as it was then. In this case, changes in all sorts of other issues have taken place – legislation has changed, spending power has changed, society has changed, tastes have changed. Customers have changed.

The unifying factor here is that customers have changed, and organizations that continue to do well are those that have changed with them.

This process of change is clearly one of great importance. It is one that must continually be in progress and it is a process that is filled with conflict. On the one hand, companies want to keep selling the same products in the same way so that they can become efficient and exploit the financial benefits of long-term mass production. On the other, they must develop the products they make to avoid losing their customers. Employees are efficient when they understand and can use effectively the existing technologies within the company. However, they must learn new skills and technologies, at which they are initially slow and inefficient, so that they can introduce new products. On the one hand, organizations find change risky, difficult to embrace and filled with uncertainty; on the other, they must change in order simply to survive.

In this chapter we shall examine the process of product development and learn why it must happen, how it can be managed and understand why it is such a difficult thing to get right. We shall start, in Section 5.2, with customers. We must understand why their needs change. We shall then see how these changing needs manifest themselves as demand for products. From this we can describe what it is that we require from the product development process. After this there are two major issues that we shall address. The first is how to go about the overall process of product development. We shall investigate models of company operation that lead to successful product development. These models describe good practice for the product development process. They are sometimes called 'product development procedures' and they are described in Section 5.3.

Once we have a procedure, we can investigate particular tools and techniques that are used to accomplish specific tasks within the process. For example, product design specifications are a particular tool in the product development process and are used for a special purpose. In Section 5.4 we shall examine eight of the most important management tools for product development.

Product development is the process by which new design ideas are brought from non-existence to ownership by customers. Clearly, this process will have an effect on all other departments of the organization. From marketing and market research where the ideas that customers will pay for are identified, through to manufacturing where products are actually made, there is no area of company activity that is not affected by it in some way. Finance is certainly affected. Product development could consume an almost unlimited amount of money and yet there may only be very limited amounts available to fund it. Even the personnel function is affected through a requirement to ensure that the company personnel offer the correct range of skills to develop the required product portfolio. Product development is therefore a theme running through an engineering organization.

Product development is the very lifeblood of an organization. Without it there can be no continued survival. However, it is a difficult process filled with conflict and so requires careful management. The required outcomes must be specified, and in achieving them we may expect to have to employ all our management skills and all sections of this textbook are therefore relevant.

The learning objectives for this chapter are as follows:

1. To understand why companies must develop new products
2. To understand three models of how the product development process may be conducted
3. To meet and understand eight important techniques in the product development process.

5.2 Customers and product development

5.2.1 The customer's need for product development

The basic forces which drive the product development process are the developing needs of customers. If we are to be successful at the product development process we must therefore understand how these needs change. These changing needs come about for many reasons, some clear, some not so clear.

Technology is an obvious cause. As developments in materials science, for example, have advanced so also have engineers' abilities to offer improved performance specifications for products such as computers, engine components, or domestic appliances. These advances in specification are clearly desirable to the customer. A customer faced with buying a new washing machine will clearly be influenced by the speed, reliability, ecology, and power consumption of the products on offer. Advances in technology are used to improve all these specification items and therefore influence customer choice.

Fashion in product development is used to cover a much wider range of products than in its colloquial usage. Fashion usually applies to clothes or records and describes the fickle behaviour of the customers of these items. In product development it applies to all products. There are 'fashions' in the behaviour of consumers of cameras, of hi-fi, of cars, of computers, even of staples like bread. These changes come in response to society's prevailing attitudes, and as society changes so do these attitudes. For example, recycled plastic packaging would have had a very negative influence on customer choice as little as ten years ago when it would have implied cheapness and poor quality. Yet today it is virtually the norm for many supermarket products. Customers will specifically buy products for their environmental friendliness.

The *economic climate* affects us all. In times of plenty consumers have more spending power and their wealth is employed in a greater range of expenditure than in times of high inflation and interest rates. Some products are very obvious in their sensitivity to this effect. Exotic holidays, for example, will clearly only sell well when people are relatively well off. Other products have more complex behaviours. DIY products, for example, sell better in recession because people cannot afford to have the work done for them. On the other hand, they have less money in any case and so are fiercely competitive about price. These two effects offset each other and the DIY market is much less dependent on economic climate than one might suppose. Companies must therefore develop products that suit the economic climate of the time. For example, custom test equipment sold in periods of economic boom may be designed to offer sophisticated methods of calibration, uncompromised use of materials and absolute accuracy, while in time of recession it is designed primarily for reduced cost, for productivity and for reliability.

A paradox of product development activity occurs during a period of recession when it is normal to find most companies increasing the amount of product development in progress. At first, this seems strange since it will involve them in more expense at a time when they can least afford it. The explanation is that when the market size reduces it becomes more and more important to increase market share. The only way to do this is to attract new customers and this requires product development. This effect works alongside another strategy, which is to reduce profit margins and so keep sales prices down. Before long, only the companies with the resources to survive a recession are left and the market is divided among a smaller number. Each company, however, has a greatly increased product range. From the customers' point of view this means that much better value for money is on offer and much greater product choice available as a consequence of all the product development. For those customers that can afford to purchase during a recession, value for money is excellent. A good example is the enormous range of new car accessories and developments that manufacturers offered during the recession of the early 1990s.

Escalating expectations are part of human nature. When one does not own a particular product at all, the purchase of even the most basic model is likely to provide satisfaction. Your first foreign holiday or new car is usually very much enjoyed. Before long, however, customers' tastes develop and they expect much more the second or third time around than the first. This escalation can be seen in cars. Twenty years ago the level of comfort provided by the cars of the time was considered amply adequate by the customers. Today, such a level is considered so basic that no manufacturer would even think of bringing such a car into the market. This has nothing to do with technology; the same specification could have been offered twenty years ago. Expectations such as these come about from increased product experience. These may be for reduced cost, increased reliability, greater safety, more accessories or improved user friendliness.

Another escalation in expectation accompanies increasing age and spending power which bring with them expectations of luxury and exclusivity. Companies wishing to meet all these needs must offer a range of products that suit. A cross-channel ferry, for example, offers a range of passages, from a foot passenger without a seat through to a luxury cabin with waiter service and a separate restaurant. The different expectations of the people in each of these categories is clear as one walks about the ship.

The *internal environment* of the company may also cause a need for product development. Often this comes from a need to reduce cost. A product development exercise that originates internally will often be transparent to the customer. For example, an electronics company that has successfully launched a particular model of video recorder and cornered a larger than expected section of the market may then redesign the internal features of the machine for reduced costs, taking advantage of manufacturing economies that have been made possible by the increased sales. The customer may not be able to tell the difference between the new and old model but if the work is successful the accountants certainly can. Reasons for product development that originate internally can include cost control as described above; manufacturing techniques where a redesign is performed to take advantage of new techniques; or elimination of unnecessary features where product attributes included at the launch of a product have proved to be unnecessary in practice. This last category might include a test socket removed from an electronics board after discovering that the failure rate in production did not warrant the expense of the test and the socket.

Having understood something of the customers' need for product development we shall now examine how we may determine when a new product is required. After this we will develop an understanding of what the ideal product development process should offer.

5.2.2 Product life cycles and GAP analysis

In this section we shall consider two concepts that assist in determining when product development is required. We start by looking at how a typical product behaves from a sales point of view. All product sales decline in the end and by characterizing the process we can recognize what is happening to a product and estimate more accurately when its sales life will be over. The second concept we shall look at is GAP analysis. This makes use of the product life cycle concept and offers a way of determining when an organization should be ready to launch a new product so that company turnover objectives may be met.

Product life cycles

One central feature of a product is its life cycle. When a product is first invented or developed the number of sales is zero, no units have been made or sold. Once the product is launched, the introduction phase is started. The number of customers wanting the product but who are as yet without it is at an all-time high. Initially, however, sales are slow since people are unfamiliar with the new product. Some customers may wait a while before purchasing, perhaps to let the price drop (as often happens after a launch) or see how others who have purchased like the product and so make a more informed purchase. Once these customers are convinced of the product, they start to purchase and sales rapidly accelerate during the growth phase.

However, after a while everybody who wants an example of the product has one and so the number of people buying levels off. The number of new customers to which sales may be made now depends on how many new customers are entering the marketplace. For example, in the teenagers' bike market, a successful product launch will start with the whole group buying a bike and then settle down to only those who became twelve or thirteen years old in the last year buying, since everyone else now has one. This is the maturity phase. As time goes by, however, other products may be launched and so even the new entrants to the market are lost as customers since they then turn to other products. These other products may be the new products of the same company or of the competition. This is the decline phase. This life cycle behaviour is shown in Figure 5.1.

Of course, the product life cycle is not the same for all products. Some are very long-lived such as beer or nails. Others are extremely short-lived, such as a chart-topping single or a particular model of electronic calculator. It is tempting to think that the long-lived ones, such as bread, have an infinite life cycle but this is not true since such products are in fact many products. Twenty years ago, for example, almost all bread sales in England were for sliced white loaves. Now there are many different bread styles for sale in supermarkets. It is true that there will always be a market for bread and the manufacturers can do almost nothing to make us eat more since our appetites are not, in the long term, responsive to advertising. However, the individual products within the bread market do have life cycles, some of which are quite short. The changing fashions in health advice, an imagined association perhaps with particular socio-economic groups or even the weather can have an important effect on the success of a new product in such a market.

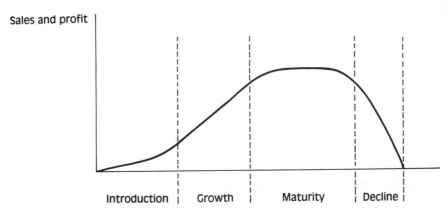

Figure 5.1 The product life cycle

GAP analysis

The product life cycle concept is a very important one when managers are trying to plan the financial activities of an organization. Naturally, they are concerned about how much money will be received from the sales of products, and since they realize that sales will eventually fall off because of the product life cycle they want to know when this will happen and what effect it will have. The analysis used is called a 'GAP analysis' and takes its name from the fact that it makes clear when a gap between desired income and income from existing products will occur. Clearly, this is an important time to identify because the organization must plan the launch of a new product for that time so that revenue from the new product sales takes over from the falling sales of the old one. Product development may be a lengthy process, often taking several years. It is therefore important that the GAP analysis extends sufficiently far into the future to chronicle all the phases of the next few products. New product development consumes a large fraction of profit and it is important to spend the profits made from one product on developing the next. In this way increasing revenue from the sales of new products which replace the decreasing sales of the old ones.

Leaving the start of the product development process too late results in insufficient time being available to develop the product properly. The product development teams then face an ultimatum: either get the new product ready in time to defend sales against the opposition or go out of business. It is sometimes said in business that there are only two ways to go bankrupt. The first is to have a poor financial management system and find one day that you owe more than you have in the bank. A healthy business can go bankrupt overnight in this way (see Chapter 4). The second way is to fail to identify when the development of a new product should be started. This bankruptcy is much more sinister. It is a process of seduction rather than a bolt from the blue. The error starts with a reduction of product development expenditure. The initial result of this is that expenditure reduces and sales are unaffected, which leads to a very profitable year. The next year, budgets are maintained and again there is a good profit, perhaps with a slight reduction in sales. The following year, extra money is spent on sales and again good profits are achieved. The next year, sales decline more rapidly, and new competition appears. The much-expanded sales force can no longer compete with the competition and what is needed is a new product. There is, however, no money available to fund a new product, and, in any case, there is no time to get it ready. The liquidation of the company is inevitable from this point. This pitfall is well known to business managers and the bankruptcies that it causes usually come about from errors in estimating the times and costs of new products rather than ignorance of the entire concept altogether.

A GAP analysis is shown in Table 5.1. The numbers in the table are in thousands of pounds and the columns are explained below.

Sales from product A are in decline and by year 5 they are completely exhausted. Sales from product B reach a peak in years 5 and 6. Product C does not start to produce sales until year 4 and from then on there is a sharp rise in

Table 5.1

Year	Desired income	Sales from product A	Sales from product B	Sales from product C	Total sales	GAP
1	230	150	90	0	240	−10
2	250	140	120	0	260	−10
3	270	90	180	0	270	0
4	290	30	190	20	240	50
5	310	0	200	50	250	60
6	330	0	200	110	310	20
7	350	0	190	170	360	−10
8	370	0	180	200	380	−10
9	390	0	170	200	370	20
10	410	0	150	200	350	60

total income which levels off in years 8, 9 and 10. When the income from all these products are added together the total sales profile is produced. When it is compared with the desired income in the GAP column we can see that there will be a shortfall in year 3 through to year 6 and again from year 9 onwards. The company can then prepare a rational plan to combat this shortfall and engage in a product development programme designed to launch two new products, one in year 3 and one in year 9. In Figure 5.2 the GAP analysis is shown graphically.

Such an analysis helps companies to plan rationally but one must remember that the predictions from such an analysis are only as good as the data that goes into them. The most important piece of data in the GAP analysis is the forecast of sales for each product. A board of directors considering such data will want to know what reasons there are for believing the forecasts, who the customers will be and what it is that will make them buy the company's products. The preparation of such data is usually undertaken by marketing personnel and is described in Section 3.4.

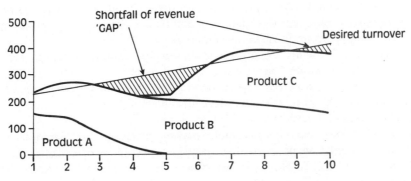

Figure 5.2   GAP analysis

5.2.3 The ideal product development process

We have seen how the needs of customers change and also how we may plan when new products should be launched. We are now in a position to set out what we require a product development process to provide.

We must start by identifying the attributes that we desire the product development process to have. The first and foremost requirement is that its action results in products being developed that meet customer needs and therefore will sell well. To be able to do this the process must have good contact with customers and be efficient at determining what they actually want. In addition, the organization must constantly review the progress of its new products under development and reject ones which show signs of not being acceptable to the customer. For this reason, good product development means having many products under development and leaving the final choice as late as possible.

Sometimes it may not be necessary to develop an entirely new product at all, sometimes product developments are based very much on existing products. For example, if a car manufacturer has identified lack of reliability as the most important failing in an existing product it makes no sense to embark on a costly product development programme for an entirely new model since the best way to ensure the desired level of reliability may be to modify and develop the existing and well-understood design for which there are many data available. This first requirement must be seen alongside the need for a business to be a self-sustaining financial concern.

It is obvious that the product development process could become a bottomless pit of expenditure. Imagine a group of engineers left to their own devices who have been asked to produce a verified design for a new hatchback car and given an unlimited budget. The result would almost certainly be financial disaster. Even if the car was an inspired design there would be no guarantee that the investment could be recovered from sales. Too much money would have been spent on product development.

In contrast, one could spend almost no money on product development and have a very aggressive selling price. After a few years, however, the market for the existing products would have disappeared and there would be no products to sell. The result again would almost certainly be financial disaster. Not enough money would have been spent on product development.

Before a development budget can be chosen the amount of money available from the total customer base must be estimated and then the product development budget selected appropriately. The second requirement of the product development process is therefore that it completes the development of new products within a predetermined budget.

In an ideal business all this would happen instantly. For example, a car manufacturer would have products that meet the expectations of today's customers already in the showrooms. However, designing and tooling up for the manufacture of a car takes time and so customer expectations are always in advance of the products available for purchase. This places great pressure on competing manufacturers to bring new designs into the market. As a measure of this, it used to be normal for a new car to take well over ten years from concept drawings through to wide availability in showrooms. This process now takes only eighteen months. When one considers that this eighteen-month period relates to a car that may have all new components, from the engine to the instruments, one can understand what a feat of organization is required to do all this and still make a profit. A third criterion for the product development process is therefore that it should take place in an appropriate timescale. In this context, appropriate is defined as a sufficiently short period of time to ensure that the delay does not cause a reduction in the desirability of the product when it is launched. This period will be very different, for example, between a singles record and a new supertanker.

In developing this picture of the ideal product development process we can see again the conflict that it poses. On the one hand, a business should spend a great deal on product development to ensure that it always has products available that are guaranteed to sell well. On the other, it should spend hardly any to ensure a good return on investment. It should also take its time over the process to ensure that the product really is an answer to customer needs, yet it should take no time at all to ensure that the needs do not change between design and launch. It is little wonder, then, that product development is difficult to manage. When one adds to this the complex technical advances that usually accompany new engineering products, it is clear that successful product development poses the ultimate challenge to engineers.

In summary, the vital requirements of the product development process are that:

- It must result in products that meet customers' needs.
- It must control budgets and return a profit.
- It must be completed within an appropriate timescale.

If an organization can meet these three criteria then it is likely that it will be successful in developing new products and will survive in the medium to long term.

5.3 Managing the product development process

Product development is a function of an organization and it is a task that all organizations have to accomplish. There are, however, many different ways of managing it. Product development, which includes design, is often thought of in conjunction with manufacturing organizations, although all creative organizations have to go through the process. Even abstract products such as computer programs or banking services have to be designed and developed, and a sensible methodology must be employed to ensure that this process is successful. We have already seen that the process of product development is filled with conflict and in this section we shall look at some of the ways it can be managed.

We shall start by looking at models that have been developed to describe the process in Section 5.3.1. We shall then examine the structures that companies use to facilitate the process in Section 5.3.2. Lastly, in Section 5.3.3, we shall examine the special role of finance in product development.

5.3.1 Models of the process

Before a successful product can be made, various stages must be completed. Product development starts with the needs of customers and the good ideas of imaginative product developers. The ideas must be expressed in a design of some sort. The market for this design must be assessed. The way the product will be made must be determined. Modifications to the original idea are almost inevitable and so product development becomes an iterative process. This is shown in Figure 5.3.

Many models of the product design and development process have been proposed to help engineers to tackle this difficult task. They all have similarities since they all describe the same process. In this section we shall look at three of

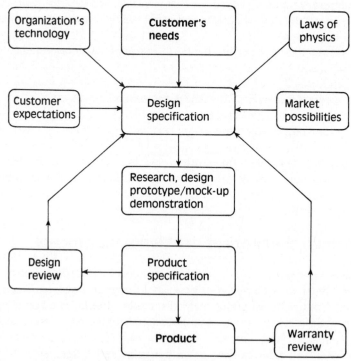

Figure 5.3 Idealized product development process

them. The first is due to Pugh,[1] the second to BS 7000,[2] and the third to Pahl and Beitz.[3]

Pugh

In his book, *Total Design*, Stuart Pugh describes a concept he called the 'design core'. In this model there are six stages involved in the product development process that must be successfully completed in order for good product development to occur. The first is called 'Market', in which customer needs and relevant market factors are determined. The second is called 'Specification', in which the product is specified. Most models of product development include a section very similar to this. In Pugh's model it is called the product design specification (PDS) and this term is in general usage, although it is often abbreviated to product specification (PS). This document is described in much more detail in Section 5.4.3. After this comes the 'Concept Design' stage, in which overall design approaches to the problem in hand are decided. For example, a product designer will have settled on what power sources will be used, the overall approximate sizes of assemblies, on how the user will interface with the product, on how energy will flow around the product, etc. The 'Detailed Design' stage follows in which the overview design of the previous section becomes detailed and specific. This phase ends with the bills of materials fully prepared and all drawings complete. There is a wealth of difference in the level to which the product is specified between these two phases. After this 'Manufacture' begins and, with it, the last phase of 'Sell'. This structure is shown diagramatically in Figure 5.4.

It is a key point in Pugh's model that the boundaries between the phases are not fixed. It is not that a group of concept designers finish their job and hand over a concept design to a group of detailed designers who then prepare detailed drawings from them. Far from it. In fact it is often the same individuals who prepare both. In the extreme, one may even find a particular individual who sees the development of a product through all stages, although in reality it is very rare even in a small company for just one person to handle a product right from inception and initial customer contacts through to manufacturing. The point of the model is that whoever does it, the new product must travel through these different phases with as much iteration between them as is necessary. Failure to do this will most likely lead to a poor product.

BS 7000 Guide to managing product design – Part 1 Section 3, 1989

So important is the successful regulation of the product design process that there is a British Standard defining good practice. Companies that operate a product development procedure in accordance with the standard may use the fact by advertising their certified design control procedures (provided that they have been assessed).

The standard uses phases through which a product development must go in order for a successful conclusion to be likely. In the standard the first phase is

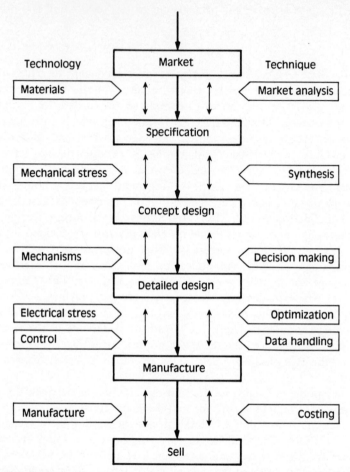

Figure 5.4 **The design core – after Stuart Pugh**[1] ***Total design—Integrated Methods for Successful Product Engineering.*** **(Addison-Wesley Publishers Ltd, Wokingham, England, 1991. ISBN 0-201-41637-5 pp 5–11, reproduced by kind permission of the publishers)**

'Motivation', in which a trigger or need for the product is identified. The second phase is 'Creation', in which all the creative design takes place. In the text of the standard, a distinction is made between the early phases of design when definition is poor and the end of the design phases when detailed drawings are finished, together with bills of materials and the product definition is very high. The third phase is 'Operation', in which the product is owned by its customer and is put to the use for which it was designed. Lastly, the product enters the 'Disposal' phase when the customer discards the product perhaps recycling it, perhaps trading it in or simply throwing it away. Figure 5.5 shows the idealized product evolution process according to BS 7000.

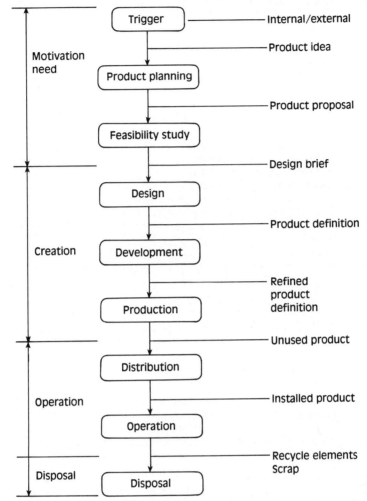

Figure 5.5 The idealized product evolution – BS 7000² Part 1 1989. Extracts from BS 7000 are reproduced with the permission of BSI. Complete copies can be obtained by post from BSI Sales, Linford Wood, Milton Keynes MK14 6LE. Orders should be sent to 389 Chiswick High Road, London W4 4AL

Pahl and Beitz

G. Pahl and W. Beitz wrote a major work on design entitled *Engineering Design, a Systematic Approach*. In it they present their own version of the design process and this is shown in Figure 5.6. The process starts with a statement of the 'task' in hand and the development of a specification. It ends with a 'solution'. The process described by Pahl and Beitz is a systematic approach to design. It is used in companies to guide the design process towards a design solution for the

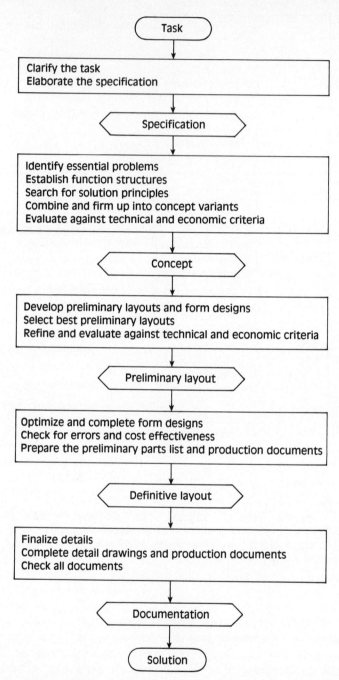

Figure 5.6 Steps of the design process from G. Pahl and W. Beitz, *Engineering Design: A Systematic Approach* **(The Design Council 1988. ISBN 0-85072-239-X, reproduced by kind permission of the publishers)**

problem posed in the marketing brief. The process is very thorough and has become the design standard of many companies.

These three models can be compared with the requirements stated above for the ideal product development system. It is clear that they all address the vital issue of ensuring that customers receive products that meet their needs. The models do not detail the way in which project plans should be prepared or resources allocated, these are matters for the individual organization. In general, if any one of the models is followed and planned effectively the organization is very likely to be successful in developing the product in question.

5.3.2 Company structures for product development

From the previous section it is clear that the product development process involves various tasks which must be performed. It makes sense therefore to choose an organizational structure that suits these tasks. An efficient organizational structure will naturally make the process flow smoothly and effectively. However, this is not as easy to do as one might think. Products are very diverse and so are companies. Some are large and have long periods of time between the launch of each new product such as Airbus or Cunard. Some are large and have a short period between each new product, such as Apple Macintosh or Canon. Others are small and have very long periods between new products such as MG cars or Fuller's beer. Still others are small with rapidly developing products such as a Formula One racing company or Quad hi-fi. With this diverse mix of product and company it is important that the product development structure suits the company's overall position as well as its product development needs.

We shall now look at four methods often used to organize the product development process in engineering organizations. Each has its own benefits and weaknesses and is appropriate to particular company situations.

Research, development, engineering and manufacturing (RDEM)
In this case the organization is divided into separate departments each having very specific responsibility for a particular aspect of the development process. One department will research the market and suggest products, another will design and produce prototypes, another will continue the development to allow products to be manufactured. Finally, another department will manufacture the product. In this functionalized approach the cross-department communication can be difficult, and often there will be no particular engineer who stays with the product through its life. The design process of such a structure is characteristically slow but is very thorough. Staff are employed in those areas in which they are most competent and there may be a high level of specialization. However, care has to be taken at the point where responsibility is transferred from one department to another to ensure continuity and to prevent duplication of effort; following transfer of the product a department is free to start on the next project. In this approach it is common to find an 'Engineering Department', although this term is

used very broadly and may cover something quite specific such as CAD/CAM or may be rather general, including manufacturing engineering, design and production control.

The RDEM model is shown in Figure 5.7. In general, the RDEM approach is suited to product development environments that must be very thorough and where the timescales are relatively long, such as major defence projects. The boundaries between the phases require the preparation of very detailed reports describing the exact state of development. These must occur since when the project moves from one phase to the next, responsibility transfers with it. The manager who takes over responsibility will clearly want to be certain that the milestones of the previous phase have been met. For this reason, the RDEM model has a burdensome level of documentation.

Project approach

The opposite approach to the RDEM model is to treat the development of a new product as a discrete project in its own right. One person is put in overall charge and they are responsible for all the necessary tasks. The continuity is good, and often individuals see the product right through from conception to manufacture. The process is characterized by speedy product development and a good match between what the customers want and what they receive. Accountability is good and control is easier to achieve than in the functional organization. The main disadvantage of this type of organization is that the engineers involved need to have many skills, and this requires that they work on tasks other than those at which they are best. With a project-based approach it can be difficult to ensure that useful work of a suitable level is always available for everyone, and it is not always possible to regulate the work to flow evenly, causing periods of frantic activity and periods of underloading. To operate effectively with this type of organization requires much more flexibility from both the organization and the individual engineers. This type of product development organization is shown in Figure 5.8.

Matrix

The matrix mode of operation involves a group of development engineers who work on various product development projects at different times. It is unlike the RDEM model in that engineers do not specialize in one particular area and it is unlike the project approach in that they do not see the project right through. Usually a development manager allocates staff for short periods as each development requires. This system offers great flexibility, since if a problem occurs in one product development programme a great deal of effort can be brought to bear very quickly. However, since no-one sees a project through and because the engineers do not have the opportunity to specialize, the results can be disappointing. This organizational structure tends to be suited to small groups of

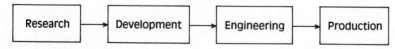

Figure 5.7   The RDEM model

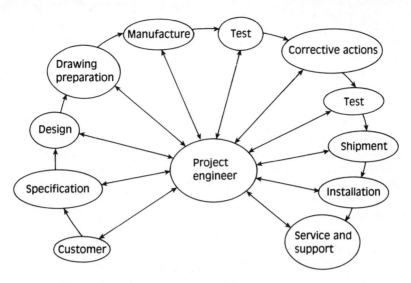

Figure 5.8   The project approach to product development

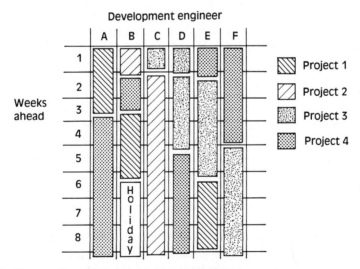

Figure 5.9   The matrix model of product development

highly motivated and gifted engineers who respond well to the constant stream of new challenges. The matrix model is shown diagrammatically in Figure 5.9.

Mixed
In some organizations it may be desirable to operate a 'mixed mode' in which one or more of the above methods operates simultaneously. For example, an electron microscope manufacturer may have two threads of product development. One

might be the development of major new machines that offer, for example, higher magnification or lower cost. Such product developments are aimed at the fundamental operation of the device and offer clear benefits to the customer. Typically, these product developments are slow, with new models coming out every few years. The other might be simultaneously developing numerous accessories for the product range. These may be one-offs for individual customers through to standard accessories that are sold with most units. Typically, these accessories are relatively cheap, take a short length of time to develop and are numerous. Clearly, the two aspects of the company's product development are very different in nature and should be handled in different ways. In such a situation it is common to find a project development team dedicated to the new instrument side and a small group of engineers operating a matrix of projects for the accessories.

5.3.3 Finance and product development

The financing of product development poses special accounting problems. This arises from the fact that while a product is under development, expenditure on it comes from the product development budget, which is a fixed sum. However, when it is in production the manufacturing department incurs the costs of manufacture on a continuous basis. A regular amount of money is consumed each month which is recovered from sales. In between is an ill-defined time. It might seem that the boundary between development and production should be clear but in practice this is not the case. Imagine a group of product development engineers putting the finishing touches to a bill of materials and correcting the last tiny error in the documentation. Clearly, by this stage the product can be made and no company will want to hold off manufacture of an important new product for the sake of some minor details of the documentation. In reality, production usually commences as soon as the development engineers are confident that the remaining development work can be done before production requires it. Development and production are, for some period, occurring simultaneously. This makes possible a very dangerous financial mistake.

The company will be anxious to determine the final accurate selling price of the product as soon as possible in order that profit may be controlled. Product development costs are paid out of profits, and fixed amounts are put forward to achieve specific product development goals. Manufacturing, by contrast, is a self-sustaining continuous process, the profits from one batch of units paying for the raw materials for the next. If the engineers performing product development assist the manufacturing department in the manufacture of the first units then some of the product development budget will actually be spent in manufacturing. The consequence of this is that the final product costings (used for final determination of sales price), which are based on manufacturing expenditure alone, will be too low. Later, when the product development support is finished,

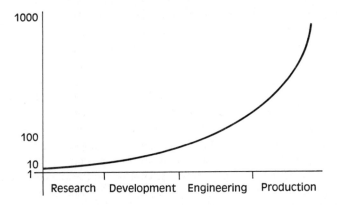

Figure 5.10 The cost of a design change

the costs of the support given by product development will have to be borne by production. Consequently, the product will cost more to make than was thought and less profit will be returned.

The lesson is that while product development engineers may sensibly be used to assist in the early phases of production they must not spend money on work that will be required continuously from the manufacturing department in the future. They must only spend money on activities that get the manufacturing department to the point where it can deal with all aspects of manufacture.

The financial importance of getting product development right is shown in Figure 5.10. The cost of making a design change in the research phase is only one-thousandth of the cost the same change would incur if performed on the product when in manufacturing. This comes about from the expensive equipment that accompanies manufacturing contrasted with the ease of respecifying a component at the design stage. Figure 5.10 vividly shows the importance of preparing an accurate PDS.

5.4 Management techniques in product development

5.4.1 Product development planning

Planning the development of a new product is a notoriously difficult thing to get right. The reason for this is that the product designers are reluctant to plan until the design is finalized, since one cannot plan when one does not know what work will have to take place. The executive management of a company, however, insist on plans since they cannot possibly commit themselves to a development programme without knowing what they will get out of it. For this reason, product development planning is usually divided into two separate phases: the development of the specification and the conversion of the specification into a product.

The preparation of the specification starts with identification of customer needs, the market analysis, and early concepts of how the product will be used. Any fundamental requirements are identified and data gathered. The end of this phase is marked by the preparation of the product development specification (PDS). Usually plans for the preparation of a PDS describe the investigations to be undertaken, the data to be gathered and the accuracy and limits on quality of the data that result. Plans do not normally extend beyond the preparation of the PDS at this stage since it is normal to review the product and decide whether or not to continue with the work. It may seem strange that a company would go as far as preparing a PDS and then choose not to proceed further. In fact it is common for companies to have many products under development and then choose the best for further development. The rejection rate can be as high as ten to one for new products at the PDS review stage. The costs associated with this phase are, however, small compared to the translation of the PDS into a product in production. Good product development companies argue that it is only by having a well-prepared portfolio of PDSs available that rational decisions may be made about which products should be selected. Certainly, it is true that companies who have only one product in the PDS stage are exposing themselves to a dangerously high risk.

A great benefit of having a PDS is that plans for the rest of the product development process can sensibly be prepared since the nature of the product (and therefore the work necessary to develop it) have been defined. Therefore, once a product design specification is ready and the product has been selected for development, a plan of the product development process through to manufacture is be prepared. Deadlines and milestones for various activities are identified. It is likely that the major deadlines will be imposed by external factors such as the need to meet a specific customer delivery or to have the product available for an exhibition. This phase can be planned in much greater detail. The plans should include the points at which design reviews will be held. Often the plans are based on the product development organizational structure that the company uses. For example, if it is the RDEM model, then the plan will detail the work to be undertaken in each of the phases. Planning techniques that can be used for this are described in Chapter 8.

5.4.2 Identifying customer needs

From the material we have covered so far it is clear that in order to develop a successful new product an organization must do more than simply have a good idea. It must be in contact with its customers and find solutions to their problems.

In traditional companies this activity used to be called 'marketing' and was performed by a department of that name. Today, however, successful organizations realize that placing the customers' needs first is so important that it must be an attitude of all departments within the business, and so all departments are involved in the product development process. Nonetheless, market investigations must occur to assist the business in understanding its customers, even if they are no longer performed by a separate marketing department. In product development such investigations are called 'market research'.

Market research is taken very seriously by product developers. For example, Sony sent a product development team from Japan to live and observe life in California. They were looking for opportunities for new products and were struck by the teenagers they saw carrying heavy hi-fi equipment while skateboarding. The need for miniaturization was clear, and this experience of the design team gave rise to the Walkman. Today, this progression in miniaturization seems only natural, but when the Walkman was first launched it was a revolution. Product developers only really understand their customers by meeting them and talking with them.

5.4.3 Product design specification (PDS)

Initial discussions with the customer result in a list of 'requirements' and 'desires'. This information is then used to develop a specification of the product. This specification is not in any way a technical description of the product, indeed, it may contain points that are very uncertain from a technical point of view. Often such a specification is described as being 'solution neutral', meaning that while it states clearly what is wanted, it says nothing about how it is to be achieved. How something is actually achieved is a design issue and should be solved by the designers. At this stage the requirement is for a specification, not a solution. The PDS describes the attributes that a product will need in order to be successful in the market, in terms of the customer needs. For example, the PDS of a new product to compete in the domestic washing machine market will include details of ideal maximum load, power consumption, size and all the other attributes upon which a decision to purchase may be based. The amount of information that the specification contains can become very large, and some companies produce a separate marketing PDS to cover the marketing issues and leave the PDS to contain only technical information.

A marketing specification will define the numbers in which it is expected that the product will be sold, the consumer groups who will purchase it and the all the other commercial information regarding the product. This is then used by the engineers responsible for converting this paper description of an imaginary product into a real, profitable product from which customer satisfaction and organizational wealth may flow. Once a specification is prepared it is used by those involved in the product development process as a working document. It is normal for it to be a controlled document with issue and revision numbers. Any factors likely to affect the specification must be examined and agreement reached. Any new information gained from further market analysis must be included as it is received.

The organization may have other requirements that have to be met by designers. There may be a requirement to use parts that are common to other products, there may be limits on the materials that can be used, or there may be certain regulatory safety or equipment standards that have to be met. These should be included in the specification.

The product development specification considers all these factors and usually consists of a series of headings with a description following each. Examples of some of the usual inclusions are as follows:

Physical principle	Size
Working environment	Weight
Performance specification	Service requirements
Reliability	Chemical effects
Product life	Appearance
Vibration sensitivity	Surface finish
Delivery rate	Noise emissions
Failure modes	Shipment technique
Sales price	Patent protection
Raw material supply	Special processes
Compliance with standards	Instrumentation

Notice that the specification contains no design calculations or justification of design principle selection. All such design work will be subsequently contained in design files.

When the product specification is prepared, design work can start in earnest. Creative design is a skilful process and it seeks to achieve a goal. People often think of designers operating in a vacuum, suddenly having an inspired idea about how something may done in a really clever way. Good design rarely happens this way. Like all creative processes, it happens best when the people involved are challenged and are motivated. The PDS and the available time and financial budgets provide the challenge to which the designers must rise. Good designers welcome an accurate PDS, seeing it as a valuable statement of what must be achieved. Poor designers see it as a limitation and an encumbrance. It is interesting to reflect on the importance design consultants place on the PDS. They are people who live by designing. Such organizations refuse to undertake design work without a complete and unambiguous PDS being agreed by both sides first. Such consultants know that unless a PDS is established one can never know when the design is complete, and for them, payment depends on doing exactly that.

5.4.4 Decision making

The product development process is filled with decision points. At every stage designers must decide which options to reject and which to accept. It is important that such decisions are made on rational bases. In fact, it is not just product development that requires sound analytical decision taking, all creative thinking does. For this reason, we have included a section on this important topic in Chapter 10. Special mention of it is made here since it is so important to the product development process.

5.4.5 CAD, IT and CAE

CAD (computer-aided design), IT (information technology) and CAE (computer-aided engineering) all have particular relevance to the product development process. They can be used as tools to greatly extend the capabilities of the company.

CAD

CAD in its minimal form is nothing more than a digital drafting board. This is how some organizations still view and use it. However, this is a gross misunderstanding of CAD. It is true that a CAD system involves people using computers to produce drawings but this is only one facet of CAD.

CAD is often introduced by a management wishing to increase the speed of drawing production. Once operational, this will almost certainly happen. In fact it is not so much the preparation of drawings that is speeded up as much as their manipulation and quality. After all, the drawings still have to be 'drawn' by entering them into the computer system and this takes time. However, once they are in the system their manipulation can be very quick. For example, modifications that involve moving large sections of the drawing around are impossible on a drawing board and trivial on a CAD system. If the drawing office staff understand the system and establish a library of modular product elements then these may be quickly recalled and changed to form new drawings more quickly. Other drafting advantages of CAD include accuracy, clarity and the ability for a draughtsperson to spend more time ensuring that the design is good and less in the act of preparing a quality drawing. In addition, the CAD system can be used to see how well parts fit together; they can be 'assembled' on the screen and the limits and fits examined. Some systems have features to examine how the build-up of tolerances will affect the assembly, clearly something no drawing board can do. Even these compelling advantages, however, do not do justice to the capabilities of a CAD system.

The major advantages are seen when the system is linked into the information system of the company. For example, a production control computer will have a library of the bills of materials for all the products in manufacture. When a new product is introduced to manufacturing it is therefore necessary to enter this lengthy and detailed data into the software. However, if it is prepared on a CAD system using a compatible format, the production control software can read it directly. The drawing office has control of the database and a modification becomes very simple to implement. In many companies modifications and maintenance of the manufacturing bills of materials consume the bulk of production engineering time. Anything that improves the speed and accuracy of this process is clearly desirable.

On top of all this, however, such a system has the advantage of shortening the period of time that it takes for a design to travel from design to manufacture, and so the new product lead time is reduced. This factor is, for most companies, more important than all the other benefits put together and they will be prepared to incur expense installing an integrated CAD system to obtain this benefit.

The CAD system can be linked into more than just the production computer. The sales office can use the same system, perhaps having their own drawings for use with customers who can then more rapidly gain an understanding of what the company has to offer and so buy more effectively. The design engineers can design using an up-to-the-minute library of materials, components and processes that are available and so produce more simply the best compromise between use of existing resources and acquisition of new ones. The accounting system can read a database of component costings and so produce more accurate costings, perhaps that even vary from batch to batch, showing how costs vary rather than having to use less accurate averages. In addition, time and materials spent can be booked directly to accounts that relate to individual products or subassemblies to any desired level of resolution. Clearly, such a company would be making use of computing technology that goes beyond the idea of CAD as a drafting tool. CAD is often the place where such developments start, but once they have gone as far as described above they have become IT and CAE.

Not all companies utilize CAD to this extent since the benefits that come from each area described above are more important to some companies than others. The central point is that to get the most out of their CAD system, a company must determine which areas of the company would benefit by being linked to the CAD system. In a highly automated, flexible manufacturing system, there will be many links; in a small, single-product line pressing company, for example, there will be fewer.

IT

Information technology is the name given to any work involving digital electronics to assist the flow of information in an organization in order to improve performance. Usually this means moving information between computer systems. This often requires a 'host computer' that interacts with the other systems, manipulating the data as it is moved between them. For example, a CAD file will contain a vast amount of data that is used to store the information necessary to draw the drawing. When the details are transferred to the production computer much of this information must be discarded and only the bill of materials data actually moved. Software is required to do this, IT software. Another example is electronic data interchange (EDI). This is using IT to replace as much as possible the routine telephone work serviced by the sales office with, for example, order progressing. It involves the company having a computer system to which its customers may log on and find information concerning their order such as delivery date, dispatch details, VAT codes, etc. The information is updated by the sales team and much time saved.

IT can be used internally or externally. For example, a DIY superstore may use software to examine the till receipts of its customers in order to optimize the layout of the store by determining the popular items. IT can also be used in the product itself. For example, the engine management system in many modern cars keeps a track of the usage history of the car so that at the next service the engine may be individually adjusted for its own typical conditions. This information is not provided to the customer, all the customer sees are services that are more effective at maintaining the car.

CAE

Companies use the term CAE to describe a complete product development process that is computer enhanced. In an ideal CAE system customers participate in the CAD preparation of their individual product, the system schedules and manufactures it for them, the lead time is small, the delivery is on time and the product information provided is correct. CAE is then a complete integration of computing into the product development and manufacture process. Not many companies operate such a complete system although many are moving in this direction. Such a system offers accurate, rapid meeting of customers' needs. However, the costs of so much equipment are high and its effective management is far from straightforward.

5.4.6 Drawings and drawing management

Engineers imaginatively create new things. They develop technology to produce new solutions to all manner of problems. Describing such solutions is not easy, even language cannot always cope. Imagine the confusion and ambiguity that would result if you tried to explain in precise detail something as simple as an electrical plug to someone who had never seen one before. Few people would depend on such a description to ensure accurate manufacture. The problem will become greatly exaggerated when one tries to describe something as complex as a car.

A car contains thousands of individual parts. Some will be bought directly from suppliers as standard components, others will be produced by other manufacturers to specific requirements, others will be produced in-house. Initially, the components are assembled together into small sections, each with its own function or integrity, called subassemblies. A gearbox, for example, or a dashboard, are both subassemblies of a car. Such subassemblies may require extensive testing and calibrating before they can be built into the final product. One would not manufacture a car, for instance, by putting all the pieces together and then testing it, only to find that a minor fault in a small subassembly caused it to fail the final test. Subassemblies are joined to other subassemblies and components to form progressively larger sections of the product. A car engine, for example, is a subassembly of the product but is itself composed of many subassemblies. In the final stage of the production process the last level of subassemblies are brought together, and the product is assembled in its entirety. Occasionally single components are also brought together in the final stage of assembly. For example, the engine of a car, containing thousands of components, is fitted on the production line, yet so is the windscreen which is clearly a single component.

So how do you describe a product as complex as a whole car? Where do you start? The description must be sufficiently accurate to allow repeated manufacture of identical cars. It must be utterly unambiguous and must deal with the very different ways in which components may be made. The answer that engineers have developed is a language of their own, which copes with this specialist communication problem. It is the language of drawings.

There are two major areas associated with drawings and we shall now investigate each in turn. The first is the use of drawings in practice in which we shall consider how drawings should be prepared well from a management point of view. The second is the drawing office itself, which has responsibility for the preparation of new drawings and the modification of existing ones.

Drawings in practice

In this section we shall not consider the technical aspects of drawing preparation. Technical drawing is a skill in its own right and engineers study it separately. There are many standards that govern the preparation of drawings and each organization has its own preferences and standards. We will, however, consider drawings from an information control point of view. We will do this by looking at some examples.

Figure 5.11 shows a drawing of a component. The drawing shows not only the engineering aspects of the component but also much management information. The important features of the drawing are as follows:

● *The details*: The majority of the area is taken up with the details of the component itself. The dimensions, tolerances, manufacturing techniques and all the other important aspects of manufacture are described in such a way that repeated manufacture of acceptably similar components can be guaranteed. Of course, no two components can ever be absolutely identical, and so the choice of tolerance bands are an important part of the drawing. The drawing does not show many properties of the component which may be important in its operation, its second moment of area perhaps, or its thermal conductivity. This is because such properties are themselves a function of the dimensions, materials and processes specified, they are 'implied' attributes and therefore stating them is a repetition.

● *The bill of materials*: After the details themselves, the bill of materials is the most important item on the drawing. The bill for the drawing in Figure 5.11 can be seen above the title block, and contains two items, a plate and a bush. From the drawing it can be seen that these two items are welded together to form the top plate. Of course, each of these components will in turn need a drawing to describe them and elsewhere in the organization the drawings MME0123 and MME0355 will be stored. These items are said to be 'called up' by the bill. Each drawing calls up the components necessary to manufacture the object it depicts. In the end, drawings will be called up that show items that are very simple and are not composed of more than one thing. Such drawings are called 'component' or 'detail' drawings.

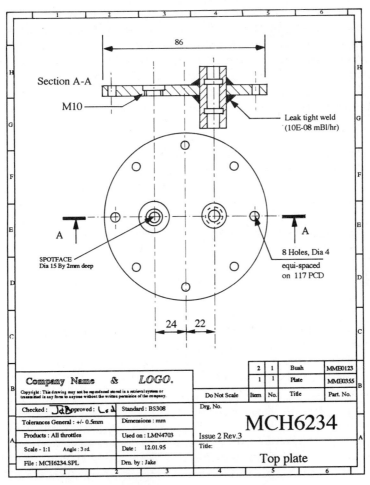

Figure 5.11   A drawing of a component

● *Border*: The drawing is surrounded by a border which not only allows the edge of the drawing to become damaged without making the details unreadable but also contains a grid reference system. Occasionally this can be useful when engineers are, for instance, talking about a drawing over the telephone and need to specify the area they mean. It is of particular help when a drawing is very complex.

● *Company name and logo*: This is usually displayed on a drawing together with a copyright reminder. By virtue of being prepared on company paper the company own the copyright of the drawing. If the organization is involved in restricted work, such as defence, the drawing may carry other warnings about reproduction.

● *Drawing number*: All drawings have to be uniquely identified. There is obviously no point in having two drawings with the same number. The simplest way of doing this is to number them sequentially and many organizations do this. As the number of drawings increases, it helps both the filing and interpretation of the drawing if a number that carries meaning is assigned to each drawing. Often the number will indicate something about the item, if it is machined, bought in, welded, or what class of component it is, for example. It is common for the engineers to become so familiar with these numbers that items are referred to by number rather than descriptions.

● *Title*: A title, as specific and descriptive as possible, is assigned to each drawing to help in understanding the overall function of the component. Care must be taken to avoid repetition of obvious but uninformative titles that get repeated easily such as 'flange' or 'bracket'.

● *Reference data*: At the bottom of the drawing on the left are ten boxes that contain more information. Two of them, 'Products' and 'Used on', are described in the next paragraph. The 'scale' box states what scale has been used and, in this case, the projection system (third angle). 'Date' of drawing records the date of first issue, the 'file' box records the filename under which the drawing is stored on the computer system and the 'Drn' box records the draughtsperson who drew it. It is common to include such matter for assistance in controlling drawing production and in rapidly locating drawings within complex filing systems. The remaining boxes are self-explanatory and they record the tolerances used throughout the details, the units that have been used for dimensions, and any drawing standard that applies.

● *Bill of material references*: As well as the bill of materials which shows those items necessary to make the component, it is normal to reference those items on which the component is used. Two boxes in the drawing refer to this, 'Used on' and 'Products'. The first box indicates the next item 'upwards' in the bill, i.e. that item whose bill of materials calls this drawing. The second indicates the ultimate destination of the component, usually a product. Of course, a component may be used on more than one item and there may be a list in the 'Used On' box. These two pieces of information are of particular use when engineers are discussing possible modifications to drawings since they can see at a glance how many other components or products are affected. There are clearly problems with the number of entries in this box if a given part is used on tens, or even hundreds, of products. For this reason, some companies ignore the reference and use the bill of materials directly to trace the usage of a part. In this way the bill of materials relates every single component that forms a product to all the others. To find out more detail about components one moves down the bill to progressively more detailed drawings and to gain more of an overview or see how components relate to each other, one moves up the bill to drawings that progressively show more and more of the product but that have less detail.

Figure 5.12 shows a complete assembly in which the component shown in Figure 5.11 is used. Such a drawing is known as a general assembly drawing. Item 6 on the bill of materials is the top plate, drawing Number MCH6234. In the details of the drawing the top plate can be seen surrounded by the other components that form the phase change and sensor assembly. However, in this drawing none of the top plate details are shown.

The system of describing physical objects outlined above is very effective and is adopted in some form by practically every engineering organization in the world. It has clear advantages of economy, thoroughness of description and ease of interpretation. However, these advantages can soon be lost if the drawings are not managed in an intelligent way. In the following we mention the most important issues that must be addressed during the management of a drawing collection:

- *Avoid repetition*: The structured approach to drawing preparation, where each drawing 'calls up' the detail drawings of the parts that compose it, is called an indented set of drawings. It is self-evident that quality manufacture of products from the drawing set cannot be guaranteed if there are inconsistencies within the drawing set. If one drawing shows a particular dimension to be one value while another shows it to have a different value, errors will be inevitable. Consequently, if a dimension is to be changed, it must be changed on all the drawings upon which it appears. If the draughtsperson is to avoid having a list of all the drawings on which every dimension appears there will have to be rules to govern the repetition of dimensions. The simplest rule is 'never repeat information'. This is also a sensible rule from the point of view of economy. An exception is the inclusion of an overall dimension on a general assembly drawing without which the reader would have no idea of size. In Figure 5.12 an overall diameter is shown.

- *Don't draw everything*: There is no upper limit to the amount of work that can be put into perfecting a drawing set. The issue at stake is not whether the drawings are perfect from an esoteric point of view but is whether the drawings are fit for their purpose. In an organization mass-producing very complex equipment the need for complete and accurate description extends to nearly every component. If, on the other hand, an organization is making just one demonstration unit for internal evaluation only, the drawings need show much less detail. At one extreme is the drawing of every component, at the other is the preparation of a few sketches, circuit diagrams, critical dimensions and photographs. Even when great detail is

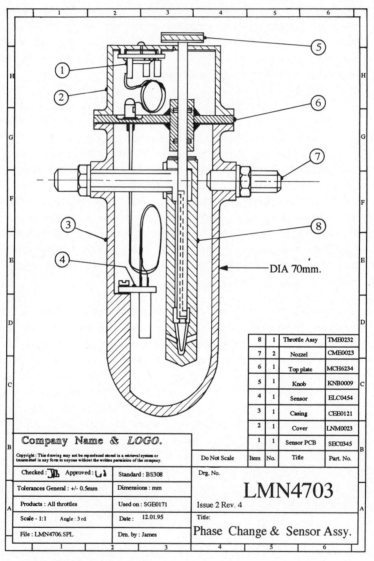

8	1	Throttle Assy	TME0232
7	2	Nozzel	CME0023
6	1	Top plate	MCH6234
5	1	Knob	KNB0009
4	1	Sensor	ELC0454
3	1	Casing	CEE0121
2	1	Cover	LNM0023
1	1	Sensor PCB	SEC0345
Item	No.	Title	Part. No.

DIA 70mm.

Company Name & *LOGO.*

Copyright: This drawing may not be reproduced stored in a retrieval system or transmitted in any form to anyone without the written permission of the company.

Checked :	Approved :	Standard : BS308
Tolerances General : +/- 0.5mm		Dimensions : mm
Products : All throttles		Used on : SGE0171
Scale - 1:1	Angle : 3 rd.	Date : 12.01.95
File : LMN4706.SPL		Drn. by : James

Do Not Scale

Drg. No.

LMN4703

Issue 2 Rev. 4

Title:

Phase Change & Sensor Assy.

Figure 5.12   A general assembly drawing

required savings can still be made. In Figure 5.12 one of the components shown in the bill of materials (item 5) is a knob. It might well be that this item is bought in from a manufacturer who mass-produces the item. There is clearly no point in drawing such an item in detail. The entry on the production control system for each part number will usually indicate whether a drawing exists or not. The important issue is that the item is completely described and in the case of a bought-in item this may be accomplished by simply listing the manufacturer's part number.

● *Support documentation*: In the preparation of products many documents other than drawings are required. These might be assembly instructions, test instructions, specification documents, packaging instructions, manuals, quality control documents, safety procedures. All these documents are important and must be obeyed to ensure fault-free production. With an indented drawing system it is easy to relate each of the documents to its assembly by the part number. A database of all the part numbers, often held on computer, states which documents apply to the part in question. This technique allows the organization to have as much or as little detail as is appropriate yet still know for each part what documentation applies. In some industries there are specifications that apply to the preparation of such documentation. For instance, some military specifications state that every component must be drawn and, whenever two components are joined together a procedure, referenced in the drawing must be followed to assemble them. If procedures such as this are used the amount of documentation that must be produced becomes enormous (for example, the documentation associated with the development of a passenger jet airliner weighs far far more than the plane itself). In such cases, the organization must see documentation as part of its business. The preparation of such documentation is clearly onerous but when managed well, it does bring excellent product quality control.

The drawing structure therefore provides a framework upon which other documentation may by referenced.

The drawing office

The aim of the drawing production system is always to be able to provide accurate drawings to meet the needs of the other areas of the company. This means that the drawings must be ready when required, that they are distributed to those who need them, and that they are maintained in a 'current' state afterwards. Often this responsibility rests with a drawing office in which a group of draughtspeople report to a drawing office manager. It is becoming more common, however, for drawing offices to be seen as extensions of 'design offices' with designers using a computer-aided design package and then producing their own drawings.

Preparation

The drawings are prepared on either CAD systems or traditional drafting boards, in accordance with the standards used by the company. A manager is usually responsible for the running of the drawing office and plans its workload according to the prevailing priorities of the other departments. The draughtspeople work closely with the design engineers to ensure that the drawings are correct from a design point of view.

During the preparation of drawings there must be a well-defined system for revision, particularly following a design review. This is because while the product design is evolving, so are the drawings. They must reflect the current state of the design and they must be controlled. It is essential that the system used to carry out this control is appropriate to the level of control required, the company and the product, and it must not inhibit the design process. The control method should ensure that the reasons for design changes are adequately documented. There must also be effective controls to ensure that drawings in preparation are not inadvertently released to the shopfloor for manufacture.

Parts lists and bills of materials may be produced by the drawing office when the final drawing set is complete, or they may be produced and updated throughout the design process, with the drawing office having responsibility for checking part numbers and format before they are entered into the production control system.

Before a drawing is finished it must be checked. It is clearly important that the drawings that are used throughout the organization are correct. There are several aspects of the drawing which must be checked: its drafting integrity, its design, its ease of manufacture and its suitability for use with other drawings. Responsibility for checking drawings from a drafting point of view is usually given to a draughtsperson other than the one who drew it and a box is provided for an approval signature, as shown in Figures 5.11 and 5.12. Responsibility for checking the technical content of the drawing is usually given to one engineer.

Drawing release

Following completion of the drawings the drawing office's responsibility is then to release, update, store and retrieve the drawings as appropriate. The release procedure is the way in which copies of the master drawing are made available to others for the purposes of manufacture, review, preparation of quotation and so on. The procedure used must ensure that the people using the drawing always have the current drawing available. This may mean that a new set of drawings is released every time a new batch of product is to be made.

Effective storage and retrieval of drawings is important because of the requirement to release drawings to the manufacturing department for manufacture, but is also important because the drawings form a record of the product at each stage of its life. The importance of this is illustrated when a customer who bought a product some time ago returns and either wants more products built to the same specification, or wants spare parts for the existing product. The manufacturer must know what was used in that product.

Drawing release means sending copies of the relevant drawings to the manufacturing areas that require them. In an organization running a formal production control system there will be an organized process to respond to incoming orders and to ensure that all the relevant departments receive the drawings they require. Exactly how this procedure operates will depend on the nature and size of the organization concerned, but it is easy to see that several departments may require the same drawing. Sometimes it happens that a drawing that needs to be released has some modifications required that have not yet been implemented. It takes time to modify drawings, and if the release of the drawing is urgently required something must be done to get round the problem. The solution is to issue 'marked-up prints'. This describes the process of taking a print and then making the modifications in longhand on the drawing, on every copy that is required. The modifications must be initialled by a draughtsperson and an engineer in order that the modification is authorized and to prevent the organization working from unofficial 'marked-up prints'. The modifications may then be made later. Clearly, there is room for error in this technique, and it is therefore normally only used in exceptional circumstances.

The purchasing department will need to refer either to the drawing itself or to the bill of material in order to purchase raw materials. The manufacturing function will clearly require the drawings in order to machine and produce the component. There may well be several different production versions of a drawing showing the different stages of manufacture. After this the inspection department will require drawings against which to inspect the components, and often these are separate 'inspection' drawings showing only the important issues requiring verification. The test department may require drawings in order to test and optimize a product. Finally, there will often be a packaging drawing showing the way in which packaging is to be carried out and what items are included for shipment.

Drawing modification

Design is never over. Even when products are in manufacture, there is almost always design work under way. This may be to reduce difficulties in the manufacturing process, it may be to develop the design, perhaps reducing cost, or it may be because a major design review is in progress. All these result in a need to modify existing drawings.

Unless it is likely that a proposed design modification or change will affect the basis of the original design, the modification will probably be carried out by the drawing office. Any modification should be checked and approved and a drawing updated before it is carried out on the shopfloor. Obviously, if there is an urgent production problem which requires a speedy modification this procedure may put a significant delay into the system, which is unacceptable and marked-up prints may be used.

In addition to the implications for the design, modifying products and parts will have implications for the business, its suppliers, and customers, and these should be considered before a modification is made. If a product specification is changed by a modification then future customers need to be aware of the new specification. It must always be the case that a customer receives a product of the specification ordered, regardless of how difficult complexities of the manufacturing system and numbers of units make it to achieve. Particular problems can arise if a customer is expecting a number of products all to the same specification and the modification would mean a change part-way through the order. Similarly, if the company has placed orders with suppliers and then needs to change the specification or number of parts then it should ascertain whether this is contractually viable and whether the supplier is prepared to agree to the change in requirements. For the company itself a design change may result in stock becoming obsolete or processes being changed, both of which can result in significant cost.

5.4.7 Design reviews

At various stages in the design process there should be design reviews. These should be carried out formally and include people who are not intimately involved in the design. The aims of design review are both organizational and technical. The organizational aims are to check that the design is going to plan, to confirm that it is correctly defined, and that the product is developing in the way that the organization desires. The technical aims are to ensure that the specification is being met, that appropriate design options have been considered, that the design is feasible for manufacture, inspection and test, and that critical calculations and decisions are correct. When changes are made to the design following review they should be documented so that the reasons for the changes are clear to everyone, even long after the decision point has passed. This ensures that the rationality of the changes is open to scrutiny and that work is not retraced.

The design review usually follows a formal meeting agenda. The designers will give a presentation describing the work completed and written material will be circulated for reading and scrutiny beforehand. The design reviews are usually held at major milestones during the product development programme which are laid down in the initial project plans.

5.4.8 Intellectual property rights

Intellectual property rights are an area of law that refer to patents and trade marks. They may be used to provide a commercial advantage by preventing competitors from using the ideas of the organization. We shall now look briefly at these areas concentrating on how they may assist product development.

Patents
Patents date back to Elizabethan times when the Crown wished to encourage the importation of good ideas from abroad. At that time you could travel abroad,

learn of some new process or idea that was not used in England then bring it back and claim it as your own. The law gave you a monopoly for twenty years during which time nobody else would be allowed to make a profit out of 'your' idea. This piece of legislation was successful in bringing new ideas and also made a few entrepreneurs very rich. When the twenty years had expired, the idea passed into the public domain and anybody could use it. This principle lasted right up until 1978, when it was changed and instead of merely bringing an idea from abroad one now has to think of something entirely new.

This twenty-year window of exclusivity makes for a very favourable commercial situation since it removes any of the business problems associated with having competitors. The organization has the entire marketplace to itself and anybody wanting to purchase the product will have to approach the organization. With this advantage, an organization that owns a patent on a product (or even just a part of a product) is virtually guaranteed success; provided, of course, that the product has a market. The important issue is owning a patent that cannot easily be infringed and which has commercial opportunities. These rewards are the legal recognition of the moral right an inventor has to a share of the profits that the idea produces.

A patent might cover a completely new invention such as the hovercraft or the transistor, or a process, for example the Pilkington process which aids greatly the manufacture of glass. It might be an add-on device that solves some problem with a standard piece of equipment. The important thing is that it is novel. The idea is then written down in a patent application. This application must contain sufficient information for someone knowledgeable in the field to be able to understand the idea. One vital condition is that the idea has not already been disclosed. Thus if a research student publishes a paper describing a new idea and subsequently wishes to patent it, a patent cannot be granted. Once published, the idea passes into the public domain. The date at which the application is lodged with the patent office is the important one, and any similar idea lodged afterwards will not be granted a patent. After the application has been received a 'patent search' is conducted in which a patent office employee specializing in the area of the patent will search the entire patent database for possible infringement of existing patents. If there are previous patents that the new application infringes then the inventor may have to modify the invention in order not to infringe them. In the worst case a previous patent is found which describes the new invention completely and, of course, no patent can be granted.

If the idea is new, the inventor must decide in which countries the idea should be protected. Patenting is an expensive process costing about £12 000 per country in which protection is sought. Furthermore, if it is contested there will be expensive litigation to pay for, which will very likely cost much more than the patent application itself. For these reasons, private individuals cannot normally afford to exploit their ideas through patents and they enlist the help of commercial organizations with the financial ability to provide protection in exchange for a share in the profits.

After twenty years the idea will be in the public domain and anybody can use it. Twenty years might appear a long period but the introduction of a completely new idea to a market is a slow process and companies often find their patents running out when the product is in full production. The important strategic achievement is therefore to use the twenty-year protection period to establish a position that effectively excludes new entrants to the market. If at the end of the period a company has established a cost-effective, efficient and dependable distribution to an overwhelming fraction of the market it will be almost impossible for new entrants to justify the investment, and a position of commercial security will have been established.

Patents are emotive issues for creative designers: they symbolize the value that society puts on creative thinking. As such, product developers often think that their companies should actively be pursuing as many patents as they can. This is not true. The goal of product development is to make the budgeted profit out of a new product. If this can be achieved without the use of costly patents, then so much the better. It is perfectly possible for a company to develop and exploit a new idea and use its skills to bring a product into the marketplace and secure an unimpeachable position without the use of patents. After all, this is exactly what a successful company must do with a product that cannot be patented.

Patents should then be used as a tool in the product development process in an appropriate way. They make excellent protection for a minority of product developments and provide a breathing space that can permit spectacular financial achievements to be made. On the other hand, they are extremely expensive and require the services of a patent attorney to administer. Lastly, the patent search can be a most useful information source. Unlike scientific papers, all patent applications have the financial backing of someone who believes the idea will be profitable. The patent search is relatively cheap and so is a valuable source not only of technical information but also of commercial information. For example, a collection of searches could be used to reveal on which areas of research the competition are concentrating.

Trade marks

A trade mark is a word or logo of some sort. Its purpose is to distinguish your product or service from someone else's. They are familiar to everyone – for example, the Coca-Cola logo, the Shell petrol sign or the Mars Bar logo. They each have an instant association. This makes them potent tools for advertising and promotion. For example, a petrol company television advertisement may show a tired motorist pulling into a very clean and well-maintained forecourt with a shop. After refuelling and buying a sandwich and a warm drink the motorist leaves refreshed. The advertisement will undoubtedly show the company logo on the illuminated sign. This message is then recalled by a driver passing a service station and seeing the illuminated sign. An entire message is conveyed by a trade mark. It is clear that the sign must appear the same in both cases, and the colours used in trade marks are usually defined very accurately. No deviation from the prescribed format is permitted by the organization since they want the association to be strong. Additionally, if the sign on each service station was a little different the customers might perceive the company as unable to control quality in their advertising and assume them the same applies to their products.

5.5 Case study – David Main; BEV Appliances

David Main works for BEV appliances. This is a large domestic appliance manufacturer with a range of nearly twenty products, although most of the income is generated from just two of them, the toaster and microwave oven.

David graduated two years ago and joined BEV straightaway. He received a company training that involved him working for short periods in many different departments. There were various business skills and awareness courses that he attended along with other graduates that had joined the group of companies, and he had a 'mentor' assigned to him to help him find his way around the company and ensure that he got the most out of his training. After about seven months he joined the product development department and spent the last five months of his training period there learning how product development was achieved within BEV.

David describes the work of the product development department in his own words:

I have four main projects on the go at the moment. The first is a development of the 'Vayerous'. This is our main toaster product which sells in vast quantities. The second is a cost-reduction exercise on another product that has been in production for some time. The third is an investigation of a completely new product concept that we are evaluating. The fourth is a review of plastics. There is a possibility that some of our components could be made in plastic and we have to look at the option. Apparently this comes up every few years and, for one reason or another, we stick with metal.

On top of these I have to support the product development issues that come up from production. For example, two months ago a supplier went out of business. They made a particular mechanical timer that we have been using in a spin dryer. The dryer has a knob on the front and the user dials up the amount of drying time required and the timer counts down. It's like an old eggtimer mechanism. We don't sell many of this product and have enough timers to last until December (five months away) but replacing a bought-in subassembly like this takes time. I have a list of potential suppliers now and I have visited three of them with my boss. We have one supplier left. Once the decision is made we will have to prepare a new mark issue for the product. This means new drawings, and there are bound to be differences in the way the subassembly fits into the product. After this, because it is a product for sale to the public and consumes electricity, we have to get legal approval for the change and have the design specification registered so that we can continue to sell the product. In a case like this where the change is small its not very difficult, but for a new product it can take ages. I'm sure there will be no problem in doing this but it all has to be organized. With a job like this you simply cannot afford to have sales stop because we didn't start the work early enough.

We get a lot of general technical queries up here as well; we were asked to help the secretaries in the front office set up a new computer-based office last year. This was obviously not much to do with developing new products but in a way it is useful to become involved with things like this. My boss said that the department had to do a similar exercise for the accounts computer, and in that case we were able to select a system that fitted into the way products are made and developed in this company. There are loads of different systems about and most of them can be configured in lots of different ways. It was only by having technical involvement from this department that the system chosen could be guaranteed to be appropriate for our system.

Anyway, back to the main projects. The development to the 'Vayerous' is the inclusion of a 'brownness' detector to tell when the toast is cooked. One or two rival products have this feature and we have them in the lab downstairs. I conducted a 'competitor appraisal report' on them. This is a standard company report that evaluates the position of competing products on a technical basis. Both systems use a photocell to measure the reflectivity of the bread and this works quite well on white bread, where the colour change is quite marked, but is not good with brown bread. Our product is near the top of the price range and many of our customers eat almost all brown bread. My job is to design and develop a system that works reliably for brown bread. My boss and I work together on this project. We have in infra-red system that works well but it is impracticably expensive. Our next project milestone is to have chosen the basis upon which the system will work. We have a brief to fulfil that states the performance criteria and an estimate of the cost of components required. The performance is compared against an internal guide we have for toast quality which deals with mechanical properties of a slice of toast made from bread of a particular moisture content, etc. We even have a machine for determining the Young's modulus and failure load of toast! There are two other possible techniques for measuring the brownness that we are trying and I have to get the tests ready for about a month's time. They are much cheaper and so far look as if they will work well; I'm afraid I'm not allowed to tell you how they work, though.

The cost-reduction project is very straightforward really. We have a deep fat fryer that has a special insulation material to keep the outside from getting too hot. Sales of this product have been slowly increasing for some years now and we have a situation where a cost saving of even fifty pence on manufacturing costs would make a design investigation lasting a few weeks well worth the effort. There is a feeling about that the increased numbers in production should make an alternative to the existing supplier favourable. In this work I have looked at alternative designs, but these are very limited. It all comes down to how much you have to pay for each unit of thermal resistance. Each manufacturer has its own brand, and to complicate matters, one option would be to make our own here in-house, buying in the much cheaper raw materials. My boss has to answer this question at the board meeting in three months' time so I have to give him all the data he needs to prepare the report by then. If we recommend the in-house option then there will be a lot of other things to sort out, like the financing and the effect on manufacturing. If this happens someone quite senior, probably a board member, will take responsibility for the project and product development will no doubt continue to be involved, but many other departments will have a role to play too.

The third project is one I particularly enjoy. We had a meeting about eight months ago. We have it every year when we all try to think of new products. There is a brainstorming session led by one of the marketing managers. There were all sorts of people there, some from other companies within the BEV group and some people with no technical experience at all. We were in groups and ours was talking about breakfast time. We were looking for the problems that people face as they get up and have to get ready for work, naturally with a view to the products they might buy. Somehow we got onto the washing up, and how you always find that either the thing you want is dirty and in the dishwasher or dirty and on the draining board, the point being that it's never clean and in the cupboard. That was how we thought of the cupboard that is a dishwasher. You simply put the things in the cupboard and the next time you open it the stuff is clean. It was great, this is what the customer actually wants. There were obviously loads of technical and cost implications and part of the oneday meeting is about all of that. Each year two or three ideas are taken further and the cupboard/dishwasher was one of them. What happens is that after the brainstorming session we have a selection procedure and the ideas that are selected are then specified in a general way. It is by no means a PDS but it is sufficient for someone to go away and start developing one. For the cupboard/dishwasher, that was me. So far, I have some market information on who might buy and in what quantities. I also have about five possible technical layouts. There are a number of limitations on the design from the word go, like supply voltages and cleanliness standards, and, for example, the size of a typical kitchen cupboard is a real problem. We have to offer something new, an advantage that will persuade the customer away from existing cupboards and into something new. Often these sessions come to nothing but sometimes they do produce good ideas and they do make you think about the products and how people actually use them and what they actually want.

Lastly, there is this review of plastics. I've only just started on it and have to say I have little enthusiasm. I'd rather be doing all the other work. I have the report from the last time this was examined and I think the simplest thing to do is to use that as a starting point, see how things have changed and then see if the changes affect the conclusions. With a bit of luck, that will be the quickest way of doing it as well.

5.6 Summary

Product development is one of the greatest challenges that a company faces. We have seen how the needs for customers change with time and examined ways of characterizing these changes. We have seen different procedures for product development that may be applied. We have also examined a collection of the most important tools that are available to the product development engineer.

Product development is naturally attractive to creative designers. Many engineers come into engineering believing that engineering actually is product development. However, product development is difficult to do effectively. Not many engineers have the skill to take an idea all the way into the market, and the journey is never easy even for those who are practiced at it. The procedures and tools described in this chapter should be used and modified to suit the situation in hand. It does not matter if the detail of the procedures is varied or if some of the tools are not appropriate. It does matter, however, if the underlying principles are ignored or if the tools are incorrectly used. It is sometimes said that there only two ways to go out of business. The first is right now by mismanaging the finances, the second way is in a few years' time. This is achieved by not developing new products that sell well when those in production at the moment will no longer sell.

Product development is like any other raw material or department of a company. One cannot do with too little or too much of it. One must buy enough of it, at a good enough price, without having any left over or buying any of the wrong sort!

5.7 Revision questions

1. Why is there a need for new products?

2. Describe how GAP analysis may be used to assist in the product development process.

3. All companies require product development. Discuss.

4. How does product development in a domestic appliance manufacturer differ from that in a machine tool manufacturer?

5. Describe three models of the product development process. What are their similarities and differences? Why should they have anything in common?

6. Write your own product development specification for a new model of camera aimed specifically at students. You must solicit the opinions of ten customers and use the data you gather to assist in the product development specification.

7. Describe how drawings are used to assist the engineering description of complex products such as a car. What advantages do drawings bring?

8. Describe the management information normally contained on a drawing.

9. What is a bill of material?

10. Explain in your own words the following terms:

    'Marked-up print'
    'Drawing release'
    'General assembly'
    'Detail drawing'

11. Describe good practice in the management of drawings. For each point you make give an example of the sort of error that might occur if the practice is not followed.

12. Who normally checks a drawing and why?

13. How can a company try to ensure that its new products stay ahead in the marketplace to the exclusion of all others?

14. Choose a new product with which you are familiar that has recently come onto the market such as a new camera or hi-fi module. Who is the product aimed at? What evidence have you for this? List the product attributes that are important to this group of consumers. Try to estimate the cost distribution within the product and see how well this agrees with the attribute list (start by dividing the product into a list of twenty or thirty subassemblies and components).

References

1. Pugh, S. (1991), *Total Design – Integrated Methods for Successful Product Engineering*. Addison-Wesley. A complete look at the product development process placing especial emphasis on the product design specification and the 'Design Core' activities of Market, Specification, Concept Design, Detailed Design, Manufacture and Selling. An excellent text that is easy to read.
2. BS 7000 (1989), Guide to managing product design – Section 3, British Standards Institution. The standard includes sections on managing at the corporate level, project planning, financial control and the design brief. All these are important product development topics and the standard makes excellent reading for aspiring product development engineers.
3. Pahl, G. and Beitz, W. (1988), *Engineering Design: A Systematic Approach*, The Design Council, London. A very thorough and rigorous presentation of a design methodology. The book explains the design process through the establishment of a specification through conceptual design, embodiment design and detail design. An important book that should be known to all designers.

6

Operations management

Overview

Engineers are inextricably linked with the process of manufacture. Some engineers will work with manufacturing systems, others will design products for subsequent manufacture. All engineers will need some knowledge of the manufacturing process and the factors that affect its operation. In this chapter we will examine the issues relating to the management of the manufacturing processes.

6.1 Introduction

The operations activity in an organization is where the processing of inputs, in the form of materials or information, takes place to produce the outputs of the organization, i.e. the products or services that meet the needs of its customers. In a manufacturing company the operations activity will include all the areas associated with manufacturing the product, in a design consultancy it is the activity associated with carrying out the design work and producing the design specifications and drawings required by the customer. An operations activity can be represented, as shown in Figure 6.1, as a number of inputs to processing, a processing activity, and outputs. Operations management is concerned with the balancing of these three. Let us consider the case of Alesales Ltd.

Alesales Ltd make home-brew beer kits. They have three kits, for stout, bitter and lager. The kits are assembled from raw materials, malt, hops, oats, etc. which have to be packaged into small containers in the quantities required for the particular recipe. The kits are supplied with ingredients and brewing instructions in a printed carton. When the company was established the kits were assembled by hand, with two staff carrying out all the operations. In order to benefit from supplier discounts all materials were bought in bulk on a monthly basis. The kits were made and held in stock at Alesales premises until they were required by small retailers or individuals who responded to a national advertizing campaign in *What's Brewing*. As the business grew, the two staff were unable to meet the output requirements and gradually the number was increased and some

Figure 6.1   The operations activity

automation was introduced, a malt extract canning line. The company continued to make for stock, buying materials and selling as previously.

Already we can see that Alesales needs to know a number of things about its operations activity in order to manufacture and trade successfully, i.e.:

1. What ingredients and other parts are required for each kit
2. How many kits of each type it is going to make so that it can make purchases and determine what the personnel are going to do
3. How much the kits cost to produce so that it can make a profit on sales.

In addition, someone must have thought about the cost of the canning line and the impact that this would have on the business, as well as the premises required, their size and layout.

After a few months of operation the company started to receive a number of complaints from the customers of the stout kits; the gravity, i.e. the alcohol content, of the beer was too low and there was little flavour. An investigation found that the complaints related to kits that had been in stock for a long time. Consequently the flavour of the hops had diminished and the age of the malt meant that it was not providing the correct sugar-to-alcohol conversion. At the same time, the company started to have cash flow problems even though it had a large number of customer orders. This meant that there were insufficient funds in the bank to meet bills for materials. This latter problem was particularly puzzling, as the company was doing well by all accounts.

What was happening at Alesales? The company was clearly prospering, as it was able to increase production and maintain a full order book, and yet kits were defective and the company had money problems. In order to examine the possible causes of these difficulties we need to know more about the production system. First, let us look at what may cause the defective kits. There is clearly a problem with the storage of the kits being excessive. There are two possible reasons for this: (1) the production was not tuned to the customers' requirements, i.e. the product mix was incorrect, with the consequence that the stout kits were stored for a long time because they were excess to requirements, or (2) the retrieval of kits from storage did not allow for the time in storage – a last in, first out system leaving the kits there that had been stored for the longest period of time.

Now, let us consider the cash flow problems. Could these be linked with the problems of the stout kits? This is probable if a lot of kits have been made and stored, and there is a delay before sale. In order to make something you should have procured the raw materials, which means paying for them, and you have to pay for the people, machines and factory which allow the manufacturing process to take place. Goods sitting in stores represent money spent and are not bringing money in. The act of storage is a non-value added activity in that it is an activity that is costing the organization money but which adds nothing to the value of the product for the customer. We also know that the materials for the kits were still being bought monthly even though it is probable that the selling pattern of the kits is different from that which existed in the early days of the company. Indeed, we do not know if this monthly ordering system is appropriate to the manufacturing lead time, which is the time required to make the product from receipt of customer order to its being available for sale. If the manufacturing lead time is only a few days, or even a few hours, then a month's stock begins to look a little excessive. It is likely that a lot of money is being spent on production and then there is a significant delay before this is being recouped from customers.

To address these problems Alesales Ltd needs to assess the management of its operations activity. The company must know what product mix is required by its customers and plan production accordingly. It must also plan purchases to meet its production schedule. It must consider how it manages its materials and how it ensures that the quality of its output is not jeopardized. The company must also recognize that its financial viability is affected by the way in which the operations activity is managed, in particular because of the costs of production which must be incurred before sales can take place. To achieve an effective operations activity operations management techniques are used to plan, schedule and control all the activities associated with the operation, and to manage the resources (materials, people, finance, time and machines) that are required to achieve the output.

Alesales is a simple case, it has three products, a few raw materials, a small workforce, and only one automated line. However, the principles of operations management apply to all organizations, irrespective of their size or product.

Operations management covers all the activities associated with performing the operation – the procurement of materials, the production scheduling, the layout and maintenance of plant, the quality assurance of the processes. In this chapter we shall look at the issues associated with the procurement and management of materials, and production planning. Quality assurance is covered in detail in Chapter 7.

The learning objectives for this chapter are as follows:

1. To look at the way in which manufacturing activities can be organized and to understand the implications of the type of organization
2. To describe the role of production planning and the data required to achieve this

3. To introduce the concept of materials management and to describe some of the methods used in industry, including materials requirements planning (MRP)
4. To explain the Just-in-Time method of production control.

6.2 Organization of manufacturing

Manufacturing takes raw materials, parts and components and transforms them into finished product. This can involve a few or many processes, and these can vary from a simple machining process to a complex assembly and test procedure. When considering the organization of manufacturing we need to know what processes are being used and the number and types of products being produced from them. This information will allow an examination of the organization of the factory, in terms of the machines and operators, and of the way in which materials are used and are moved around the manufacturing facility. This examination is important, as the way in which the manufacturing is organized will have implications for process utilization, stock levels, cash flow, planning and scheduling complexity, and manufacturing lead time. It is also necessary to be able to identify non-value added activities. These are activities that will incur a cost to the organization but which cannot be passed on to the customer as they do not cause the value of the product to be increased. Examples of non-value added activity will include storage, handling and inspection. In addition to the cost implications, non-value added activities will also affect lead time and cash flow.

Irrespective of specific process there are four production methods used in manufacturing: job, batch, flow and group. Each of these is appropriate to a particular set of circumstances according to the demand for the product, the production volume required, the range of product to be produced in a single facility, and the degree of standardization of the product. The four methods are described below.

6.2.1 Job production

Job production is used when products are made singly or in small batches, i.e. when a small number of products are processed at each stage before going on to the next stage. A typical example would be a pattern maker producing a pattern for a new casting; one pattern maker works on one pattern and does the whole job. When the job is finished another can be started, as shown in Figure 6.2. Another example would be in a construction company where a large team will work to produce the one product, the building or works, before going on to the next. Job production can operate effectively for low- and high-technology products, and for either simple or complex processes.

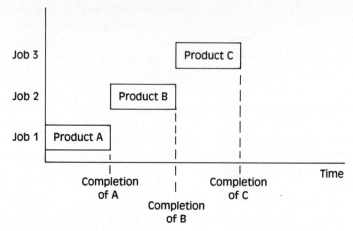

Figure 6.2   Schedule for job production

When one operator is responsible for the whole job the planning is simple and the manufacturing lead time short, as one job is processed at a time. The main requirements are that the operator has the correct tools and training. The time to produce the product is governed by the rate at which the operator can work and during the process of production value is being added all the time. A great advantage of this method is the ability to modify the product; making changes after each product has been produced is not a complex operation and thus this method is very appropriate to companies who are just testing the market and producing prototypes, and who need to be able to react with speed to changes. It is also appropriate for organizations producing to individual customer specifications. It does mean, however, that the tools and equipment required for a particular job have to be available when needed, otherwise the operator will be idle.

When this type of production is used for a large project, such as a construction activity or shipbuilding, then it will require a great deal of planning and a lot of organizational flexibility.

6.2.2 Batch production

In batch production the product is built in a series of stages. At each stage a number of units of product, the batch, are processed before the whole batch moves on to the next stage, as shown in Figure 6.3. Modifications can be incorporated into product at the next batch, but are difficult to introduce part-way through a batch because of the difficulty in tracking what has happened when each unit of product does not have unique identification; it is the batch which is identified.

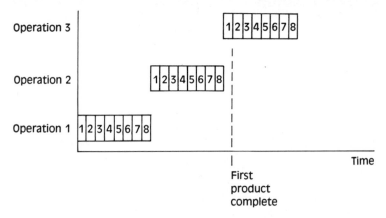

Figure 6.3 Schedule for a product requiring three operations, produced in a batch of eight

In this type of production one operator or machine can be dedicated to a particular process but is able to work on a number of different products requiring that process. For example, in a company producing a range of electronic goods a flow solder machine can be kept fully utilized processing batches of different products one after another. This method of production is particularly suitable if the manufacturer has to support and produce a wide range of products. It is also appropriate when there are limitations on the equipment and skills available such that these have to be shared, or when specialized skills or equipment are required that could not justify being kept for infrequent work. Batch production can be used when the organization produces a variety of products and when the demand for products is variable, as batch sizes can be varied to suit requirements.

Companies using batch production typically use a functional factory layout in which similar processes and functions are grouped together. All jobs, irrespective of product, requiring the processes in a particular area are processed in the area on any of the available facilities. An example of a functional layout for batch manufacture is shown in Figure 6.4.

At a machine in a process area there may be a batch of Product A in progress and a batch of Product B waiting to be processed. In practice, this means that there will be a number of units of Product A waiting, having been processed, one unit of Product A being processed, and the remainder of the batch of Product A and the whole batch of Product B waiting to be processed. So we can see that this type of production requires very careful planning in order to minimize stock on the shopfloor, since it means that there will be a lot of part-processed material on the shopfloor waiting for further processing. This material is called work-in-progress. Batch production means that all processes have to be scheduled as well as the batches of product, making planning complex. Control is also complex because of the requirement to keep track of all batches of all products. However, this is being made easier by the use of barcoding and automated shopfloor data collection.

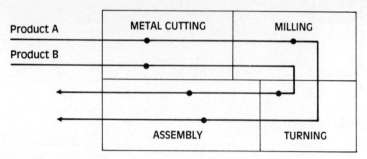

Figure 6.4 Functional layout

Batch production is used when the availability of skills and/or equipment is limited or high levels of utilization are required. It invariably means a longer manufacturing lead time than in job production, as seen in Figure 6.3, and will involve having many more materials waiting to be processed with the consequent cost implications of holding stock and having non-value added activity.

6.2.3 Flow production

Flow production is the process of manufacturing the product in a number of discrete steps but with no waiting at each stage for a batch of components to be produced. It combines the advantages of both job and batch production for high-volume production. In this case the products pass through a production line, as if they were batches of one (see Figure 6.5).

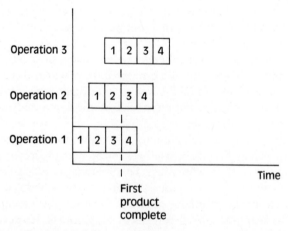

Figure 6.5 Schedule for flow production of four products through three operations

In order for the flow production system to operate effectively there should be no waiting time between each stage in the process, requiring that each operation should take the same amount of time. The method used to calculate the times for successive operations and determine the production rate is called 'line balancing'. In addition, because one product will be following after another if the line breaks down at any point there will rapidly be a build-up of product at the break and this will bring the whole line to a standstill. Thus, preventative maintenance, and its planning, plays an important role in organizations using flow production methods.

This type of production can be used where there is a fairly constant demand for the product, and the product is standardized, i.e. there is little variation in the product specification over a period of time. It is thus typical of high-volume and highly automated production lines such as those found in food processing and consumer goods manufacture. All operations for the processing of parts must be clearly defined and the materials for the stages of manufacture must be available promptly at all times. Companies using flow production have a factory layout based upon the product where all the processes required for the product are grouped around a central line along which the product travels, as illustrated in Figure 6.6. Although there will be a requirement to hold some stocks of parts at the side of the line for assembly work-in-progress stock will be minimized because work will be continuous.

Flow production allows for the shortest possible lead time, as shown in Figure 6.5. However, it does require that the production process be broken down into small, similar-size operations which usually involves a level of de-skilling that can make the production line jobs somewhat boring for the operators and this can significantly limit the personnel who can be recruited to such work. Flow production is not suitable for products that may require modification or when there are many variants. In addition, because it requires dedicated plant then there must be a consistent and stable demand.

6.2.4 Group technology

Group technology is an extension of batch production in which the factory is organized with the aim of putting all the processes required for a particular

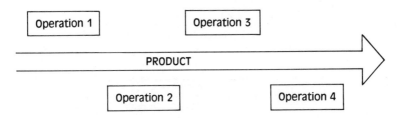

Figure 6.6   Factory layout for flow production

product, or family of products, into one area. This area is called a cell. Within the cell a small flow line operates, with machines and processes dedicated to the types of product being produced in that area. The product is produced in small batches. The principle of group technology is that it achieves the benefits of flow production and high standardization, that is, low work-in-progress and short lead times, for a situation where there is small-scale batch production and a range of products. This is illustrated in Figure 6.7.

A comparison of Figures 6.7 and 6.4 shows the benefits of group technology over standard batch production when different products are being produced. In the cellular layout the process route will be much shorter, and because only one type of product will be put on a machine in a particular cell then the scheduling of the machine is simplified and the requirement for resetting from one product to another is reduced. The scheduling of the cell is also simple, as only one type of product will be processed there. These benefits result in reduced manufacturing lead time and improved cost effectiveness by reducing non-value added activity, such as queuing for machine time and transport.

The capital costs of group technology can be high because of low machine utilization, therefore it is important that the groups of parts to be produced in each cell are selected carefully and that people are deployed effectively. Multi-skilling of the workforce is a feature of group technology which enables the workforce to be used to operate many different processes.

The key features of each type of manufacturing organization, relative to each other, are summarized in Table 6.1. In any one factory there is usually a mix of two or more of these types, for example the car production line which has a batch manufacture system at the line side for producing gearboxes.

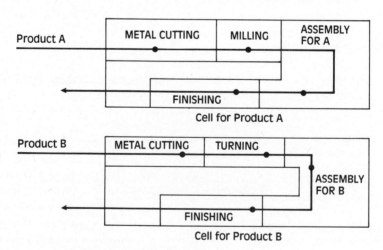

Figure 6.7   Cell layout for group technology

Table 6.1 Key features of the four methods of manufacturing organization

Manufacturing organization	Key feature
Job	Short lead time
	Low work-in-progress
	High level of value-added activity
	One product processed at a time
	Can react quickly to different requirements
	Low machine utilization
Batch	Long lead time
	High work-in-progress
	Allows for variable demand
	Allows processing of a variety of products
	High process utilization
	Complex scheduling and shopfloor control
Flow	Minimum lead time
	Low work-in-progress
	High-volume production
	Requires stable demand
	Requires standardized product
	Scheduling simple after line has been balanced
	Requires preventative maintenance programme
Group	Short lead time
	Low work-in-progress
	Low level of process utilization
	Simplified planning and control
	High level of value-added activity

6.3 Production planning and control

Production planning involves the planning of people, machines and materials in order to ensure that the output targets set for the operations area can be met. The factors that affect this are the lead times for manufacturing and purchasing, the way in which the manufacturing is organized and its level of complexity, the size of the operations function, and the objectives set for the operations area.

The activities included in production planning will vary from long-term forecasting, as required for business planning, to daily or weekly shopfloor scheduling. However, before production can be planned there needs to be data on which the plans can be based. The data needed falls into two categories: operational data, including stock data, and product data. In each of these categories there will be data that is set up when the production system is established and data developed during operation that has to be continually monitored and fed back to the user. The monitoring of the system allows potential problems and improvements to be identified, thus allowing control of the system and the opportunity to manage the individual parts of it effectively.

6.3.1 Operational data

This is the data concerned with the resources available for manufacture, the capacity of the manufacturing plant, and the stock of materials and parts.

Capacity

Capacity defines the amount of work that can be done on a particular machine, by a particular process, or within a particular facility. Capacity data is needed by manufacturing organizations so that they can make realistic assessments of what can be achieved in the first instance, and subsequently it is used to manage the manufacturing process so that plans are achieved. In order to determine capacity a catalogue of the types of resource available is required. A resource may be a single machine, a process, or a group of machines or processes that all supply the same function. When the catalogue has been compiled the capacity of each resource can be determined. There will be a theoretical capacity which is based on process specifications and the number of hours for which the process is available. For example, if a machine can produce 25 units of part B12 per hour and it is used for 40 hours per week then the theoretical capacity is $25 \times 40 = 1000$ units of B12 per week.

Unfortunately, there will usually be something that prevents a process operating at its theoretical capacity. Time is required for maintenance, changeovers or resetting when a new part is to be produced, or the operator may not be available at all times. All these will reduce the capacity of the process. By carrying out observations of the processes the effective capacity, i.e. the real capacity, of the process can be evaluated.

This data allows the production controller to know the capacity of the manufacturing facility, thereby allowing realistic plans to be made. It also allows opportunities for improvement in resource utilization to be identified which would increase the cost-effectiveness of the operation. Any resource that is working at less than full capacity can be seen as a charge on the organization since it still has to be funded while it is idle.

Collection of capacity data during the production process is still carried out in many companies by using manual recording methods that are then fed back to a central processing station. Barcoding linked to shopfloor data collection points is taking over from this but the principles are still the same. Typically, in a machine shop operating a batch production process a ticket will be attached to the batch of parts. This ticket will indicate what the parts are and what processes are to be carried out. When the batch of parts has been processed, the operator who carried out the work will complete the ticket, indicating the amount of parts processed, the amount that met specification, and how long the processing took. The ticket will be passed to the finance department who can use the information to cost the job. The production control department will use the process times indicated on the ticket to update process capacity information.

Part specifications
These are required for every part used in the manufacturing process. The parts are usually given a part code, or number, which is then used to identify each one uniquely. Against this part code there will be part definition information. This will include a clear description of the part, which may be achieved by reference to a drawing number. There will also be supplier information for those parts that are bought from external suppliers, including the supplier's/manufacturer's part code, the cost, and the delivery lead time. Part specifications will have to be set up at the start of any production control system. They will then have to be monitored and adjusted accordingly, to reflect changes in cost and lead time, as well as any regrading of the specification. A complete list of all part specifications is called a parts list.

Stock
This is the name given to the materials and parts held for manufacturing. The correct materials in the correct quantities must be available for processing to take place and therefore throughout the manufacturing process stock must be monitored and controlled to ensure that this is the case. The production controller needs to know when parts will be available, what parts have been used and what parts are needed. This information comes from an effective stock control system, as described in Section 6.4. The stock control system will use the information contained in the part specification for determining order details. There will also be further data recorded against the part number, live data indicating the number of units in stock, those allocated to a particular job, units that are due in, and the current cost of parts.

6.3.2 Product data

This is the information that is used to define the product and the processes that are used in its manufacture.

Bills of material
These are sometimes referred to as product structures. The bill of material defines the parts, and quantities, that go into any subassembly or product, and usually reflects the way in which the product is manufactured. Bills of material are usually multi-level. For example, consider a kettle, as shown in Figure 6.8. To understand how the product is made you work up from the bottom. The level 2 parts have to be assembled before level 1 parts. It is usual to describe any part that has its own bill of material as a subassembly, unless it is also sold in its own right

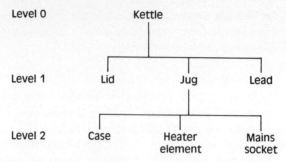

Figure 6.8   Product structure for a kettle

as a product. It is also the case that when looking at subassemblies the level numbers will change from those indicated in the product structure. The top level of any bill of material is usually level 0.

Bills of material can be prepared on the basis of the time that it will take to process the product, so, for example, each level could relate to all the parts that you would use in one week. The chosen period is known as a time bucket.

The bill of material for a jug is shown in Figure 6.9. The full specification and stock details for each part will be taken from the parts list. The bill shown in Figure 6.9 shows only specification information and not details that are subject to continuous updating, i.e. cost and stock information.

Because they define the product, bills of material need to be controlled to ensure that any changes to the physical item are reflected in an amended and authorized bill. Bills of materials and their relationship with drawings are discussed in Section 5.4.6.

Routings

These define the processes that are used in the manufacture of the product and the sequence in which they occur. For parts that are made regularly the routings might be given to the manufacturing department in the form of a standard part schedule. For other parts special route cards may be made up. The route card will indicate the processing times for each stage of production, given in standard

Level	Part No.	Part	Quantity	Supplier Code	Lead Time
0	J100	Jug	1		1
1	J239	Case	1	D32	24
1	EP210	Heater Element	1	E65	3
1	EP554	Mains Socket	1	E65	3

Figure 6.9   Bill of material

minutes (SMs), and will thus allow the manufacturing lead time to be calculated. An example of a route card is shown in Figure 6.10.

The data required for the manufacturing operation and the way in which it is used is shown in Figure 6.11. For all data used in the manufacturing process effective control is imperative, this is discussed further in Section 7.2.5 under 'Document and data control'.

6.3.3 Scheduling

With the manufacturing data, shown in Figure 6.11, it is possible to assess the resources required for the manufacturing activity and then to schedule them to meet the output requirements. There are, of course, two further issues that need clarification before a realistic schedule can be produced. The first is that the output requirements have to be defined, and second, there has to be a review of the inputs and required outputs so that the resources can be managed to achieve a match.

The output requirements may be defined directly by customer demand. In this case as orders are received a new product is scheduled into production. This would be the case with job production systems and also with some flow production systems where Just-in-Time production methods are being used (see Section 6.5). In the other situation forecasting methods are used to give an indication of demand over a particular period and this is used to define output. This method will invariably lead to some production of product which will then go into stock as finished goods and be held until required by a customer.

Where batches of product are processed the size of the batch to be processed at a time has to be determined. Clearly, the batch size can have serious

Part No. DL9003				Issue No. 1		
Description: Base Flange				Batch size: 50		
Operation No.	Description	Dept	Machine			Drawing No.
			No.	Type	S.Ms.	
10	Grind face	M	305		15	DL9003
20	Drill ten 5mm holes	M	306		8	DL9003
30	Transfer to stores	SI				

Figure 6.10   Route card

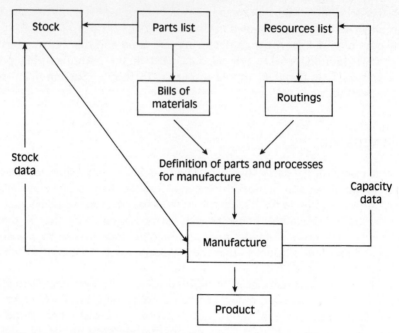

Figure 6.11   Use of manufacturing data

implications for an organization. Too large a batch will lead to excessive spending on parts, their storage and processing, as well as tying up money in work-in-progress and finished goods stock. It will also increase lead time. A large batch will have some advantages when a functional layout is used because it will reduce the numbers of changeovers required at a particular process. However, this has to be countered by the delay on the process availability for other products. Conversely, too small a batch can lead to problems with not having enough stock to meet customer demand, and the costs of production may be much higher due to the increased number of changeovers and resetting of machines.

Economic batch quantity

If you consider a factory that needs to produce an output of 1000 units per year, you might think that this could be achieved by producing one batch of 1000, two batches of 500, five batches of 200, and so on, the quantity in the batch defining the frequency with which a batch of products needs to be produced. The number of batches made and the quantities in the batch will have implications for the company in terms of storage requirements, both in-progress and in the Stores area. It will also have implications for the amount of control required over the particular work areas, suppliers and subcontractors, and for the costs involved. A method that is used to optimize these parameters is called economic batch quantity (EBQ).

The costs associated with batch production can be categorized as direct and ancillary. The direct costs are those that are associated directly with the production of the product and would include parts costs and process costs. The ancillary costs are those that exist independently of the numbers of product. These will include the costs of equipment maintenance and set-up costs for internally supplied goods, and the purchasing costs for those goods provided by external suppliers. It is usual to divide the ancillary costs for a batch evenly over the number in the batch. Consequently, because of the stability of the ancillary cost the proportion allocated to each product is lower, the larger the number of products in a batch. However, the problem of producing goods in large batches is that they require you to have a large stockholding, the cost of which increases with the number of units in the batch. The use of the EBQ method is designed to analyze these problems and provide an optimum solution. The EBQ can be shown graphically as in Figure 6.12.

The EBQ can be calculated according to the formula:

$$EBQ = \sqrt{\frac{2SD}{IC}}$$

where

S = ancillary cost per batch
D = annual usage
I = annual holding cost as a fraction of the stock value
C = unit cost of the item

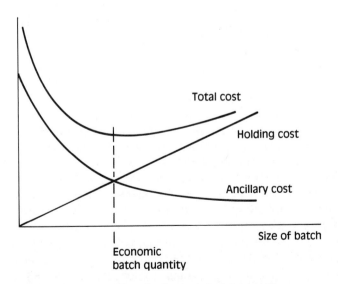

Figure 6.12   Graphical representation of economic batch quantity

In this case it is assumed that neither the unit cost nor ancillary cost vary with batch size. The calculation of the EBQ also assumes that stockholding costs, unit cost, and demand are constant, and that the quantity ordered is provided in one delivery.

EXAMPLE

A company makes telephone booths, for a stable market where the demand is 450 units per month. The booths are made in batches, with all the units in each batch being completed at the same time. Given the following information, calculate the economic batch quantity:

Machine set-up cost per batch = £150
Stockholding cost = 10% of stock value per annum
Unit cost = £37.50

$$EBQ = \sqrt{\frac{2 \times 150 \times 450 \times 12}{37.50 \times 0.1}}$$

EBQ = 657

The main advantage of using the EBQ is that it gives a value which is optimized for a certain data set. There are, however, a number of problems associated with its use that have to be considered. The main problems relate to the assumptions that have to be made, i.e. that unit price and ancillary cost remain constant throughout the year. There is also the problem that the ancillary costs, and more particularly the stockholding costs per batch, can be very difficult to assess. Finally, the EBQ will not often produce a number that is consistent with supply systems.

The schedules

A number of schedules are required to operate a manufacturing facility effectively. The primary schedule is the master production schedule (MPS). This is the list of all the demands made on the production system; in its most basic form it will record what products are required, in what quantity, and when. An MPS will be updated at regular intervals, appropriate to the product lead time. An example of an MPS is given in Figure 6.13.

For any of the schedules to be realistic and achievable they must take into account the availability of materials and processes. There is consequently some iteration involved in assessing available resources and managing these in order to produce an acceptable schedule. This process is shown in Figure 6.14.

In spite of this, a number of organizations still do their scheduling on the basis of infinite capacity, the assumption being that there is no restriction on the capacity available to meet the requirements. Clearly, this can lead to overloading with consequent delays. An improvement on this is rough-cut capacity planning.

Master Production Schedule at 10 March			
Part Number	Product	Quantity	Required by
B214	LW receiver	50	30 March
A911	SW receiver	100	5 April
B629	FM receiver	50	8 April
B730	FM/SW receiver	75	15 April
A004	headphone set	250	1 May

Figure 6.13   Master production schedule

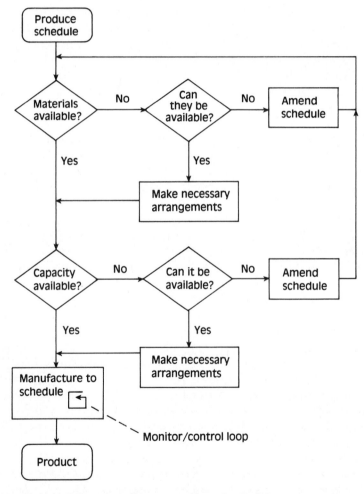

Figure 6.14   Review of resources for preparation of schedule

This is not detailed planning but looks broadly at overall capacity requirements and available capacity for that period. It is a recognition that there is finite capacity but may still lead to bottlenecks at specific highly utilized processes. Rough-cut capacity planning may be used to amend an MPS. More detailed and accurate scheduling can only follow from a realistic MPS and will incorporate full capacity planning, which takes account of the effective capacities of all processes.

The schedules are required to give instruction to process operators and those responsible for procurement. They must be easy to understand and their review against actual performance is an everyday management task if the manufacturing programme is to stay on target.

There are various production planning techniques, such as Just-in-Time and materials requirements planning, explained later in this chapter, that define the way in which the information can be used to produce the subsequent work schedules for ordering parts and then for loading processes. However, fairly simple scheduling can be done, if lead times are known, using Gantt charts and critical path analysis, as explained in Chapter 8.

6.3.4 Capacity planning

Capacity planning leads to accurate scheduling of resources and allows available capacity to be matched to requirements. It requires an assessment of whether or not there is available capacity. This follows from an assessment of the MPS, and subsequent action to meet the capacity requirements. Insufficient capacity is clearly a problem and must be tackled immediately. However, overcapacity also brings problems for an organization:

● *Undercapacity*: An organization may have insufficient capacity because it is fully committed already or a resource is unexpectedly unavailable, such as a machine breaking down. Equally, an organization may decide that it wants to exceed its own capacity in order to avoid missing a market opportunity. In any case, if the situation is not remedied then it will invariably lead to dissatisfied customers. If the problem is a short-term one then there may be opportunities to achieve the extra capacity in-house. Typically, operators can be given more work, such as by offering overtime, or machines may be more heavily loaded. These are short-term options as they will increase the cost of the operation and may have effects on quality and operator morale.

A further option would be to increase the workforce temporarily by employing agency staff who can then be laid off when the demand has been met. This option can be costly, as there will be agency fees and employment costs. It may also give rise to a training need which could lead to further loss of capacity if fully trained people have to be taken from their work to provide this. Quality may be compromised if personnel are used who are not fully trained. The other option in this situation is to subcontract work. In this case, work is sent out to other companies who will then carry it out, to the required specifications, as suppliers. This is an effective means of increasing capacity as it also relieves the difficulty of trying to shuffle schedules and conduct negotiations with personnel. It carries the advantage that goods produced should meet specification, as payment will depend on this. However, this also means that the organization is contributing to the overheads of the subcontractors rather than to their own overheads, and some control of the process is lost as the subcontractors will also have other customers to satisfy. It is quite common to find companies using subcontractors on an almost permanent basis if the subcontractors can offer a service or process for which the organization has only limited use and thus saves the expense of purchasing and operating the process.

If the undercapacity situation is to be long term then the organization will have to take more permanent measures to deal with it. These steps may include increasing the workforce, purchasing new processes or machines or increasing the availability of processes, for example by introducing shift work. There is clearly significant cost associated with these options and it is thus important to identify the time for which the undercapacity situation is likely to endure.

● *Overcapacity*: In certain manufacturing environments, for example in Just-in-Time production, a certain amount of overcapacity is not seen as a problem (see Section 6.5). However, for many companies overcapacity will rapidly lead to a fall in profits which cannot be sustained. A long-term problem must therefore be tackled and two approaches are possible. The first method is to get rid of the extra capacity so that it is no longer a financial burden. The second is to use the extra capacity, for example by no longer subcontracting work or by taking on work under contract for other companies.

By using the methods mentioned above the capacity requirements and available capacity can be matched.

6.4 Materials management

Materials are essential for manufacture and their cost is likely to be a significant proportion of the total product cost. Therefore, materials have to be managed so that production schedules can be met, and money used effectively, and to ensure that quality of product is not compromised by poor-quality materials. The primary function of a materials management system is to ensure that the right parts are available for manufacture at the right time, at the right price. It therefore involves identifying requirements, purchasing materials, and making them available for manufacture.

6.4.1 Stores

The heart of most materials management systems is the Stores, where stock is received, held, and issued, and where the records of these transactions are generated and maintained. There are three types of stock that are commonly defined:

1. Raw materials, parts and components
2. Work-in-progress
3. Finished goods.

Raw materials, parts and components are all the materials that are to be used for manufacture of product, in original form. This might include items such as sheet steel, parts made to the company's design, and proprietary parts, i.e. those made to someone else's specification. These parts are usually bought according to a planned purchase scheme and are kept in the Stores until required for processing. Assemblies that have been manufactured and then returned to Stores prior to further processing will also be included in this category.

Work-in-progress is the material that is on the shopfloor either being processed or awaiting processing. This will typically include parts that have been released from Stores to make assemblies and part-built assemblies.

Finished good are those that have been completely processed and are in a state fit for sale. They are then stored in this state, until despatch, in a finished goods store (FGS).

The purpose of the Stores is to provide a safe and secure place for holding goods and to make those goods available when required by the appropriate company departments. Security is necessary in order to prevent loss due to pilferage or misuse. The Stores should also preserve the quality of the goods by providing adequate packaging, space and handling, as well as having established procedures for stock rotation and use to avoid deterioration of stock held.

The layout of the Stores must consider the types of parts that are to be stored and the access that has to be provided. It should also take into account the way in which parts can be stored, for example whether or not they can be stacked. In addition, consideration must be given to any packaging or special storage requirements. Given that the parts must be easily accessible in Stores, some companies design the layout based on part number, on groups or families of parts, the frequency with which access is required, or on the size of the parts. Irrespective of how the Stores area is laid out, it is essential that all parts within it are identifiable and that their inspection status is clear, that is, whether or not they are available for manufacture or sale.

Some Stores require specially defined areas, sometimes with restricted access. These are called bonded and quarantine areas. The bonded area is used to hold parts that have high value or that are required for special jobs. The quarantine area is the place where goods that cannot be used are held pending decisions being made on their disposition, and this would include parts that had failed inspection. Depending upon the company and on the types of parts used, there may also be other special areas defined such as static-free for electronic components and a lockable store for flammable materials.

bonded area so that they cannot be used. The Stores stock is manually counted and checked against the records.

Continuous stocktaking can be done as an alternative to the annual stocktake and does not require the Stores to be closed. Before being able to use this as a method for accounting it has to be approved by the financial auditors. The technique requires small amounts of stock to be checked on a sample basis with the requirement that all the goods must be checked so many times in a given period.

6.4.2 Purchasing

Goods must be purchased to meet manufacturing requirements. However, the way in which purchases are made will be either dependent or independent of the demand placed on the manufacturing system. In the dependent demand situation goods are ordered according to real customer demand of product. The intention of this method is to minimize stockholding. This method is used in the technique called materials requirements planning (MRP), which is described in Section 6.4.3, and in Just-in-Time production systems, discussed in Section 6.5.

Independent demand
With this method the organization must make some assessment of probable use of goods over a period and will then decide what goods and quantities to hold in stock. Stock items can then be withdrawn from Stores for use as required. The main issue in this method is deciding how much stock to hold, as the holding of stock clearly has implications for the costs of operating the manufacturing facility. The organization therefore has to calculate, for every item to be stocked, a reorder level which is the level to which stock is allowed to fall before being replenished, and a reorder quantity, the predetermined amount by which the stock will be replenished.

● *Reorder level*: Figure 6.15 shows how stock may be reduced over a period of time and then replenished. The decrease in stock may not be linear, as illustrated, but the diagram shows that there is a step increase in stock which is then allowed to decline to a defined level. If the minimum stock level is zero then there will not be sufficient stock for the period while waiting for the new stock to arrive. The reorder level therefore has to take into account the average use of stock in the lead time.

In the independent demand situation there is also the risk that stock will not be available to meet increased demand and there is no insurance for unexpected demand such as may be caused by breakage or fault. For this reason, it is usual to hold safety stock. The amount of safety stock will reflect the risk of having an out-of-stock situation. Factors to be assessed might include the consequences of not meeting customer deadlines, with the possibility of penalties, or the costs of having idle processes. The level of safety stock must be considered carefully. It

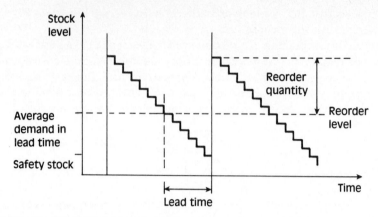

Figure 6.15   Stock depletion and reorder

represents an immediate charge to the organization because it must be paid for and stored. In the long term there may also be an additional cost if it is not used.

The reorder level (ROL) is calculated according to the formula:

$$ROL = gL + SS$$

where

g = average weekly demand
L = lead time in weeks
SS = safety stock

● *Reorder quantity*: The reorder quantity chosen must reflect the usage of the item, its lead time and its cost. In addition to these, account should be taken of the cost of both ordering and holding stock, and the ordering rules imposed by suppliers. For example, there may be minimum sale quantities and there will be defined buying units, such as electronic components which will be packaged in tens or hundreds.

The economic batch quantity, as described in Section 6.3.3, can be used to provide an indication of the size of order to be placed. When used in this situation the ancillary cost will be the cost of placing the order with the supplier. The resulting EBQ will have to be assessed and amended to take into account buying restrictions.

Price breaks will also affect the reorder quantity chosen. Many suppliers will offer a discount for bulk purchases, effectively lowering the unit price. Since the EBQ calculation assumes a constant unit price there must be a way to make an objective assessment of the effect of the price discounts offered. This can be done by either calculating the EBQ for each of the prices offered or by looking at the total cost of operating the system with the quantities where the price breaks just become available. For example, an organization has demand for 250 cast aluminium parts per year and estimates that holding stock of the parts costs 10% of the stock value. A foundry offers the following prices for bulk orders of these parts:

> 1–9 parts £8.50 per unit
> 10–99 parts £8.25 per unit
> 100 + parts £8.00 per unit

The cost of placing an order is £5.00.

1. Using the equation for EBQ calculate the EBQ for each of the prices offered.

$$EBQ = \sqrt{\frac{2SD}{IC}}$$

where

> S = ancillary cost per batch
> D = annual usage
> I = annual holding cost as a fraction of the stock value
> C = unit cost of the item

At £8.50 per unit: $EBQ = \sqrt{\dfrac{2 \times 5.00 \times 250}{0.1 \times 8.50}} = 54$ units

At £8.25 per unit: $EBQ = \sqrt{\dfrac{2 \times 5.00 \times 250}{0.1 \times 8.25}} = 55$ units

At £8.00 per unit: $EBQ = \sqrt{\dfrac{2 \times 5.00 \times 250}{0.1 \times 8.00}} = 56$ units

From these calculations we can see that the only feasible answer is for 55 units at £8.25 per unit. In both the other cases the EBQ cannot be obtained at the price which led to its calculation. We also see that the EBQ is not particularly sensitive to the price breaks offered.

In order to verify the optimum order quantity we can calculate the total cost of buying using the EBQ and the quantities at which price breaks are available.

2. Calculate the total cost of operating the system.

> Total cost = purchase cost + reorder cost + stockholding cost

$$\text{Total cost} = D \times C + \frac{D \times S}{Q} + \frac{Q}{2} \times C \times I$$

where

D = annual usage
C = unit cost of the item
S = ancillary cost per batch
Q = quantity ordered
I = annual holding cost as a fraction of the stock value

Note that the stockholding cost is calculated on the basis of the average stock held over the period, i.e. half of the order quantity.

For the feasible EBQ, 55 units, the total cost is

$$250 \times 8.25 + \frac{250 \times 5.00}{55} + \frac{55}{2} \times 8.25 \times 0.1 = \pounds2107.91$$

If we now look at the total cost where the £8.25 price just becomes available, i.e. at 10 units, we have:

$$250 \times 8.25 + \frac{250 \times 5.00}{10} + \frac{10}{2} \times 8.25 \times 0.1 = \pounds2191.63$$

At the point at which the £8.00 price can be realized, i.e. 100 units, the total cost is:

$$250 \times 8.00 + \frac{250 \times 5.00}{100} + \frac{100}{2} \times 8.00 \times 0.1 = \pounds2052.50$$

From this we can see that although the EBQ calculation will give a feasible solution when price breaks are offered it will not necessarily provide the optimum solution and must therefore be tested by calculating the total operating cost.

Finding a supplier

Finding a supplier for an item is known as sourcing, i.e. finding a source for supply. In more traditional organizations goods are multi-sourced whenever possible. This means that there are a number of possible suppliers for any item. The reason for this is that traditional purchasing is adversarial, the philosophy being that in order to get the best deal the suppliers have to compete fiercely for trade. Single-sourcing and a move to partnership sourcing is a significant move away from the traditional systems and is usually associated with Just-in-Time manufacturing. With this method of sourcing the aim is to find a supplier that provides the best value rather than the best price and then to develop the supplier/customer relationship so that it is mutually supportive. The concept of value extends beyond price because it considers all the attributes of the goods and their supply. For example, it may be possible to buy a component at a low price. However, if it fails, or is delivered late, or there is no after-sales service then the value of that component to the customer is lower than one that may have a higher price but is more reliable and is delivered promptly. When a supplier has been found that can supply good value then the aim is to build a partnership with the supplier which will ensure that the value is sustained and subsequently improved. This relationship must be a two-way one and such 'partnership' sourcing invariably leads to benefits for the supplier, such as prompt payment of invoices and a commitment to purchase, and benefits for the customer of consistent supply and quality.

In addition to whether multi- or single-sourcing is used there will be other factors that affect the choice of supplier. Some organizations, such as those belonging to a group of companies, are required to purchase from other companies in the group. A number of organizations and industries have special requirements that affect the way in which they procure goods. For example, companies supplying to the nuclear industry have to be able to trace all parts back to source, which means that they must use suppliers that can also do this. Increasingly, companies also require suppliers to have independently certified quality systems.

There are a number of trading rules and pieces of legislation that will affect purchasing. For example, in order to say that something has been made in the European Union a defined percentage of the parts have to have been sourced there. There is legislation, following from other European directives, which defines cases when purchase requirements have to be put out for tender. This includes public supplies above a certain value. In this case the customer has to notify a number of potential suppliers that goods are required and then the suppliers provide a tender to this offer, i.e. they indicate what they can supply and under what circumstances. The methods of review of tender are defined in the legislation, and following review, the order is given to the supplier with the best tender.

A number of stockists (companies that trade in a large number of stock parts such as electronic component suppliers) will maintain stock of parts within a company without the need for internal stock control and purchasing. For example, there are fastener stockists who will maintain stocks of parts in special stores areas in customer organizations, visiting on a regular basis to refill bins and then invoicing afterwards.

Placing orders

Having determined the quantity to be ordered and when the order has to be placed on the supplier, the next step is to place the order. Orders for work to be done in-house, for example parts to be assembled, are called works orders. Works orders will be released to the shopfloor and should be accompanied by any necessary documentation such as assembly drawings or route cards. The works orders will be used to carry out detailed scheduling and loading of work centres.

Purchase orders are the orders placed with outside suppliers and the actual selection of suppliers is covered in detail in Chapter 7. However, having ascertained that the supplier can meet the specification required the order must ensure that the purchase requirements are fully specified. A typical purchase order is shown in Figure 6.16.

A purchase order must contain all the information required to ensure that the correct parts are supplied in the correct quantity at the time that they are required.

Date ....   Purchase Order No. 12345

Requisitioned
from ....   **ZEB ZEB
MANUFACTURING LTD**

A/C No.:

Post Code:

Bridgend Road
Whitecross
WC54 7GB

Tel: 0492 12648
Fax: 0492 78534

This Purchase Order is subject to the terms and conditions of purchase overleaf. Invoices and enquiries relating to this Purchase Order should quote the Purchase Order number.
All prices specified in Sterling unless otherwise stated.

Qty	Description	Finance Code	Unit Price	Total Price
DATE REQUIRED:			VAT	

VAT No: X437 9990   TOTAL

Deliver to:
....

Contact:
....

Signed on behalf of ZEB ZEB
Manufacturing Ltd:

....

Figure 6.16   A typical purchase order

The order must therefore contain:

- *Full supplier details* – name and address.
- *Full delivery details* – address for delivery and date required by. There may also be special instructions (for example, if a particular carrier is to be used or if the goods are to be delivered to a particular person).
- *Specification of goods* – this may be by reference to a part number, may be given in full, or may be achieved by reference to a drawing. When a drawing or other documentation is used the issue information should be provided on the purchase order to ensure that no obsolete or unauthorized documents are used.
- *Order details* – these will include the quantity to be delivered and perhaps payment information. Specific requirements such as testing to be carried out or quality assurance procedures to be followed must be stated.

When an order is placed it may be for a quantity of goods to be delivered in one batch or a number of batches. When the delivery date for the batches is not identified specifically but there is an indication of when it is likely that goods will be required this is called a 'call-off' order. The company can 'call' for predetermined goods to be delivered at short notice. The advantage to the company is that administration is reduced significantly and yet the parts are delivered as needed. There is no need to hold stock of one large order.

After placement of the order some progress chasing may be done by the company to ensure that it will be delivered on time. Delivery of the order will be accompanied by some form of delivery note which is used to verify that the order has been received by the customer. Following this, the supplying organization will send an invoice which demands payment for the goods. Invoices are often 30 or 60 days, which means that they should be paid within those time periods. The consequence of this is that an organization can have access to and use goods for a considerable period before having to pay for them. This is a very significant factor in the financial operation of the company, and one of the more common reasons for small firms going out of business is the lack of cash flow resulting from their inability to recover monies owing on invoices within the time that they have to pay for goods (see Chapter 4).

The actual cost of placing an order can be significant and should be considered when orders are raised for small purchases. The cost of the order will include the documentation associated with it, any phone, postage or fax charges, and the time required to place the order, progress chase, receive the goods, and invoice for them. In addition, many suppliers impose a minimum delivery charge. It is quite feasible that an order can be placed for goods costing £5.00 but the cost of placing the order and the minimum delivery charge amount to £30.00.

6.4.3 Materials requirements planning (MRP)

MRP is a dependent demand ordering system, i.e. it uses the real customer demands on the organization to calculate when orders should be placed with suppliers. It is used for both external and internal ordering.

MRP looks at the materials that are required in order to meet the production requirements. It uses lead-time information to calculate when parts are required on the factory floor. It thus allows effective management of stock and work-in-progress, and provides information for cash flow planning. Figure 6.17 shows the operation of MRP.

The input to the MRP system is the master production schedule (MPS). When the MPS is produced, some rough-cut capacity planning is done and the MPS may be amended following this level of planning. With this list of products the bills of material can be interrogated to provide a list of all the parts required to meet the production schedule.

The bill of material processing will look at all the bills of material for every product required to meet the MPS. It will analyze the bills in order to determine all the parts that are needed to make the products, calculate the quantities of parts, and determine when the parts are required. The time that parts are required will be calculated by working backwards from the time that the product is required using the manufacturing and supplier lead times: the lead time is the time from placing an order to having it delivered, i.e. ready for the next

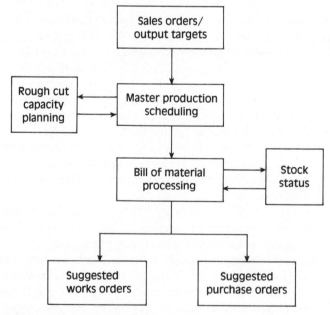

Figure 6.17   Materials requirements planning

process. This first stage of the bill of material processing determines the gross requirements.

When all this information is calculated a check will be made against the stock status file in order to determine what parts are already available or what will be available according to orders already placed. This information is used to refine the materials requirements, giving the net requirements. The output of the processing will be in the form of a list of suggested purchase and works orders.

To show how an MRP operates we will consider the indoor television aerial in Figure 6.18. The bill of materials for the aerial with stock data, as at 10 March, is given in Figure 6.19. Notice that the level numbers are indented to show clearly the relationship between assemblies and subassemblies; this is called an indented bill of material. The stock data is on-hand stock only; it is assumed that there are no orders due in and that no parts are allocated.

There is a requirement to produce 50 aerials by Friday 23 March and a further 20 by Friday 30 March; this information is provided by the MPS. Using this information and the bill of materials we can calculate the gross requirements, i.e. a list of all the parts that are needed and when. We know the quantities because the bill of material gives that quantity of each part required to build one unit of product. We know when the parts are required because we have the lead times. Note, however, that lead times are given in working days and do not

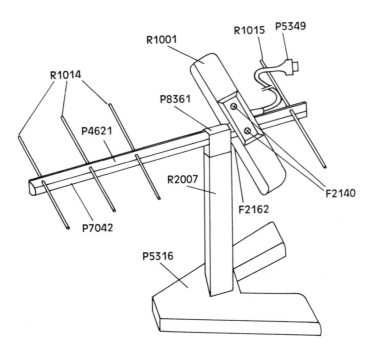

Figure 6.18 Television aerial, part no. A3521

Level	Part No.	Description	Qty	Unit	Lead time days	Stock on hand
0	A3521	Aerial	1	Ea.	1	5
1	A3522	Receiver assembly	1	Ea.	1	15
2	P7042	Support arm	1	Ea.	5	7
2	P8361	Mounting clip	1	Ea.	5	7
2	R1001	Al strip 25mm×1mm	0.45	m	3	1.5
2	F2162	M4×10 Al	1	Ea.	2	100
2	F2140	M2×6 Al	2	Ea.	2	75
2	P5349	Connector & cable assembly	1	Ea.	4	10
2	P4621	Main arm	1	Ea.	6	2
2	R1014	Al bar 3×2×14mm	3	Ea.	3	24
2	R1015	Al bar 3×2×20mm	1	Ea.	3	0
1	R2007	Nylon 15mm sq tube	0.15	m	2	0.45
1	P5316	Stand base	1	Ea.	8	30

Figure 6.19   Bill of material for television aerial

include weekends. Thus, for the aerials required on Friday the 23rd the orders for the stand base must be placed nine working days earlier, i.e. Monday the 12th. Note that the lead time for the base is eight days but a further one day is required to manufacture the aerial after the parts are available. This is easier to see if a picture is drawn as shown in Figure 6.20.

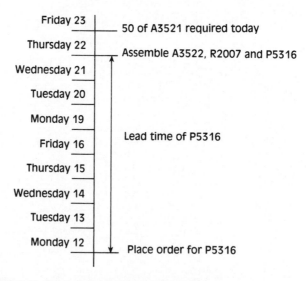

Figure 6.20   Sketch to check order dates and lead times

The gross requirements for the television aerials are given in Table 6.2. The gross requirements are used in conjunction with stock data, the quantities of stock on hand and those due in, to calculate the net requirements. Where a safety stock is usually held this must also be allowed for so that a reorder is triggered if the stock will fall below this level. The net requirements are calculated as:

$$NR = GR - OH - DI + SS$$

where

NR = net requirements
GR = gross requirements
OH = stock on hand
DI = stock due in from orders already placed (not allocated)
SS = safety stock

Table 6.2   Gross requirements

Part no.	Date required	Lead time	Date for order	Quantity
A3521	30 March	1	29 March	20
A3522	29 March	1	28 March	20
R2007	29 March	2	27 March	3
F2140	28 March	2	26 March	40
F2162	28 March	2	26 March	20
R1014	28 March	3	23 March	60
R1001	28 March	3	23 March	9
R1015	28 March	3	23 March	20
P5349	28 March	4	22 March	20
A3521	23 March	1	22 March	50
P7042	28 March	5	21 March	20
P8361	28 March	5	21 March	20
A3522	22 March	1	21 March	50
P4621	28 March	6	20 March	20
R2007	22 March	2	20 March	7.5
F2140	21 March	2	19 March	100
P5316	29 March	8	19 March	20
F2162	21 March	2	19 March	50
R1014	21 March	3	16 March	150
R1001	21 March	3	16 March	22.5
R1015	21 March	3	16 March	50
P5349	21 March	4	15 March	50
P7042	21 March	5	14 March	50
P8361	21 March	5	14 March	50
P4621	21 March	6	13 March	50
P5316	22 March	8	12 March	50

In this example we are assuming that there is no safety stock and that there are no orders due in. The net requirements for the week 12–16 March are given in Figure 6.21 as a series of suggested orders. Note that in this case the requirement for P5349, the connector and cable assembly, is given as a works order as this is produced in-house. All purchase orders are externally sourced. Net requirements must be calculated chronologically, earliest first, as the effects of a change in stock level must be carried forward.

In this way MRP provides information which can be used to schedule purchases and allows detailed scheduling of shopfloor processes on the basis of parts requirements. It is not a shopfloor scheduling system as more detailed capacity planning and scheduling may still need to be done, but it is widely used and there are many computer packages available on the market. It is not a procedure that is performed manually as the number of calculations would be ridiculously large in a real situation with a variety of products, and where a number of parts are used on different products.

The orders placed for this first week will be used to update stock information. This may have a bearing on subsequent calculations, for example if there is a minimum order quantity for a part that will provide excess stock that can be used on other jobs. For this reason, MRP processing must be done on a regular basis.

MRP can be used to allow analysis of 'what if' situations, allowing speculation of the consequences if the manufacturing system is loaded differently, or to assess if it is feasible to make something within a given time. For example, in the case of the television aerials, the stand base P5316 had to be ordered by 12 March in order to achieve a product completion date of 23 March. If the order for the aerials had been received on 14 March then the MRP would have highlighted that the completion date could not be met.

Advantages and disadvantages of MRP
MRP has as many advocates as it does opponents who say that it is not sufficient for today's complex manufacturing systems. Thus we shall briefly review the

Part number	Works orders		Purchase orders	
	Quantity	Required by	Quantity	Required by
R1014			126	16 March
R1001			21	16 March
R1015			50	16 March
P5349	40	15 March		
P7042			43	14 March
P8361			43	14 March
P4621			48	13 March
P5316			20	12 March

Figure 6.21 Suggested orders

advantages and disadvantages of MRP:

● *Advantages*: The main asset that MRP has is that it has been widely used for a very long time. Therefore there has been a lot written about it, it is a proven technique and there is much expertise that an organization can call upon. The technique is designed to allow production in minimum lead time while maintaining minimum stock levels. It can do this successfully with basic information on product structures and lead time, and it can be used with any type of manufacturing organization (batch, flow, etc.).

● *Disadvantages*: The information supplied for the MRP processing must be accurate, and clearly, with the amount of information required and the fact that lead times can change, this can be problematic. The system is not tuned for detailed capacity planning which can lead to problems with meeting schedules. Finally, MRP is a 'push' system. This means that work is pushed onto the shopfloor irrespective of whether there is real demand, rather than that used to feed the MRP processing, and also with no account of whether the work centres have available capacity. This means that use of the technique can lead to materials waiting at work centres and processing taking place when there is no requirement for it.

For many organizations the disadvantages of MRP are far outweighed by its advantages. However, for those that do suffer its limitations, or who see the Japanese style manufacturing systems as the systems of tomorrow, then the answer is to move from MRP to JIT.

6.5 Just-in-Time

Just-in-Time (JIT) was developed in Japan. It is a philosophy of meeting customer needs when required and with minimum waste. In the JIT environment any non-value-added activity is deemed to be waste. These activities are costly to the company but do not increase the value of the product, that is, the amount the customer will be prepared to pay. There are many non-value-added activities to be found in a manufacturing organization using more traditional production methods, and these will include inspection of parts and holding stock of parts and finished goods. Elimination of the waste produced by these activities can be achieved by eliminating defects and by not overproducing, i.e. by making to order.

The aim of a JIT system is to ensure that goods are available at the time that they are required, whether these be parts, subassemblies, or products. The consequence of this is increased productivity and flexibility. Increased productivity means that product can be made in the shortest possible time, with minimum resources. Flexibility means the company's ability to react to changing circumstances, whether this be a change in customer order or a modification to the product design.

The main difference between a JIT system and an MRP one is that JIT is a 'pull' system whereas MRP is a 'push' one. In the JIT system parts are ordered and work is done when there is both a requirement for work and the facility for doing it. In the ideal JIT system there should be no queuing of material at work centres and no stock of parts. When there is a real customer order work will begin at a work centre but no work from a preceding work centre will be passed on until it is 'pulled' by operators because they are free to work on that job immediately. The system can lead to perceived idleness of operators and work centres. However, carrying out work that is not required just to keep people and machines busy is seen as a waste and this time can be used for other productive activities such as maintenance, problem solving, etc. JIT manufacture is typified by a make-to-order, single-unit flow line production system in which there is a steady rate of production. Essentially product is made in batches of one.

6.5.1 The principles of JIT

The JIT aim is achieved by a disciplined approach which involves three principles applied to the organization:

1.   The elimination of waste
2.   Total quality control
3.   Total employee involvement.

Waste in a system can perpetuate errors and problems and so the principle is to eliminate the waste, highlight the problems and then resolve them. An example is the holding of stock. If stock is held and a fault is identified in a component or something is not made correctly then the stock can be used to replace the faulty item with very little difficulty. The fact that the component was faulty can be overlooked, the fact that an operator is not working to specification can go unnoticed. Take away the stock and immediately the problem becomes visible and action can be identified and taken to ensure that the problem does not recur.

The usual analogy for this phenomenon is the one shown in Figure 6.22. As the waste (the water level) is reduced more problems (the rocks under the water) are exposed and therefore to maintain production (the ship's progress) the problems must be resolved.

Total quality control leads to the elimination of waste by eliminating defects. However, within the JIT environment the aim is not to detect defects but to prevent them occurring in the first place by tracing any problems back to their source. This involves the whole organization from product development, ensuring that new products can be manufactured to specification, through to purchasing who must ensure that bought-in parts achieve the specifications required for

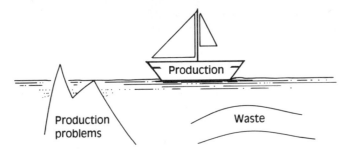

Figure 6.22   The JIT/production ship analogy

manufacture. The emphasis within the manufacturing area is on statistical process control and in-process testing rather than on inspection after processing. This ensures that processes can achieve specification instead of finding out if they have achieved it. Quality control and total quality management (TQM) are covered in Chapter 7.

The third principle of JIT is total employee involvement, which recognizes that the greatest asset of any organization is its workforce. This principle requires that management provide leadership which leads to employees wanting to be involved in what is happening. They must also give opportunity and encouragement which includes providing education and training, and using teams at work. Multi-skilling the workforce is of great importance in the JIT system, and means that workers are trained in a number of tasks and skills. Multi-skilling provides increased production flexibility by being able to redeploy workers, when required by changes in product or demand, with minimum disruption to the production line and with minimum losses in productivity. It also has a great advantage for the workforce. Workers can contribute more effectively to problem solving because they understand the wider implications of what they suggest, and this also leads to increased motivation.

6.5.2 JIT techniques

A number of specific techniques are used in JIT to achieve increased productivity and flexibility. All these follow from the three JIT principles. Five of these techniques are described here, i.e. work-in-progress reduction, set-up reduction, Kanban, supplier integration and balanced scheduling. Other JIT techniques are group technology (described in Section 6.2.4), multi-skilling the workforce (mentioned above) and preventative maintenance.

Work-in-progress reduction
Work-in-progress reduction is achieved by reducing batch size, the target being a batch size of one. As batch size reduces, the amount of work waiting to be

processed is reduced, as is the time to process any batch. This leads to advantages for the company because less space is required and less stock. Lead time for a product is also reduced, thus providing improved customer service. In addition to these benefits the non-value-added activity of handling and administration are reduced. Flexibility is obviously increased. If there is a small amount of work-in-progress then changes do not have to wait a long time before being introduced.

Set-up reduction

Set-up time is the time taken to effect a changeover from one product to another on the same machine. There are two components in set-up, the activities that can be carried out while the machine is still producing the previous product and those that can be carried out only when the machine is not producing.

If the set-up time is long then this can represent a significant cost. The machine is out of action for that time and the operator is unproductive. If this cost has to be absorbed by a small number of products then the product cost may be disproportionately high. For example, if it takes 1 hour to set a machine which then produces product for 1 hour before being reset then the cost of that batch of product must include 2 hours of machine and operator time. Some complex machines can take in excess of 8 hours to set up and so the economics of the situation lead to the production of large batches with all their disadvantages – tying up the process for other products, long lead time, high work-in-progress. In order to avoid this 'requirement' for large batches the aim in the JIT system is to reduce the set-up time so that small batches become economically viable.

The most effective ways of reducing set-up are targeted at reducing the activities that have to be done while the machine is out of action. This can be achieved by looking at methods to allow these tasks to be carried out while the machine is still operating, for example by fetching tools or removing the requirement for this to happen by having tools next to the machine. Standardization of parts will also help to decrease set-up time by reducing the requirement for tool changes.

Kanban

Kanban is a system for 'pulling' materials through the manufacturing process. It is a signal that work is required and can be done, and therefore that materials are needed. Kanban is the Japanese word for card, and many companies do use cards to signal work requirements. However, any clearly visible method will do. Some companies use boxes or bins, others have areas marked out on the floor.

The Kanban system is activated by a final assembly schedule, indicating that products are required. The operator at the final assembly thus has a Kanban to pull work from the previous workstation. This operator will then use a Kanban to indicate that he or she requires work from a prior station, etc. In practice, there will be a very small amount of buffer stock between workstations and the reduction of this will provide the Kanban signal.

Supplier integration

The role of suppliers in the JIT system has major significance. If suppliers do not supply on time or meet the specifications set then the system cannot work. Therefore a company using JIT must work closely with its suppliers. The aims of building a good relationship are:

1. To improve the design and quality of goods and services
2. To improve logistics, i.e. the movement of material from the supplier to customer
3. To reduce the requirement for high levels of stock.

The strategy used for achieving these aims requires the building of a relationship which benefits both supplier and customer. The companies work together with the concept of team working extending beyond the internal production system and both companies will be involved in design and supply decisions from the start of a project. This type of partnership requires reducing the supplier base to a minimum, and using local suppliers where possible. It requires the supplier to provide quality assured goods with reliable and frequent deliveries. The supplier should not be used as a store for the JIT organization but should be able to produce goods to a predetermined schedule of expected requirements. The benefits for the supplier are assured trading, prompt payment, and a reduction in transactions.

Balanced scheduling

Balanced scheduling is the equating of the production rate to the sales rate. This may seem an impossible task, particularly when companies offer a wide range of products. However, by modularizing and standardizing subassemblies and components it is often possible to predict sales demand with some accuracy. The variation is then for final assembly which is customer-specific. If the lead time of the product can be reduced to be within the normal customer demand time then there is no problem and the scheduled can be balanced. This will mean that production rates for critical resources, i.e. those that set the lead time, may be set per hour, day or week.

The JIT techniques cannot be introduced piecemeal if results are to be obtained. The successful implementation of JIT follows from a thorough examination of the manufacturing facility and the products being produced. It requires the involvement of design engineers as well as manufacturing engineers, as so much of its success can be attributed to standardized and modular design and the ability to manufacture in a flow line.

JIT has many benefits, and companies that employ the JIT techniques are happy to talk of the savings in time, costs, space, that have resulted from the use of the system. These companies have also found that worker motivation has improved. The difficulty that some companies experience with JIT is that in order to work effectively it requires a culture change, particularly in the way that management and workers work together. Some companies would have difficulty finding benefits from changing to a JIT operation. In companies where there is a very stable demand or a limited product range it is unlikely that the costs of change could be recouped. JIT also requires excellent relationships with suppliers. Failure of supply will be very disruptive in a JIT environment.

The KAB Seating case, below, describes some of the benefits realized by the implementation of a JIT system. In addition, the Edwards High Vacuum case study in Chapter 8 describes how the company changed from a traditional batch manufacturing system to a JIT one and mentions some of the benefits that were achieved.

6.6　Case study: KAB Seating Ltd

The following case study shows why a company chose to implement a JIT system to manage its manufacturing facility. It gives real examples of what can be achieved by JIT, and it describes some of the issues that are involved in the implementation of such a system. (Reproduced, with permission, from *Just in Time: An executive guide to JIT*, produced by the DTI as part of the Enterprise Initiative 'Managing into the 90s' series).

KAB Seating business is driver comfort, safety and health. It achieves this by producing ergonomically designed seats to give the correct postural support and seat suspensions designed to eliminate harmful vibrations. The principal market is work vehicles (such as trucks, tractor, off-highway vehicles). Approximately 65% of the main Northampton plant's output is for export. The customer base is worldwide including Ford, Iveco, Case, Komatsu, Caterpillar, Scania, JCB and Massey Ferguson.

The main factory in Northampton employs approximately 450 people. The product range is diverse with production varying from very small to medium volumes. The company has a high degree of vertical integration producing its own pressings, foams, covers, etc., in addition to the obvious assembly of the product.

In late 1984 the parent company Bostrom was subject to a management buy-out by the incumbent directors. Since then the company has grown by 300% and in late 1988 floated on the UK stock market.

Why Just-in-Time?

The management identified a number of external and internal forces that would occur, and recognized that Just-in-Time could be used to meet these. They included:

- The need to be extremely flexible in reaction to customer schedule changes

- Movements towards daily/JIT deliveries to customers
- The need to reduce financial gearing by reduced inventory
- The need to establish distinctive strengths in the market, which could be used to improve efficiencies in any subsequent company acquisitions
- The need to keep a highly competent management motivated by applying new ideas
- The need to involve factory workers in the running of the company and enable them to work in small teams using job rotation
- The pressure to reduce cost by improved methods and improved material supplies
- The opportunity to create space which could subsequently be used for growth.

The question was whether JIT techniques could be used in a low- to medium-volume production environment. It was immediately recognized that the project would be large-scale, encompassing virtually all aspects of the business.

Performance targets
In the planning and evaluation phase, a number of specific targets were set:

- *Inventory*: To increase the number of times inventory was turned over each year by 200%. To reduce total inventory by 40%. Although the target turn rate does not look ambitious, it should be noted that only 40% of the material supplies were considered suitable for JIT, with a large element of the balance coming from Continental sources.
- *Lead times*: The in-house manufacturing lead time for a complete product should be reduced from six to eight weeks to less than 10 days – press batch sizes of an average of five days, and two to three days of work-in-progress on welding/fabrication/paint/sub-and final assembly.
- *Set-ups*: Reduce average press set-up times by a factor of six times to reduce press shop batch sizes to one week's production without loss of efficiency.
- *Timing*: Major savings must be realized within six months.

Implementation
There were four strands to the implementation of the project:

- Supplying management, organization of the supply base, JIT delivery arrangements, etc.
- Identifying and installing integrated flexible production cells by the use of group technology
- Reducing in-house set-up times using work study techniques
- Modifications to the information systems, both computer and manual.

Here we focus on the key second and third strands.

(1) Integrated production cells: *implementation phase*   In order to promote the fastest passage of components through the factory, work-in-progress and the movement of it must be kept to a minimum. The answer was to create a series of manufacturing cells for the eight product groups that integrated all perations after pressing but prior to final assembly. These operations include anual and robotic welding, drilling, light pressing, spot and projection welding and subassembly. Figure 6.23 shows the material flow prior to reorganization and the large amount of material handling necessary to support it.

A large amount of analytical work and detailed planning was required in designing the cells to balance utilizations, work flow and layout, materials handling, and standardization of components. The programmes for all operations to final assembly rely on the pulling effect created by usage on final assembly. No production in the factory is dependent on computer-generated gross requirements. Each section is triggering the previous section as stocks are used up. The labour force is required to be flexible and has been involved in the various stages of the project. Despite a great deal of scepticism, there has been a complete cooperation in the radical changes to operators' working areas. Each work area – final assembly line, manufacturing cell and so on – has its own hourly paid direct operator as a group leader. The factory is now run with 300 direct operators but requires only three supervisors and no progress chasers.

(2) Set-up time reduction: *implementation phase*   It was obvious from the beginning that the major cause of internal stock was coming from our own press shop and that this was quite simply due to the traditional calculations of set-up

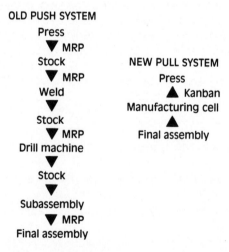

Figure 6.23   Material flow prior to reorganization

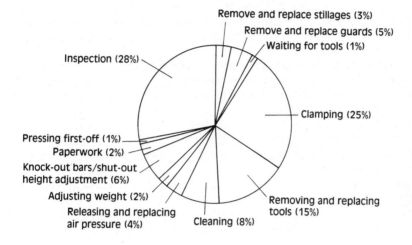

Example of original multi-tool
press set-up on 300-ton press

Remove and replace stillages (3%)
Remove and replace guards (5%)
Waiting for tools (1%)

Inspection (28%)

Clamping (25%)

Pressing first-off (1%)
Paperwork (2%)
Knock-out bars/shut-out
height adjustment (6%)
Adjusting weight (2%)
Releasing and replacing
air pressure (4%)

Cleaning (8%)

Removing and replacing
tools (15%)

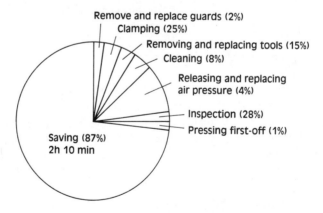

Example of proposed press set-up
times for 300-ton press

Remove and replace guards (2%)
Clamping (25%)
Removing and replacing tools (15%)
Cleaning (8%)
Releasing and replacing
air pressure (4%)
Inspection (28%)
Pressing first-off (1%)

Saving (87%)
2h 10 min

Figure 6.24 Work elements for the 300-ton press

time against running time. Batches of between one and three months' usage were being run. Therefore, by setting the objective of maintaining the current productivity, and setting the target of one week's batch sizes, this gave the objective of a reduction of six to eight times in set-up times. Therefore, the major area to aim for in set-up time reductions was the press shop. Initially a number of video tapes were taken of press tool set-ups on 300- and 100-ton presses.

From the small sample it was apparent that set-up times for the 300-ton press could run as high as 3.5 hours. An analysis of the typical work elements is in Figure 6.24. The objective was to reduce the set-up time to a maximum of half and hour.

To achieve this would need total reorganization of the way the press shop was managed. Some of the reorganization required capital investment, but most of it was the commonsense application of work study and good housekeeping. The actions included:

- Totally new layout of presses to give improved space utilization
- Creating detailed operation sheets for each piece part operation giving tooling details, etc.
- Dedicating each piece part operation to a specific machine, colour coding the tooling and storing it adjacent to the press
- Manufacturing a kit of spacers and air pins for each tool, stamping them with the piece part and operation number and storing adjacent to the press
- Designing and manufacturing tool kits containing standard spanners, nuts, bolts and so on for each press
- Buying a tool-handling trolley
- Establishing standard clamping centres and tool shut heights.

6.7 Summary

In this chapter we have examined some of the issues that affect manufacturing. We have described a number of techniques that can be used to manage the activities that must take place in the manufacturing facility in order to balance the three elements of any operations activity, i.e. the inputs, the processes, and the outputs.

6.8 Revision question

1. What are the key features of job, batch and flow production methods of organizing manufacturing?

2. Describe what is meant by group technology.

3. What are the implications for a company changing from batch production with functional layout to group technology? What are the likely advantages of such a change?

4. Under what circumstances would a change as in (3) above not be appropriate?

5. What data are needed to operate a materials requirements planning system? How can this be generated?

6. What strategies can be used by companies to deal with overcapacity? Describe the implications of each strategy.

7. What strategies can be used by companies to deal with undercapacity? Describe the implications of each strategy.

8. What is the role of a Store? Describe the main activities undertaken by a storekeeper.

9. Explain the terms:

   Safety stock
   Reorder level
   Reorder quantity

10. What factors affect the values used for reorder level and reorder quantity?

11. Bidol Ltd manufactures filing cabinets. It has two products, a four-drawer cabinet and a two-drawer cabinet. It makes 3000 and 4000 of each, respectively, per year. It currently buys in drawer handles in batches of 1500 pairs at 24p per pair. However, a new supplier is offering the following prices:

    1–2000 pairs at 25p per pair
    2001–5000 pairs at 23p per pair
    5001 + pairs at 20p per pair

    The cost of holding stock of parts is taken to be 10% of the stock value and the cost of placing a purchase order is £7.50. Should Bidol change their supplier? If so, what quantities should they buy in to achieve maximum savings?

12. Describe the operation of an MRP system, in particular defining 'master production schedule' and 'bill of material'.

13. The demand for the phase change and sensor assembly, shown in Figure 5.12, for the next four weeks is given as:

Week	1	2	3	4
Demand	400	700	1000	1000

    These quantities must be available at the end of each week. Using the stock information below, produce an MRP for the next four weeks. There is no safety stock and no parts are allocated. The first day of the first week is day 1, the factory works a

5-day week.

Part	Stock on hand	Quantity due in	Day due in	Lead time (days)
LMN4703	250	200	7	5
TME0232	1100	240	4	7
CME0023	2000	1000	11	2
MCH6234	600	—	—	10
KNB0009	1500	800	10	3
ELC0454	750	1000	3	15
CEE0121	750	1000	5	10
LMN0023	750	1000	8	10
SEC0345	410	500	10	6

Comment on the results of the MRP.

14. Explain how set-up reduction and Kanban are used to achieve Just-in-Time production.

Further reading

Adam, E. E. and Ebert, R. J. (1989), *Production and Operations Management: Concepts, Models and Behavior*, 4th edition, Prentice Hall International.

Baily, P. and Farmer, D. (1990), *Purchasing Principles and Management*, 6th edition, Pitman.

Harrison, M. (1990), *Advanced Manufacturing Technology Management*, Pitman.

Hill, T. (1991), *Production/Operations Management, Text and Cases*, 2nd edition, Prentice Hall International.

Muhlemann, A., Oakland, J. and Lockyer, K. (1992), *Production and Operations Management*, 6th edition, Pitman.

Wild, R. (1991), *Production and Operations Management*, 4th edition, Cassell.

7

Quality management

Overview

Customers are the lifeblood of an organization. Without customers the organization will not exist. Yet customers are choosy, they know what they want and they expect their needs to be met. However, in return for their requirements being satisfied customers will stay loyal to their supplier, and will tell their friends and colleagues about a good supplier. A company that does not provide the required product quality or level of service will struggle to find new customers, without the benefit of repeat custom. In addition, it will spend money on activities that do not benefit its customers and which do not lead to profit. Thus, to maintain a good and profitable customer base a company must understand and meet its customers' requirements.

The definition of customer requirement is the definition of quality, and the way to ensure that this is met is by quality management. In this chapter we will examine some of the principles and techniques associated with quality management.

7.1 Introduction

'Quality' is a very emotive word and one that is regularly used and misused. 'Look at the *quality* of that!' is a phrase that may be heard in any pub, street or shop. The speaker could be talking about a vehicle, a computer, a piece of jewellery, or even another person. What can the speaker possibly mean? Let's look at another example. A burger bar offers a 'quality meal'. As a customer would you expect something that tastes good, something cheap, or just something that is edible? This wide use of the word 'quality' tends to take people increasingly further away from actually understanding what it is and how it is achieved.

In a way it is right that 'quality' can be interpreted so widely. It is synonymous with customer needs and certainly these vary. However, this doesn't help producers who must provide 'quality' goods and services in order to compete in today's markets. For them a more precise definition is required. Various people working in the field of quality management have coined definitions for it, such as 'fitness for purpose' and 'satisfying customer needs'. A more precise definition is that provided by ISO 8402:

> Quality: Totality of characteristics of an entity that bear on its ability to satisfy stated or implied needs.

This means that quality enters all aspects of the customer–supplier relationship. When you buy a new car you expect it to be fault-free. You also expect it to be delivered on time, to receive an accurate bill (or at least not to be overcharged), and that there is an effective after-sales service. The way in which these expectations are met, or not, will affect your perception of the quality of the car, its manufacturer, and the supplier. Clearly, a supplier cannot meet the individual requirements of all customers. One person's perception of an effective after-sales service will be different from that of someone else. What a supplier can do is define the quality that will be supplied. Even though it will not be possible to meet all the customers' detailed requirements at least they will know what to expect if they buy the product, and they will know whether this has been achieved.

This establishment of a defined quality, with the opportunity to review whether or not quality has been achieved, is very important. Customers are essential to a company's wealth and its ability to trade, and while there are many customers there are also many suppliers. Life is very competitive. With the prospect of a number of suppliers customers can afford to look at what is being offered in some detail, and expect to find both a product and a service that meet, if not exceed, their requirements at a fair price. The implications of this for a company that makes a strategic decision to meet customer needs are significant. The approach that it adopts may affect its internal organization, its culture, its position in the marketplace and its ability to compete and, inevitably, its profitability.

Defining the quality of goods and services is not a simple task as it means giving absolutes that are realistic. If a supplier tries to use comparative or qualitative statements – good, better, reasonable – then it's back to the problem of understanding. However, defining quality is not only for the benefit of customers. Both the under- and overachievement of quality can be expensive for a supplier. If a manufacturer can offer a product with a delivery lead time of 5 days and is not explicit that this is the standard then the effects of supplying people who expect 3 days may be (1) dissatisfied customers who will probably also pass on to their friends and colleagues the news that the company does not deliver on time or (2) extra efforts are made to deliver in 3 days in order to keep customers happy, with the consequence of an increase in cost.

So the first rule of quality management is that the 'quality' of the goods or services being supplied must be defined. The problems of ensuring that this is the right level of quality to meet customer need are discussed in more detail in Chapter 5.

When there is something that can be measured life becomes a little easier. A supplier has to now meet the standard that has been set. There are various approaches that can be taken and these fall into two categories: (1) detection of product that does not meet specification, and (2) preventing the production of product that does not meet specification. In the first category we have the techniques of inspection, test, and quality control. In the second we have the quality management principles of quality assurance and total quality management. We shall now look at what each of these mean for a manufacturing company and how they contribute to the achievement of quality.

7.1.1 Inspection and test

The idea of this approach is that products are checked after production so that faulty product can be detected and rejected. It is a method that has been used for a very long time.

A company produces plastic drainage pipes. The specification for the pipes states that they will be made of specific materials, of a certain size, and will have certain properties, such as being non-porous. In the inspection system the pipes will be made and then inspected, i.e. someone (an inspector) will assess and measure the pipes against the specification for them. This system should certainly have the effect of preventing faulty goods being supplied (as long as the inspector knows what he or she is doing and is very thorough, and all the pipes are inspected...), but at what cost?

An inspector does not add value to a product. The post should be superfluous if the product is made correctly. The problem with inspection is that even if all the pipes are perfect the company still has to employ the inspector and provide the inspection equipment. The other aspect of this is what happens when the inspector finds that the product is faulty. It may be possible to repair or rework, but if not, then the product will have to be scrapped. In the case of scrap then the company has lost the product, having paid for its materials, manufacture and inspection. In the case of rework or repair then the company may be able to retain something but at the cost of either supporting a repair/rework section or submitting to further loss by putting the product back through the production system. This last method is particularly insidious because of the lost capacity to produce new product.

Inspection and test can be used on manufactured goods but is not an effective means of ensuring quality for the customer due to its cost and the difficulty of inspecting and detecting everything. It can, and often does, lead to a culture of blame in which people may try to hide faults rather then have them detected. It definitely results in the belief that 'faults can and will occur, but all is not lost because we'll find them in the end'.

Services are very difficult to inspect because there is often no tangible product that can be measured.

7.1.2. Quality control

Quality control is a step up from inspection, the idea being that processes are monitored and controlled to ensure that they are capable of meeting requirements. In effect, we now have inspection of goods and processes with the consequent costs of finding, or not, faulty goods.

7.1.3 Quality assurance

Quality assurance (QA) is a method of managing all the activities that affect the quality of goods or services in order to prevent faults. If we consider the manufacture of car alarms, this will involve some assembly work carried out using automated assembly lines and some assembly work done by people, and it will involve using bought-in components. With a QA system the aim would be to prevent faulty components being bought in by ensuring that purchase specifications are clear and that suppliers are assessed to confirm that they are capable of meeting the specification. Operators will be trained and processes controlled to ensure that they are capable of meeting the specifications for the processes.

The operation of a quality assurance system will clearly affect many areas of the organization, including operations, some aspects of personnel, such as training, and purchasing. Most importantly, in a company that designs product quality assurance activities should ensure that the design process is managed to ensure that designs meet customer need and that the ensuing product is capable of being produced to specification.

7.1.4 Total quality management

Total quality management (TQM) is the next step up from QA. Its emphasis is on meeting all customer needs and expectations, moving beyond the product focus. It also goes further than QA in that it requires a commitment to continuous improvement in quality. One of the main principles of TQM is that the customer is defined differently. In QA the customer is the one who actually provides the money in return for the goods, the external customer. With TQM everyone in the organization is a customer of a process and in order to achieve quality for the external customer it is necessary to achieve quality for all the internal customers. Thus, it involves quality management principles in all aspects of the business.

In this chapter we will examine the principles, use, and assessment of quality assurance systems, as defined by ISO 9000, and total quality management systems. We will look at tools that can be used in the quality control and continuous improvement programmes that should form part of these quality systems. We shall examine the factors that influence the use of quality management systems. Finally, we shall consider the role of quality costing and the methods that can be used to assess quality costs.

The learning objectives for this chapter are:

1. To examine the role and application of the ISO 9000 quality system standards
2. To understand the principles of total quality management and review its application
3. To be able to use a number of quality management tools
4. To examine the factors that influence the use of quality management
5. To examine models that can be used for determining quality costs and to consider their application.

7.2 Quality assurance and ISO 9000

Quality assurance is a means of ensuring that a given system will provide assurance of the quality of the output from the system. It is about managing to achieve quality and preventing defects. For example, if the quality of a turned component is to be assured then there are a number of management actions required. The drawing for the component must be correct, and the operator should have a copy of it. The materials used must be of the correct standard, the machine must be able to achieve the tolerances given on the drawing, the operator must be capable of using the machine to achieve those tolerances. ISO 8402 defines quality assurance as:

> All the planned and systematic activities implemented within the quality system, and demonstrated as needed to provide adequate confidence that an entity will fulfil requirements for quality.

Thus, quality assurance requires a formal structured approach to the management of quality. When an organization wants to use a system for quality assurance it has two options: either it can develop a system that is unique to its operation or it can use a model that has been developed by another organization. It is this latter route that the vast majority of companies take, choosing the ISO 9000 quality system standards.

7.2.1 What is a standard?

In the UK the British Standards Institution (BSI) is charged with the production and maintenance of standards. These take a number of forms, including guidance notes, system standards, and product standards. Their aim is to produce consistency in products, in particular with reference to safety, and to provide general models for systems which can be applied to particular circumstances. The systems standards to a large extent define good working practice and common sense. All the standards are produced with the help and advice of a large number of industrial and academic experts and therefore benefit from a broad view, which may be beyond the usual scope of a company working in a particular field.

The standards can be used by companies to provide a model against which to work. For example, a company implementing a product development system may want to benefit from the standards describing product development procedures. Companies may use product standards as inputs to their design process so that they can claim compliance with a standard which defines safety or operating features for a particular product.

7.2.2 ISO 9000

The ISO 9000 standards started life in 1979 as BS 5750, a quality system standard developed in the UK and published by the BSI. The aim of the standard was to define a model for a quality assurance system that could be used in a contractual situation between a supplier and a customer. This followed an increasing requirement from customers for quality assurance which meant that suppliers were pressurized into taking on specific quality activities to meet the needs of particular customers. Clearly, an intolerable situation could have arisen if a supplier had a number of customers each with differing requirements. The standard was revised in 1987, at which point it was also adopted as both an international standard (through the International Standards Organization, ISO) and a European standard (with a Euro-Norm 'EN' designation). Throughout the world the standard is known as ISO 9000, although it contains several parts. Within each country that uses the standard it will also carry a standard designation for that country (for example, an NF number in France or a DIN number in Germany). It also has a general European designation, EN ISO 9000.

In 1994 some parts of the standard were revised. In the UK these new parts now carry the designation BS EN ISO 9000, the new labelling reflecting its formal adoption as an international and European standard. However, the old name of BS 5750 is likely to remain in general use for some time to come. The aim of ISO 9000 is to provide a model for a quality management system in any of the following cases:

1. As guidance for a quality management system
2. As a contractual document between two parties, a supplier and a customer
3. As a measure against which a supplier's quality management system can be approved or registered by a second party, i.e. a customer
4. As a measure against which a supplier's quality management system can be certified or registered by a third party, i.e. an independent certification body.

In all cases the emphasis is on satisfying customer needs.

The model is provided by means of a series of clauses each defining a specific requirement. The clauses each relate either to a particular aspect of the business, such as design, or a business or system activity such as management responsibility. The requirements are written in a way that allows an organization to define a working practice that meets the requirement and that is suitable for its own business activity. Thus, ISO 9000 is equally applicable to a firm of five solicitors as it is to a manufacturing company employing 1000 people. Taken as a whole, a company which meets all the requirements will be operating an effective quality management system.

The use of ISO 9000 is very widespread throughout the world. In the UK alone some 30 000 organizations have a certified quality management system meeting its requirements. Thus, the use of this quality management model is well recognized in industry. It has the advantage that the quality system developed by an organization based on the model, while meeting recognized requirements, can also be developed to fulfil exactly the organization's own requirements and objectives.

ISO 9000 contains nine parts, but only three define models for quality management systems. The remaining parts contain guidance notes on the selection and use of the standards, as well as on quality systems in general. The three parts defining quality management models are 9001, 9002 and 9003. They are each written in a style that makes them applicable to both products and services. However, the scope of each is different, with each being applicable in organizations carrying out various functions, as defined in the title. The most comprehensive of the quality management systems is that defined by ISO 9001, and it is this which we will consider in detail. A full list of the parts of the ISO 9000 standard, and its UK designation is given in Table 7.1.

None of the three quality management standards contain references to cost or to administration support such as personnel or finance. Some organizations see this as being a limitation in the applicability of the standards. However, the standards are given as defining minimum requirements, and it is for organizations to determine how they wish to use the principles of quality management.

7.2.3 ISO 9001

ISO 9001 is the most wide-ranging of the three quality management models in the ISO 9000 series. It is intended for use in organizations which carry out design activities as well as production. Its scope also covers servicing and installation.

Each of the three standards, 9001, 9002 and 9003, is written as a series of specific requirements, contained in clauses, that must be met in order to achieve a quality management system. The requirements provide the objectives but not the means. It is for each organization applying the standards to decide on the methods that will be used to meet the requirements. ISO 9001 has 20 clauses. The other standards have fewer but there is a great deal of overlap, with many clauses being

Table 7.1 The parts of ISO 9000, Quality systems

Standard	Date	Title	Comments
BS EN ISO 9000–1	1994	Quality management and quality assurance standards Part 1. Guidelines for selection and use	
ISO 9000–3	1991	BS 5750 Quality systems Part 13: 1991 Guide to the application of BS 5750: Part 1 to the development, supply and maintenance of software	These ISO and BS standards are identical. BS 5750 Part 1 is now BS EN ISO 9001
ISO 9000–4	1993	BS 5750 Quality systems Part 14: 1993 Guide to dependability programme management	These ISO and BS standards are identical
BS EN ISO 9001	1994	Quality systems – Specification for design, development, production, installation and servicing	A standard which can be used for certification purposes by organizations encompassing a design activity
BS EN ISO 9002	1994	Quality systems – Specification for production, installation and servicing	A standard which can be used for certification purposes by organizations encompassing a production activity
BS EN ISO 9003	1994	Quality systems – Specification for final inspection and test	A standard which can be used for certification purposes by organizations having no design or production activity
BS EN ISO 9004–1	1994	Quality management and quality system elements Part 1. Guidelines	
ISO 9004–2	1991	BS 5750 Quality systems Part 8: 1994 Guide to quality management and quality systems elements for services	
ISO 9004–4	1993	BS 7850 Total quality management Part 2: 1992 Guidelines for quality improvement	

the same in each of the standards. Table 7.2 lists the titles of the clauses and indicates in which standards they are required. Where the requirement is indicated by an 'L' it is less onerous than the requirement given in ISO 9001.

Table 7.2   Clauses in ISO 9001, 9002 and 9003

Clause in ISO 9001		Appears in ISO 9002	Appears in ISO 9003
4.1	Management responsibility	Yes	L
4.2	Quality system	Yes	L
4.3	Contract review	Yes	Yes
4.4	Design control	No	No
4.5	Document and data control	Yes	Yes
4.6	Purchasing	Yes	No
4.7	Control of customer-supplied product	Yes	Yes
4.8	Product identification and traceability	Yes	L
4.9	Process control	Yes	No
4.10	Inspection and testing	Yes	L
4.11	Control of inspection, measuring and test equipment	Yes	Yes
4.12	Inspection and test status	Yes	Yes
4.13	Control of nonconforming product	Yes	L
4.14	Corrective and preventative action	Yes	L
4.15	Handling, storage, packaging, preservation and delivery	Yes	Yes
4.16	Control of quality records	Yes	L
4.17	Internal quality audits	Yes	L
4.18	Training	Yes	L
4.19	Servicing	Yes	No
4.20	Statistical techniques	Yes	L

Each specific requirement gives rise to a need for a documented procedure, defining the actions that will be taken to ensure that the requirement will be met. The requirements will be discussed in more detail in Section 7.2.4, focusing on the clauses relating to operational areas of the organization, and in Section 7.2.5 looking at the general quality system requirements.

7.2.4 The requirements of ISO 9001 for operational activities

If a company is designing and manufacturing alarm clocks, for example, and wants to provide assurance to its customers of the quality of the product it needs to have effective working practices, defined by procedures in the operational areas that affect the design and manufacture of the clocks. These would be:

- *Sales* – to ensure that customer requirements are known and specified
- *Design* – to ensure that the product meets the customer requirements

- *Material control* – to ensure that materials used in the manufacture meet specification
- *Operations* – to ensure that the product is produced correctly and consistently.

ISO 9001 defines requirements for each of these areas that will, if met, provide assurance of quality. In addition, it covers servicing activities for companies that include this as part of their contractual agreement with their customer. In this section we will look at the requirements for each area and consider why they are required for the achievement of quality.

Sales

The clause that affects the sales activity is 4.3 Contract review. It is designed to ensure that an organization can meet its customer requirements and that these are fully specified, that there be a method for implementing authorized amendments to contracts. To meet the requirement may require some sort of review meeting before tenders or quotations are sent out, or perhaps just an approval of delivery time from the Operations Manager for a catalogue item.

The reason for this clause is that quality cannot be achieved if both parties do not agree to the same definition, thus the need for a specification. The requirement for authorizing changes follows from the fact that there will be occasions when a customer and supplier will agree a change in specification after an order has been agreed. The purpose here is to ensure again that there is full knowledge and agreement, and also that there is a system for letting people know of the change. There would be some question of quality if a company supplied a fixed satellite system for receiving Astra after the sales department and customer had agreed on a change in the order, and price, for a system incorporating a motorized unit to allow receipt of Astra and Eutelsat. It would not matter to the customer if it was delivered on time, it was not what was ordered.

Design (product development)

The clause relating to design control, 4.4 Design control, is what distinguishes ISO 9001 from ISO 9002. This is concerned with ensuring that product design meets customer requirements and that quality is defined. It has specific requirements relating to the planning of the design process, specification of design input and output, design review and measures for design verification and validation, and the system for design changes.

Design control ensures that when design is carried out the requirements of the design are fully specified, and that when the design is complete it is reviewed to ensure that the specification has been met. A major aspect of design control is the design review, which should be undertaken by someone other than the person responsible for the design. The design review should ensure that all aspects of the design have been considered. It should be an on-going process, and not just take place at the completion of the design. It is also important that a recording system is operated for recording assumptions made in design, recording criteria for component selection, and noting any critical calculations.

There are a number of techniques which can be used by engineers to improve the quality of design and product development work out these techniques are beyond the scope of this book. However, *Managing Quality*, edited by Barrie Dale, provides a good introduction to Taguchi, failure mode and effect analysis (FMEA) and quality function deployment (QFD).

Design can be a long and costly process. Errors made at the design stage can be rectified fairly easily and cheaply. However, if errors continue through to production their rectification is far from simple and may incur considerable expense, added to which there may be a problem with meeting agreed customer requirements in the agreed time. The procedures for design required by ISO 9001 follow from the procedures for good design practice described in Chapter 5.

Material control

The aim of the material control requirements is to ensure that all materials used in the production process meet specification. There are four clauses that relate to material control: 4.6 Purchasing, 4.7 Control of customer-supplied product, 4.8 Product identification ..., and 4.15 Handling ....

Purchasing is concerned with obtaining goods and services that meet specification. The requirements relate to ensuring that suppliers can meet the quality requirements specified for the goods and services, and that these are adequately stated. This means that it will be necessary to clearly define what is required, including any contractual specifications such as delivery time. However, there is no point in defining a tight specification for a product and then placing the order with a company whose abilities and probability of successfully meeting the requirements are unknown. There must be within the purchasing activity a method for assessing suppliers and determining what specifications they can achieve, in order to allow you to decide whether or not these are appropriate to your company's requirements. This is called supplier appraisal and it involves an initial assessment of potential suppliers to ensure that they can meet your requirements. It then becomes an on-going activity of monitoring the quality of supplier service and goods. Supplier appraisal may involve visits to suppliers and audits of their systems or you might accept suppliers on the basis of having a quality system either approved by another of their customers, or certified by an independent certification body. The purchasing activity is discussed in detail in Section 6.4.2.

Customer-supplied product should be controlled to ensure that its integrity is maintained prior to use.

Product identification ensures that at any time you know the inspection status of material and can therefore prevent non-specification material from entering the manufacture process or being sold. Traceability was discussed in Section 6.4.1, and involves being able to trace parts and products to their source.

Methods for assuring the preservation of product quality are required at all stages of manufacture and shipment, until the product is no longer the liability of the manufacturer. This might involve anything from ensuring that delicate goods are not stacked on top of one another to defining special packaging that will protect the equipment as it is being loaded into an aircraft.

Operations

We have seen in Chapter 6 that Operations is concerned with processing inputs to meet output requirements. The clauses in ISO 9001 relating to operations are 4.9 Process control, 4.10 Inspection and testing, and 4.11 Control of inspection, measuring and test equipment. These ensure that processes are capable of meeting output requirements and require quality control to provide assurance of continuing capability.

Process control ensures provision of the instructions, equipment and environment that are needed to carry out particular processes. These will include work instructions indicating the level of workmanship that is acceptable for any given process on a specific product. In many engineering companies these take the form of drawings. In addition, procedures for the approval of processes and process monitoring to ensure continued capability are also required. In order to control processes you need to identify the actual processes used, determine the process specifications and institute a procedure for monitoring to ensure that processes are carried out within the specification. Some processes are defined as special processes and these cannot be monitored while they are being done; it is only possible to carry out monitoring of quality after completion. Welding is a typical special process. For special processes the way in which you achieve control is by ensuring that equipment and operators can meet the required standards, and this may require special training and qualification for operators.

Inspection and test procedures are required to define the quality control that is to be carried out on the manufactured product in order to ensure that it meets the design specification. They have to define the actual form of the inspection or test, the results expected including tolerances, and what should happen to inspected product. They should also define the level of qualification required for the inspector/tester.

There is no point in inspecting and testing unless you have confidence that the equipment being used to carry out the inspection or test is capable of achieving the required level of accuracy, and that at the time it is used will achieve that level of accuracy. It is therefore essential that in any quality system where there is a requirement to carry out either inspection or test there is a procedure that, if followed, ensures that equipment is appropriately calibrated and maintained.

Servicing

Clause 4.19 Servicing requires that there is a system for ensuring that servicing is carried out to specification when it is part of the customer requirements.

7.2.5 The requirements of ISO 9001 for quality system activities

In order to hold the quality management system together there are a number of requirements that relate to the operation and functioning of the quality management system itself. These will apply in all the operational areas covered by the quality management system. They are management responsibility, definition of the quality system, document and data control, corrective action, quality records, audit and review, training, and statistical techniques.

Management responsibility
The clause on management responsibility requires an organization to demonstrate its commitment to quality by having a documented quality policy and developing an organization structure to allow the achievement of that policy. It should also have a means of reviewing the quality system for its continued suitability and effectiveness in meeting the requirements of both the organization (as defined by the quality policy and the objectives for the quality system) and the requirements of the standard.

A quality policy is a statement of aims and objectives for the quality system. It may be associated with a catchy phrase, such as 'it is our policy to be the best', but it must have real targets set in it. In the same way that the quality of a product must be defined, so must the quality of the quality system. It is a requirement that everyone in the organization is aware of and understands the quality policy.

The organization structure requires the definition of responsibility and authority for all personnel who affect quality, a commitment to the provision of adequate resources, and the appointment of a 'management representative'. It is the management representative who will have responsibility for the quality system. This person is often given the title of quality manager.

Quality system
The requirement is to document the activities of the organization that pertain to quality by means of a quality manual and quality system procedures. In addition, there is a requirement that the organization should carry out quality planning to ensure that quality requirements will be met during any project or product development. The aims of the requirements are to ensure that the quality system is clearly defined and is accessible to all relevant personnel, and that quality activities are considered at the project planning stage.

Within the quality system there will be four levels of documentation, as shown in Figure 7.1. Each of these levels are as follows:

(1) *The quality manual*: The quality manual is the operating document for the system. It may contain all the procedures that are operated in order to meet the requirements of the quality system, but is intended to give an overview of the complete system.

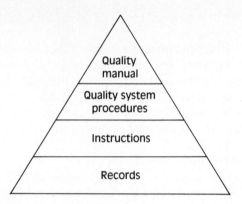

Figure 7.1   Quality system documentation

(2) *Quality system procedures*: These are the actual documents that define the tasks to be carried out in order to meet the requirements of the quality system. Any procedure must give the job title of the person responsible for carrying out the task and the job title of the person who has authority for making decisions about the work being undertaken. Procedures clearly define the way in which the task is to be undertaken and the records that are to be kept in order to verify that the task has been carried out correctly.

Procedure writing requires much care and practice to ensure that ambiguities are avoided and that the steps involved are complete and appropriate to the intended user. There are certain 'rules' for procedure writing, and these include:

- The aim of the procedure is to break down the task into a series of small, logical steps.
- It must define who is responsible for carrying out the procedure.
- It must define the scope of the activity.
- It must state how the operation of the procedure will be observed, i.e. what records will be made or what will happen. The intended user should know what the aim of the task is and when it is complete. Any further action that may be required should also be defined.
- Any supporting documentation must be clearly referenced.
- The language should be unambiguous.
- The language and detail of the procedure should recognize the level of skill and competence of the persons for whom the procedure is written. The correct technical terms should be used to avoid potential misunderstanding.

Procedure writing should be approached with the care and thought given to any written communication (see Chapter 12).

(3) *Instructions*: These are like procedures in that they define tasks to be carried out and the same rules apply when writing. The difference in the labelling of the two is that procedures relate to the quality system and instructions will relate to particular operations activities. For example, there may be a procedure that specifies that calibration of measuring and test equipment will be carried out. It will identify who is responsible for ensuring that calibration happens, but it will not give the detailed instructions for calibrating a particular piece of equipment. These instructions will be referenced in the procedure. There will be an instruction for each type of equipment to be calibrated and it will identify who is responsible for the calibration, when and how it is to be done, how the results are to be recorded, how to identify a piece of equipment after calibration, and what to do with non-conforming equipment.

(4) *Quality records*: These will follow from people following instructions, for example completing a calibration record card showing when the calibration was carried out, by whom, and the results. The records form the documented evidence that instructions have been followed and show the results obtained. They therefore provide evidence of the achievement of quality.

Document and data control
Document control is the way in which documentation is managed to ensure that it is appropriate, accurate, and available where it is needed. It is particularly important in the manufacturing area where it is essential that up-to-date drawings and part specifications are used. It is the means by which companies ensure that the correct documents are available where they are needed and that obsolete documents are removed from use. It also allows changes in documentation to be monitored and controlled. Document control is necessary business practice in all companies, but particularly those carrying out manufacturing, whether or not they use formal quality management systems. The requirement of ISO 9001 is for the company to have a system for controlling all documents that relate to the quality system, including a system for approving and issuing documents, and a method for controlling document changes.

A document control procedure should cover all the documents that define either the product or the quality system, although it may be extended beyond this if a company feels it appropriate. The types of documents that should be included are:

Quality manual
Quality system procedures
Bills of material
Part and product specifications
Drawings
Assembly, test and work instructions
Process instructions and specifications
Instruction manuals
Operating procedures
Inspection instructions.

The way in which these documents are controlled is first, by putting control information on the document and then by regulating modifications and the distribution of the document. Figure 7.2 shows a document with control information.

The unique identifier for each document usually follows a particular house format, as in Figure 7.2 using QP to indicate Quality (system) Procedure. It ensures that there can be no ambiguity when referring to documents.

The issue number shows how many times the document has been issued after being amended in full. Initially the issue number will be 1, as this is the original document. At this stage the revision number of the document will be 0. When an amendment is made the document will be redistributed with an increased revision number and the amendment clearly shown. However, it is usual after a number of small revisions or after one or two large revisions to reissue the document so that the issue number increases and the individual modifications are incorporated fully into the document. After this process the revision number reverts to 0. At this stage the document control system should ensure that all obsolete documents are withdrawn from use.

The number of pages should always be shown on a document so that the recipient can be sure that they are in receipt of the complete document.

The approval signature and date shows that one person has authority for that document and has taken responsibility for the accuracy and completeness of its

---

> Procedure: QP1
> Issue no. 2 Rev: 0
> Page 1 of 1
>
> **Controlled Document Register**
>
> All controlled documents will be logged in a central register. This register will be maintained by the document controller.
>
> The register will contain the following information:
>   Title of document
>   Current issue and revision numbers
>   Approval authority
>   Location of master copy
>   Register of controlled copies, names of holders and locations
>   Location of old issue master copies (for archive)
>
> This register is maintained on a computer database and can be accessed using the DC terminal.
> Approved by   Date: 1 May 1994
>
> Issued to: Document controller
>   Engineering manager
>   Administration supervisor

Figure 7.2   Document with control information

contents. Subsequent modifications to the document should not be made without further approval from this authority. This is particularly important when changes are being made to design or specification because the approval authority will have access to information about why a design was made in a certain way and will be able to evaluate the effects of the proposed change. The distribution list indicates to whom the document has been copied.

In order to control the document a master list of all documents has to be kept. This may be called a log or a register but should contain the information, defined below, relating to all controlled documents:

> The titles and identifiers
> The current issue and revision numbers
> The approval authority
> The location of the master copy
> The distribution list for holders of controlled copies of the document
> The location of old issue master copies.

The document log ensures that at any time it is possible to identify the current version of a document and who has copies of it. When a document is to be redistributed the people on the distribution list can be contacted and the old copies of the document withdrawn. This is particularly important if one is to avoid old versions being used inadvertently. New copies can then be issued.

In some circumstances a company might not wish to have all copies of all documents controlled, for example if a drawing has been sent to a workshop for quotation or if a product manual has been supplied as a sample. In these cases the copies should be clearly labelled as 'uncontrolled' and the recipient should understand that these will not be updated and should not be used for manufacture. Document control procedures should be as simple to use as possible and avoid excessive bureaucracy. Only in this way will they be operated effectively. The control of drawings is described in Chapter 5.

Corrective action

There are two clauses that relate to corrective action, i.e. putting things right. These are 4.14 Corrective and preventative action, and 4.13 Control of non-conforming product.

The corrective and preventative action clause acknowledges that things can go wrong, leading to 'type' faults, such as documents not being withdrawn from use. It requires systematic and objective reviews of areas of the company to identify any problems with the operation of the quality system or product in order that they can be resolved. It also defines the way in which problems will be followed up to check the effectiveness of the solution.

The requirement relating to non-conforming product disposition acknowledges that sometimes things go wrong with individual products. Non-conforming products are those parts, or products, that do not meet specification, for whatever reason. A non-conforming product procedure must provide a method for identifying the parts, taking them out of the manufacture process, and then dealing with them in the most cost-effective way. Some options are to reject and scrap parts, to reject and rework parts, or to ask for a concession from the customer. The concession is an agreement from the customer to take the product even though it does not meet the original specification. Each of these options has a cost and when determining the option to be used this cost and the effect on the business have to be carefully weighed.

Quality records

Records will be required throughout the quality system and provide documented proof that procedures have been followed. You need to ascertain what records are required, what information is needed and the most appropriate method for keeping the records, and there should be a system for maintaining them. Examples of records would include calibration records showing that equipment has been calibrated, training records indicating what training people have required and what has been undertaken, etc.

Audit and review

ISO 9001 has two requirements relating to audit and review. The first is part of the Management Responsibility requirement to have a regular review of the system to ensure continued suitability and effectiveness. The second requirement is given in clause 4.17 Internal quality audits. These audits are used to confirm that the system that is defined by the quality manual, procedures and instructions is the one that is actually being used.

An audit is a systematic check, by looking for objective evidence, that is carried out by someone who is independent of the area being audited. This latter requirement is to maintain impartiality and aid objectivity. Being objective can be very difficult for these auditors and thus there is strong emphasis placed on the training of internal auditors. In the UK there is a registration scheme that is administered by the Registration Board for Assessors, and many certification bodies will want to see that a company's auditors are registered before certifying its quality management system.

Training

Everyone should be trained to do the job that is expected of them. In order to ensure that this happens ISO 9001 requires that a company should have a system for identifying training needs and providing appropriate training. This requirement would lead to a training policy and a procedure that ensures that training is reviewed at regular intervals. The subject of training is dealt with in detail in Section 9.6.

Statistical techniques

There is a requirement to ensure that there is a method for using statistical techniques when these have been identified as appropriate for controlling and verifying processes and product. This is particularly pertinent in manufacturing where statistical process control is used routinely. The aim of the requirement is to ensure that the techniques are selected and used correctly.

7.2.6 Certification

Certification is the term used to describe the process in which a system or product is objectively assessed against a specification, by an independent body, and a certificate is awarded when the certifying body is satisfied that the specification has been met. The certificate is thus the public acknowledgement of the compliance.

There are four types of certification commonly found in manufacturing: quality management certification, product conformity certification, product approval, and CE marking. Each is described briefly below.

Quality management certification

Quality management certification is the process by which an organization's quality management system is assessed against the relevant part of ISO 9000 by an independent third party, i.e. someone who is independent of both supplier and customer organizations. The assessment involves an objective examination of the organization's working practices and procedures to confirm that the requirements of ISO 9000 are being met. A positive result leads to the issue of a certificate and subsequent registration in the Department of Trade and Industry's register of certified suppliers. A certified company will be subject to regular surveillance audits following certification and these are used to confirm continued operation of the quality system.

Certification is carried out by a certification body. In the UK any organization can set itself up as a certification body and there is no control over this. However, there is a system by which organizations can be sure that they use a reputable certification body and this follows from UKAS accreditation.

The United Kingdom Accreditation Service (UKAS) Accreditation was established to introduce some regulation of certification bodies. The UKAS will assess a certification body against a set of criteria and then recommend it to the Secretary of State for Trade and Industry for accreditation. The accreditation will be for specific activities and may not cover all the activities of the certification body. In the same way, an organization may have certification for a specific range of activities, products, or sites. Thus, an organization seeking a supplier with a certified quality management system should check first, that the certification body which has issued the certificate is accredited for the activities that are of interest and second, that the certificate covers the activities that are of interest.

If the certification is of the quality management system then it is this alone that will be assessed. There will be no assessment of the viability or reliability, or indeed any other characteristic, of the product. Assessment of product is carried out using the process of product conformity certification or product approval.

Product conformity certification

This type of certification will involve an assessment, as above, of the manufacturing system against ISO 9000 to ensure that it is capable of producing consistent product. In addition, samples of the product will be assessed to confirm conformance with appropriate product standards. The BSI kitemark is a sign of product conformance.

Product approval

Again the manufacturing system is objectively examined to ensure that it is capable of producing consistent product, but in addition the product will be assessed against criteria relating to its fitness for purpose for intended use. A product approval certificate will identify the limitations on use of the product.

CE marking

The CE mark is used to indicate product conformance with one or more of the EC 'New Approach' Directives. These directives define essential health and safety requirements for products so that they may be freely traded in Europe. A number of products are affected, ranging from toys to agricultural machinery, and the range will eventually be increased to cover all manufactured items. Any product that falls within the scope of the directives must carry a CE mark if it is to be traded.

Costs and benefits of certification

The costs involved in achieving certification can be significant, particularly for small companies. In addition to the costs of implementing the quality system there will be fees for the certificate, registration and surveillance. The level of fee will vary with certification body, as they all operate (in the UK) as profit-making organizations. The size of the company to be assessed and the part of ISO 9000 for which certification is sought will also affect the fee, as these factors influence the amount of time that will be involved in carrying out the assessment.

A registration fee is payable to the certification body which is to carry out the assessment. This fee will cover an initial assessment to determine compliance (or not) with the quality system standard. However, following certification there will be annual certification and surveillance fees as the retention of the certificate relies upon the company's ability to continue to satisfy the requirements at subsequent assessments.

The benefits that an organization can expect from certification are significant. Increasingly, customers are demanding quality management certification as a pre-qualification to tender. Thus the certificate may mean either that new market opportunities are opened or that a company can remain in a market that has imposed certification requirements. Organizations can also achieve higher profits following from reductions in failures and warranty, improved worker morale, and improved sales.

7.3 Total quality management

Total quality management (TQM) is about meeting the needs of all customers, both internal and external to the organization. Rather than a set of strict requirements it is a philosophy that quality is obtained by preventing the errors that give rise to failure and waste, and that the actions required to achieve this should be both management-led and company-wide. Failure is the inability to meet the specification and waste the resource lost to the company due to this failure. For example, if goods are supplied incorrectly this is a failure. The resulting waste is the cost of receiving the incorrect goods, the time involved in identifying and rectifying the problem, and the cost of returning them. Thus the organization should be concerned with managing all processes within it so that they are carried out correctly every time, thus avoiding failure and waste. TQM naturally affects all areas of the organization and all its employees and it is usually associated with a change in the culture of the organization.

There is a British Standard for TQM, BS 7850, from which the following definition of total quality management is taken:

> Management philosophy and company practices that aim to harness the human and material resources of an organisation in the most effective way to achieve the objectives of the organisation.

Notice that the definition does not mention quality specifically; that quality is synonymous with the objectives of the organization is implicit.

TQM follows from the theories of a number of 'quality gurus', such as the Americans W. Edwards Deming and Arnaud V. Feigenbaum (see Further reading). The key point from the work of the quality gurus is that TQM is a philosophy requiring:

- Top management commitment
- Management leadership
- Company-wide involvement
- Employee awareness and motivation
- Everyone having a responsibility for preventing errors.

7.3.1 The main elements of TQM

To consider how TQM works in an organization the processes that occur in it have to be identified. Key processes that directly affect the achievement of the organization's goals must be defined. The needs and expectations of the customers of each process must be established so that the process can be designed and managed to meet these. The main elements of the system are (1) the customer–supplier relationships, (2) the management of the processes to provide assurance of quality, and (3) the continuous improvement of the processes.

Quality chains

The customer–supplier relationships both within and without the organization are described as quality chains. For example, Company X, a customer of AB Design Services Ltd, is to be billed by the finance department for some design work that was carried out by the software services department. Company X is a customer external to the organization, for whom the organization has carried out some work. This is a classic example showing the supplier – customer relationship in which money and goods are exchanged.

We can look in more detail at what this process involves. Company X wants the design work done to its specification, and this would be an explicit requirement. However, they will also want to receive a bill that is correct; they would expect this. Thus, the suppliers to Company X are not just the design engineers but the people in finance who deal with invoicing. In the same way that the quality of the design service will be influenced by the quality of the goods and services supplied to it, the quality of the service provided by finance will be influenced by the quality of the information supplied to it. Thus the design engineers have two customers, Company X and their own finance department. Finance are both supplier to Company X and customer to the engineers.

All interactions within an organization can be viewed in this way with every one being both a customer and a supplier. The link is the process. This is shown in Figure 7.3.

These customer–supplier chains operate throughout the organization and the quality of supply to the external customer will be dependent on all the links holding together. If there is a break in the chain it will have a knock-on effect. For example, a printing company produces warranty cards for a microwave oven manufacturer. It uses an ink that is supplied by another company, which has some quality problems. In fact, the incorrect mixing of a batch of ink means that after a year the ink fades and any printing is unreadable. Naturally, the problem is not discovered for a while. In the meantime you buy a microwave oven with a two-year warranty. After 15 months there is a malfunction and you search for the warranty card. It's strange but you can only find a blank piece of paper! Quality relies on quality throughout the supply chain.

One reason some companies do not receive the customer joy that they expect

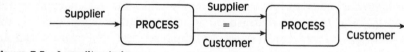

Figure 7.3   A quality chain

when they implement initiatives to answer the phone quickly is that there is nothing to be gained from having a prompt reply if the person answering is unable to deal with your enquiry.

This principle applies to both external and internal suppliers and customers. The invoice clerk needs accurate figures in order to prepare an invoice and therefore must know what was spent on the job; the clerk is relying on accurate time sheets, accurate stock lists, and accurate pricing of materials.

Everyone in the TQM organization has a role as both supplier and customer. For TQM to work, to provide the quality of product demanded by the organization's external customers, all employees must be accountable for their own performance and want to achieve quality in their own work and that of the organization. The underlying requirement here is that customers' needs are known and that it is possible for the organization to meet these needs. These follow from effective marketing and product development.

Process management

A process is the way in which inputs are converted to outputs. The inputs may be materials or information. The output will be the product that meets the needs of the customers of the process. For example, in manufacturing an input may be a pile of plastic granules and the output a garden table. The process that has converted the input to the output is plastic moulding. In the service industry the inputs are more usually information. In the example of AB Design Services the finance department would have been taking cost information and time sheets as inputs, and these would then be processed to produce an output, the invoice.

In TQM the aim is to understand the processes that are involved in meeting the customers' needs so that they can be managed to ensure that these needs are met consistently. All personnel will be involved with processes and process management, which is the method of establishing control of the processes. They will thus be concerned with assuring the quality of supply. The first stage in process management is to identify and define the processes involved. For each supplier–customer link there will be a process. To define the process we must know what the process is required to achieve, i.e. the specification for the process, and what the inputs and outputs for the process are. For example, a photographic unit is required to produce photographs for a new sales brochure. The input to the process is a list of photographs required – quantity, size, subject, colour. The output will be the photographs that meet the input specification. In between, the process will entail a number of steps, including taking the photographs, developing and printing, verifying that the images are clear, etc.

An easy way to show the steps involved in a process is to use a flow chart which provides a graphical representation of the steps by using a series of symbols. The symbols are shown in Figure 7.4. These can be used to represent the above process as shown in Figure 7.5.

We can now use this process definition to determine the controls that are required to allow the process to be managed. Controls may take the form of ensuring appropriate training of personnel, that they have equipment that can meet the specification that has been set, that they have enough time, and they

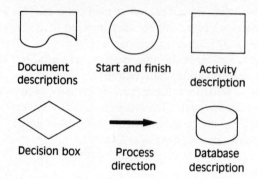

Document descriptions   Start and finish   Activity description

Decision box   Process direction   Database description

Figure 7.4   Flow diagram symbols

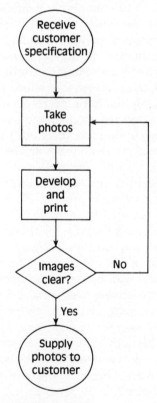

Figure 7.5   Flow chart of photo supply process

have the resources to do the job satisfactorily. Taking the above example, we can control the materials, the equipment and the processes that are used in the photo process. In order to achieve an acceptable output the film has to be of the correct type and grade, it should have been stored correctly, and used before its expiry date. The cameras must be capable of taking the type of photos specified, with the correct lenses, and these must be clean. The photographer must be capable of taking good photographs, balancing the picture and ensuring that the lighting is correct.

All these controls are concerned with quality assurance, making sure that the process consistently achieves the level of quality that has been set for it. Because quality assurance is such a fundamental aspect of TQM it is no surprise to see that the ISO 9000 series of quality system standards sits very well as an integral part of a TQM system.

The two most important aspects of process management are making sure that (1) the output meets the customers' requirements and (2) that this happens consistently. Consistency is an important issue. If the process sometimes fails there will be dissatisfied customers. It is also important for process management. If something is consistently wrong the chances are that the problem is due to one source, therefore a systematic check of the process should identify the problem and allow its resolution. On the other hand, if the process produces variable output then the problems could be many and there would be great difficulty in isolating the sources. For example, if a CNC machine always punches holes with a certain offset then there are two possible problems: either there is an error in the program or it has been incorrectly set up. It would be easy to determine which of these was the culprit and then to rectify the fault. On the other hand, if the offset is variable, so that it is variously to the left, right, 2 mm, 3 cm.... then there are many causes of the problems. It would be a difficult and time consuming job to investigate all the possible problems and aim to rectify them.

We can therefore see that an invariable output is easier to deal with than a variable one, whether we are interested in rectifying a fault or making a general improvement. Thus, one of the aims of process management must be to reduce the variability in process output.

Continuous improvement

The establishment of consistent processes is not enough in the TQM environment, it requires that processes are continuously reviewed and improved. From the management point of view there should be a system for review and improvement, providing an opportunity to do something more effectively, to reduce the variability in a process, or to save money. What is required in TQM is a method of continuous review and appraisal, leading to change when necessary, as shown in Figure 7.6.

The keys to continuous improvement are (1) knowing that a change is required, (2) making sure that any prescribed change will be effective, and (3) ensuring that the most knowledgeable people are involved in the improvement

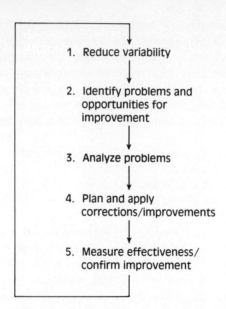

1. Reduce variability

2. Identify problems and opportunities for improvement

3. Analyze problems

4. Plan and apply corrections/improvements

5. Measure effectiveness/ confirm improvement

Figure 7.6 Continuous improvement process

process. To know what to improve, and whether a real improvement has been effected, you must have objective evidence of what is happening, so there has to be some measure of performance which can be assessed. The measure, or measures, for a process must be meaningful to that process, the data required must be readily available, and the analysis of the data must be simple and easily understood. For example, consider the process of replacing memory boards in computers returned under warranty. A number of characteristics of the process could be measured, for example the number of machines processed in a week. However, what the customer wants is to have the machine back as soon as possible, and does not want to know if the company can repair five or 5000 machines in a period. Therefore, a useful indicator for the process given the customer requirement would be to measure the time taken to effect a repair and have the machine available to the customer once more. This data will be easily recoverable from fault report sheets and shipping records and can be recorded on a graph, as shown in Figure 7.7. From this the time taken for a variety of cases can be seen, as can the trend, i.e. whether the time taken is increasing, decreasing or staying the same. In this case the trend is upwards, meaning that the time taken to effect repair is increasing. This type of presentation allows the information to be reviewed quickly and it can prompt further action (for example, initiatives to reduce the time taken). By plotting the time taken after these initiatives have been implemented the real effects can be ascertained.

The most appropriate people to monitor a process and suggest improvement will be those carrying out the process. This is where culture change comes into

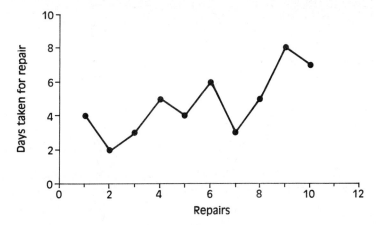

Figure 7.7   Graph showing time taken to repair faults

TQM. The people who work with the product have to be empowered, i.e. given the responsibility and authority to monitor and review what is happening, and to make changes. These people must have responsibility for their own work and an interest in both maintaining and improving the quality of this work. The significant consequence of this is that people must be trained to use tools and techniques that are available to help them in this. There are many of these and a number of them are described in Section 7.4.

Employees engaged in continuous improvement may work alone but will more often work in teams. The nature of customer–supplier chains means that inevitably there will be groups of people formed who are customers of a particular process, suppliers to it, or are process operators. Groups of people are affected by the process and thus they, as a group, should be the ones to review it. Working in teams rather than as individuals brings many benefits. There will be a broader skill base, and a variety of experience and views. In addition, people tend to work more effectively in a team, where tasks can be shared, and people can take the team roles best suited to them. Team working will also improve communications and trust in the organization, leading to an improvement in the working environment. For more information on this important aspect of working see Chapter 10.

Management commitment

To make changes in the way the organization operates, alter the focus of the business in line with TQM philosophy, and provide the training and support needed by the employees demands the highest level of management commitment. TQM will take a long time to implement, and there will be difficulties with persuading people to take a more proactive role in their work and to put quality first. It will demand large resources, in both time and money, to review and improve processes and to train people. There will be a return on the investment but this will be long term, during which the effort must be sustained. For these reasons, a TQM initiative will only work if it has full management commitment and the people at the highest level believe in it.

Management commitment does not automatically follow from a decision to implement a system like TQM. For example, if the system is being imposed from another source, perhaps a supplier or a parent company, then this may provoke resentment of the system. It is also the case that managers feel threatened by change in the same way as lower-level workers. This can lead to an outward show of approval but with practices that may actually subvert what is to be achieved. However, to make TQM work, all management must be committed and be seen to be committed. It is for this reason that companies implementing TQM spend a long time training their managers before moving onto the rest of the personnel.

7.3.2 Benchmarking

In the same way that certification is used as a measure of achievement in ISO 9000 systems benchmarking can be used to measure achievement in TQM systems. It is also a powerful tool for continuous improvement.

Benchmarking is the old idea of measuring performance against a standard. The standard may be set by performance indicators, it may be set by criteria for an award, such as the European Quality Award, or it may be assessing your performance against someone else working in the same field.

Benchmarking is a self-assessment exercise and is very important to both the achievement of quality and to quality improvement. If you do not measure you do not know if you are achieving the specification set. Similarly, you cannot know if an improvement has been made if you did not know what the starting point was. Thus benchmarking provides the essential control and measure feedback loop necessary for quality improvement.

The aim of benchmarking, and the reason it is widely used by many leading companies, is that it aims to help companies become as good or better than others operating in their field. The term 'Best Practice Benchmarking' is used when the benchmarking is done against multinationals who have been deemed 'best in the world'.

There are four key requirements in benchmarking:

1. The organization must establish what makes the difference between a good and an excellent supplier as far as its customers are concerned.
2. For each area of activity to be benchmarked standards must be set according to best practice.
3. The organization must find out how the best companies reach these high standards, in order to profit from their experience rather than 're-inventing the wheel'.

4.   The ideas and experience of the 'best' must be applied to meet or exceed the standards.

Benchmarking is clearly not a one-off. If it is to be used to help development then it must be a continuous process of review and improvement. It does, however, necessitate significant work for an organization and therefore the activities to be benchmarked must be chosen with care to ensure that they will be meaningful and useful. For example, a company might wish to benchmark its response time for dealing with customer enquiries or how much time it loses due to accidents.

Quality awards

Quality awards are gaining in number and popularity and increasingly provide standards against which to benchmark. There are national and international quality awards. In the UK there is the UK Quality Award, in Europe, as a whole, there is the European Quality Award. The winning of one of these awards leads to a great deal of prestige and publicity for an organization. However, their main value is not as a prize but as a motivator and a system for monitoring and measuring progress in a TQM environment, for which certification is not an option.

The awards function by providing a model for self-assessment. The model for the UK Quality Award is shown in Figure 7.8. The results part of the model focuses on what the organization has achieved, the enablers on the way in which the results are achieved. The model is used so that organizations can give a score for each of the award criteria. The scores, as a percentage of the maximum available, are shown on the model. The British Quality Foundation provides guidance on scoring and it is the scores that are used by assessors to determine the organizations that are shortlisted for the award, and they can also be used to benchmark.

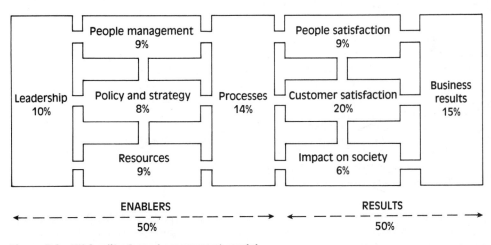

Figure 7.8   UK Quality Award assessment model

7.3.3 TQM – An illustrative case study

Mistakes should be one-offs
By NILS GRIMSMO
MANAGING DIRECTOR OF ERICSSON LTD

TQM depends greatly on learning from experience – but getting people to question their current practices is no easy task

As a world leader in telecommunications, Ericsson has always viewed quality as an integral part of its business strategy. Our market is characterized by fierce competition and the need to adapt to new market challenges is a pressing concern for any telecoms supplier. Indeed, one of the challenges inherent in operating in a high-tech environment is that the market is changing all the time and lessons have to be learnt quickly because repeated mistakes are not tolerated.

Today's customers take quality and performance for granted and therefore, while aggressive policies on price and speed of delivery are vital to secure orders, we must never lose sight of the importance of quality and performance in our own business.

The application of TQM principles plays an indispensable role in helping a company learn quickly from its experiences by building on its success and improving on its weakness. In this way new business opportunities are created which, in turn, enable the company to expand.

Our first involvement in total quality dates from the early 1980s when we introduced an internal quality assurance scheme, known as the Ericsson Quality Programme. This was further enhanced by obtaining ISO 9001 certification.

In order to become more competitive and improve our overall business performance, however, we recognized a need for further development of our quality programme and to this end we have introduced TQM. Our strategic approach encompasses:

- TQM as an integral part of the business
- Leading by example
- Empowerment/involvement
- Local programmes to meet the needs of local markets
- Having 'quality champions'
- Building TQM into work practice and business objectives
- Measurement
- Applying TQM to all parts of our business
- Producing short-term results to promote TQM benefits
- Effective training support
- Continuous evaluation of new tools, etc.
- Follow-up action by senior personnel

Within Ericsson in the UK, we have with the assistance of TQM International Ltd's consultants, introduced TQM into all our core businesses based on the 10 TQM principles (Figure 1).

Common goals have been set for all Ericsson businesses. For example, we specified that by 1995 Ericsson companies should be capable of winning international/national awards and achieving a major breakthrough or improvement in a strategic area.

All improvement activities are controlled by steering groups within each core business and follow this model (Figure 2). This represents a threefold approach aimed at:

- Improving our efficiency, mainly through specific quality improvement projects
- Improving our overall effectiveness through reviewing our processes
- Changing our management culture to a more open style, encouraging empowerment and involvement

To be effective, we have found these projects must be supported by a clear communications policy and training infrastructure. One of our most important discoveries as a company has been that learning from experience is an asset which we can use to our advantage over our competitors. This has demanded significant shifts in attitude among our staff.

Getting people to question and challenge their existing work processes is not an easy task, but it is vital to create an environment in which change is viewed as normal and staying the same is seen as abnormal.

In 1993 we decided to apply for the European Quality Award, which has been well received. Broken down into several smaller components, the award model helps managers answer questions such as 'What does a TQM model manager look like?' 'What should my process look like?' 'What should I focus on?'

The award model takes a similar approach to resolving any major problem. With a complex issue one can spend hours, days or even months talking around the problem and not necessarily finding a solution. By 'deconstructing' the problem we find it becomes easier to manage.

The model also provides a national benchmark. I'm pleased to report that in 1993 Ericsson was overall winner of BT's Network Product Quality Awards, an

---

THE 10 PRINCIPLES

- Agree customer requirements
- Understand customer–supplier chains
- Do the right things
- Do things right first time
- Measure for success
- Continuous improvement is the goal
- Management must lead
- Training is essential
- Communicate like never before
- Recognition for success

Figure 1

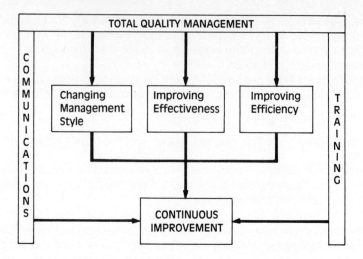

Figure 2

award which related not only to our equipment quality but also to our standards of manufacture, customers response and delivery.

Reprinted from *FIRST Magazine*, 77 Oxford Street, London W1R 2RB © *FIRST Magazine* 1994

7.4 Quality management tools and techniques

The quality system frameworks of ISO 9000 and TQM require supporting methods to aid quality improvement, assess the impact of change, and to sustain the enthusiasm for quality over a long period of time. There are a number of tools and techniques that can be used to achieve these ends. Many of these exist to allow objective gathering and analysis of data, to highlight areas for improvement or to show the effects of quality initiatives.

In this section we will examine a number of tools and techniques that facilitate the gathering and analysis of data, and help to examine problems so that solutions can be proposed.

Each of the tools and techniques described in this section can be used on its own. However, without the framework of a quality system it is unlikely to provide a sustained improvement in quality. As part of a quality system each technique used should have a clearly defined purpose and this should be accompanied by the objectives for the technique, i.e. what it is to achieve and how it ties in with other techniques being used. The use of the technique should be supported by any organizational changes that are necessary, such as allowing access to information or providing opportunities for interdepartmental meetings. Resources required to support the technique must also be provided, such as training, information, equipment, and time.

7.4.1 Quality circles

Quality circles provide a structured means of involving personnel in improving quality in their work area. They encourage team working within the workplace. Organizations using quality circles often use their own names to identify the operation (for example, they may be called quality improvement groups), but the principles are still as described here.

A quality circle is a group of people, usually four to twelve, who meet regularly to discuss issues, analyze problems, and propose solutions. These people, the circle members, are from the same work group and may be led in the circle by their work group leader, i.e. their foreman or supervisor. The circle members should be there voluntarily and the circle should have autonomy in deciding which problems it is going to tackle. However, problems should be work-related and it should be possible for action to be taken following the circle's recommendations. Thus statutory or negotiated factors, such as pay, should not be open for discussion within the circle.

Quality circles have been working successfully for many years in Japan, the USA and France. Increasingly they are being used successfully in the UK. Their great advantage is that they bring people together forming a cohesive working group, engendering strong group loyalty, and because the people looking at the problems have to face them every day they are very committed to making effective change.

To make quality circles work a number of resources will be required. The key people in the circle will be the circle members, a circle leader, and a facilitator. The role of the facilitator is to help the circle initially, providing guidance and training, and subsequently to monitor and review the way in which the circle works. The facilitator may have responsibility for a number of circles in an organization. Circle members must receive training so that they can work effectively together, managing time, holding meetings and planning projects. In addition, they will usually need some specific communication and presentation skills training. As well as these general skills the circle members may need some specific training on problem analysis methods and issues such as costing. After training the circle will need somewhere to meet and time for meetings. They may also require other expenditure for testing solutions.

Quality circles meet regularly to tackle a chosen problem. Using techniques such as brainstorming and cause-and-effect diagrams they explore the possible causes of, and solutions to, the problem. They then agree actions to be taken, perhaps to gather data to understand the problem more fully, or to run some tests on possible solutions. When a solution to the problem is agreed by the circle this will then have to be presented to management to take further action. It is at this stage that management commitment to the circles can be fully demonstrated by taking the circle's recommendations on-board and effecting the solutions proposed.

Quality circles directly aid quality improvement by motivating personnel and allowing the circle members to tackle quality problems.

Data is gathered and a problem, or an opportunity for improvement, is identified. Recognizing that there is a problem or that an improvement could be made is not sufficient in itself. What is causing the problem? How might it be resolved? Two techniques that can be used to help generate ideas and encourage analysis are brainstorming and the use of cause-and-effect diagrams.

Brainstorming

Brainstorming is a technique that uses the combined efforts of a group of people in order to generate ideas. It is commonly used by quality circles as it ideally requires between eight and twelve people to work effectively. It is described in Section 9.4.2.

Cause-and-effect diagrams

Cause-and-effect diagrams, also called fishbone diagrams because of their shape or Ishikawa diagrams after their originator, are used to build up a picture of the possible causes leading to a particular effect. Notice that the accent is on effect and not problem. The aim is to aid the process of identifying the problem, thus helping the quality improvement process by ensuring that the major problems are tackled.

The diagram is based on the skeleton shown in Figure 7.9. The categories used are usually MEN, METHOD, MATERIALS, MACHINES, ENVIRON-MENT. These categories can be varied according to the problem being dealt with but should not exceed six. There is no prescription for maximum number of causes.

The seven steps in producing the diagram are:

1.   Define the effect.
2.   Define the major categories of possible causes.

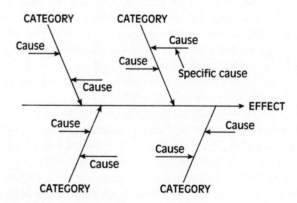

Figure 7.9   Basic cause-and-effect diagram

3. Construct the outline diagram.
4. Brainstorm possible causes in each category and mark these on the diagram.
5. Analyse each cause to focus on a more specific cause – the problem field is being narrowed – and mark these on the diagram.
6. Identify and circle the most likely and actionable root causes.
7. Gather information to verify the most likely root cause.

For example, consider an organization that is having problems with sealing vacuum vessels. The effect is that the vessel is not vacuum-tight. A quality circle can consider the possible causes of this effect and show them on a cause-and-effect diagram, as shown in Figure 7.10.

7.4.2 Data gathering and analysis

There are many ways in which data can be gathered and analyzed, allowing an objective decision about changes to be made, a review of the effects of change, or to monitor performance, thus aiding the quality improvement process. Four of the most common methods are described here. These are histograms and graphs, Pareto analysis, and control charts.

Histograms and graphs
A histogram, or bar chart, is a graphical method of keeping a count of how often something happens. For example, the histogram shown in Figure 7.11 is used to record the number of each type of incident causing the photocopier to be unavailable. The chart can be left next to the photocopier and added to by the person finding the problem. After a defined period the chart can be removed and analyzed to see the most common incident causing copier unavailability, as well as proving an overall indication of how often the copier is unavailable.

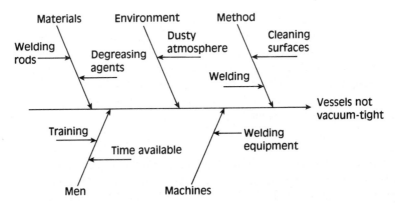

Figure 7.10 Cause-and-effect diagram to investigate why vessels are not vacuum-tight

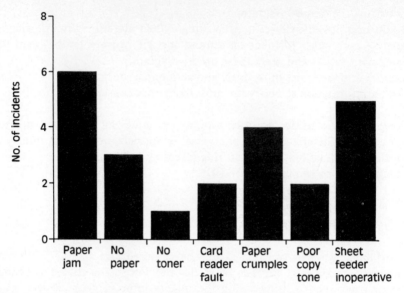

Figure 7.11   Histogram showing tally of photocopier problems

The advantages of this method for recording data is that it is easily visible and understood. It can also be added to without a need for redrawing, thus allowing simple data gathering over a period of time. Graphs also allow data to be reviewed quickly and trends observed.

Pareto analysis

Pareto analysis follows from the Pareto principle (also known as the 80 : 20 rule) that a few causes give rise to the majority of effects, i.e. in general, 80% of all the failures arise from 20% of all the faults. The idea is then to identify the significant minority of faults and thus resolve most of the problems. This type of analysis helps to focus on the main issues and to prioritize work effort.

For example, there have been a number of minor faults reported in shipments of product to an OEM customer. These faults include not having manuals for the equipment, having the wrong manuals, not having enough screws for assembly, damaged packaging, scratches on the casing, and so on. Because the list is long it may be difficult for the supplier to take action on every type of fault immediately. Thus, the main problems have to be identified. So you need to know the number of times each problem has given rise to a fault and then record them on a Pareto diagram which will show, at a glance, the most significant problems.

Unlike the histogram, the Pareto graph is produced after the data has been gathered. It requires the data to be ranked in order, highest number of occurrences to lowest number. Using the data shown in Table 7.3 the Pareto

Table 7.3   Faults with supply of equipment to OEM customers

Category	Incident	Occurrences
A	Manuals not included	18
B	Incorrect manuals included	7
C	No assembly screws	2
D	Insufficient assembly screws	14
E	Packaging damaged	5
F	Scratches on casing	9
G	Power cables not included	4
H	Equipment labelled with wrong customer name	1
I	Wrong equipment	2

diagram is drawn using the following steps:

1. Draw a histogram with the types of incident in rank order, with the highest number of occurrences on the left.
2. Over the histogram draw a graph showing the cumulative amount of occurrences.
3. Draw in a scale of 0–100% on the right-hand side of the graph.

For the above information the graph produced is as shown in Figure 7.12.

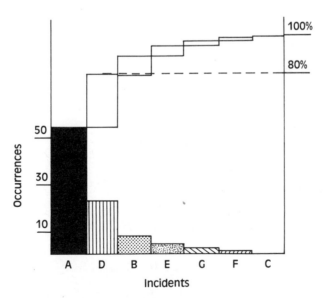

Figure 7.12   Pareto diagram of faults with supply of equipment to OEM customer

We can easily see from the diagram that by addressing only two of the nine types of incident observed we have the potential to resolve 80% of the problems. However, the Pareto diagram, as with all data analysis techniques, must be used with care. The data to be analyzed must be chosen for its significance. In the above example the most commonly occurring faults were identified. However there were two instances in which the equipment carried the wrong OEM customer name and one when the wrong equipment was supplied. These faults are clearly more significant than the customer not receiving enough assembly screws.

Control charts

Control charts monitor the performance of a process that has regular outputs. They serve a similar purpose to a simple graph in that they allow a quick glance to provide a view of what has been happening over a period. However, they are more refined in that they are used to show limits within which the process should operate. Therefore they allow a potential out-of-control situation to be detected at an early stage.

The charts follow from the theory that there will be some fluctuation in the output of any process that is in control. However, in a controlled process these fluctuations should not go beyond statistical limits. These limits can be calculated after the process has been operating for a short while and some data has been collected. There are a number of different types of control charts, and the chart described below is known as the c chart, which is based on a Poisson distribution. This is used in cases when the number of defects in a batch or item are to be monitored. Two other control charts, mean and range, are described in Section 7.4.3.

For example, a ball-bearing manufacturer records the number of ball bearings in each batch of 1000 that are out of specification, and these are shown in Table 7.4. Using this data we can calculate the mean number of defects as:

$$\bar{X} = \frac{\Sigma X}{n}$$

Table 7.4 Number of defects per batch of 1000 ball bearings

Batch no.	1	2	3	4	5	6	7	8
Defects	12	10	15	8	10	13	6	12

where

X = number of defects
$\bar{X}$ = mean number of defects
n = number of samples

Therefore

$$\bar{X} = \frac{12 + 10 + 15 + 8 + 10 + 13 + 6 + 12}{8}$$

$$= 10.75$$

This mean is used to calculate the control limits for the process. These are the upper control limit (UCL) and the lower control limit (LCL). They are calculated as follows:

$$\text{Control limits} = \bar{X} \pm 3\sigma$$

where σ = standard deviation and is calculated as the square root of the mean for a Poisson distribution. Thus

$$\sigma = 3.28$$

and

UCL = 20.59
LCL = 0.91

This information is then shown on a control chart as shown in Figure 7.13.

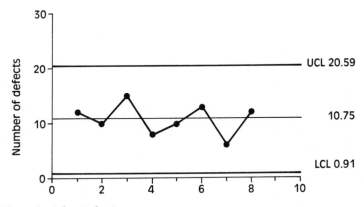

Figure 7.13   c chart for defects

The chart shows us that the samples measured are within the control limits. What this means is that the samples all contain a number of defects that are not statistically too extreme. On the basis of the data represented here there should be a warning if the number of defects starts to move towards the upper control limit. A move towards the lower control limit represents an increase in quality, i.e. a lowering of the number of defects, and, of course, this should be the aim.

7.4.3 Statistical process control

Statistical process control (SPC) is widely used to control processes, ensuring continued process capability and ability to meet quality requirements. It uses control charts to monitor what is happening so that timely action can be taken. The theory is that a process that is in statistical control will provide a predictable output and this is what is required. Two control charts are commonly used, the mean chart and the range chart. The mean chart shows values and their variation around the mean. A range chart is used to show the range of values found in a sample and then charts the variation in range for a number of samples. Limits are then applied to these, i.e. warning limits to indicate that the process is moving towards an out-of-control situation and action limits which indicate that action needs to be taken immediately to bring the process back into control. The charts are produced during the process so that there is continuous monitoring.

The techniques and use of statistical process control are beyond the scope of this book. However there are many excellent books dealing with this topic and some of these are listed in the Further reading.

7.5 Factors influencing the use of quality management

There are many reasons why organizations use quality management, and why some pursue certification to ISO 9000. Customer pressure is the reason most commonly cited by organizations, particularly where the customers have already achieved certification and want their suppliers similarly certified. Another factor is the campaigns run by government. In the UK the government's National Quality Campaign, which started in 1984, encourages organizations to use quality management principles and helps them directly by providing resources such as help for consultancy and training materials. Similar schemes operate in other countries although these have generally started much later (for example, the French government initiative started in 1993). Other reasons for using quality management include marketing, seeking to improve profits, requirements for product conformity, and product liability.

Marketing

Many organizations now require their suppliers to have a formal quality management system in order to be able to supply. In some cases large companies will take suppliers without quality systems and will then help with their development, but this is by no means widespread. Thus, in a large number of cases a quality system is a requirement to supply. This also means that the achievement of a quality system can open up previously closed markets. The effective operation of the system can also be used as a marketing tool, demonstrating the organization's commitment to meeting the requirements of its customers and establishing its reputation as a supplier of quality goods.

Improvement in profits

Any money that is spent doing things that are unnecessary, such as making faulty goods, is money that is lost to the organization. It follows then that, all things being equal, if this loss of money is stopped then the profits of the company will rise. Thus quality systems help in the attainment of this by reducing waste. The calculation of how much is lost, and how much it costs to prevent this loss, is complex and is discussed in Section 7.7.

Product conformity

Many industries and customers now require a demonstration of product conformance with agreed industry standards. This will be true in all spheres of manufacture as the requirements for CE marking under the 'New Approach' directives come into force (see Section 7.2.2). The use of quality management allows consistent output from a manufacturing system. Thus if the product design conforms to specifications then the quality system ensures that the product is made consistently to those specifications.

Product liability

Under the European Commission directive on product liability (1985) a producer will be legally liable for damage or injury resulting from defective product. This directive was written into English law as part of the Consumer Protection Act 1987. Defects can arise due to poor manufacture and design, insufficient warnings or instructions on use being supplied. It is thus incumbent on a producer to ensure that product is free from defect and that it carries appropriate warnings and instructions. The use of a quality management system should demonstrate that a producer has taken all reasonable steps to prevent the production of defective product by establishing controls for the design and manufacture of product. It is envisaged that this may give rise to a defence in a product liability case, although at the time of writing this has not been proven.

7.6 Issues surrounding the implementation of quality systems

The implementation of a quality management system will take a long time and a great deal of resource, thus the primary requirement for any implementation is complete management commitment to the system. Executives must lead by example and provide resources where necessary to maintain progress and worker motivation. This is particularly important because of the culture changes that are usually required to implement quality management effectively. These come about when an organization moves away from a 'blame' culture associated with a traditional inspection system, or an inward-looking culture in which everyone has their own job and gets on with it and there are barriers to communication.

The use of a formal quality management system naturally implies the formalization of organization structures, definition of responsibilities and definition and documentation of working practices. It is likely to also require some changes in working practices which can lead to conflict and personnel problems in the workplace. Very few people like change. With ISO 9000 there is also the risk that an organization may start documenting procedures and forget to stop. A complaint often heard about the standard is that it leads to a high level of bureaucracy and unnecessary documentation. This does not have to be the case, but in order to avoid this, care has to be taken in the design and development of the system, as with any product development.

The costs of implementing a quality system are likely to be high, and will include personnel costs. Someone must be available to implement the system and coordinate the various activities. The training bill will also be very high as employees will have to be trained in the use of the quality system, and there may also be a need for specific technical training. Training costs are often high because you have to allow not only the direct costs of paying for training and expenses but also for time that is not being used for manufacture. Documentation costs can be high as they involve the time to produce the documentation, as well as the costs of stationery, printing and secretarial services. In addition, there may have costs because equipment is required for inspection or test, or current equipment has to be calibrated. If a company decides to have its quality system assessed with the aim of certification then there will be costs for this.

Careful planning of the implementation is vital to its successful completion. It can take some years to complete a quality system implementation. Very rarely companies can achieve it in less time, but even then it will be closer to a year than six months. During this time resources must be provided, worker motivation and enthusiasm must be maintained, and the organization must still continue to operate its business. Providing targets and opportunities for achievement throughout the project is vital to maintaining impetus and is particularly important in a TQM implementation in which it is very difficult to define an end point and there are no road signs as there are with the specific requirements of ISO 9000.

Consultants are often used to aid implementation. These people can give a more objective view of the organization as well as bringing different skills and experience into it. Consultants can also spend a lot of time and money with little to show for it if they have not been well chosen and well briefed. They should be used as any other subcontractor and supplied with a clear specification of what is required.

The reasons behind a quality initiative can have a great effect upon its outcome, whether it is successful or not. One of the problems with ISO 9000 often cited by companies is that it sets in stone working practices that are not as effective as they might be in achieving quality for the customer. This is an unfortunate effect, following the indiscriminate requirements of some customers for an ISO 9000 badge, or else! Naturally, suppliers will not react well when forced to take on something as all-encompassing as a quality system. On the other hand, suppliers who see quality as strategically important for their future and their customers, and are consequently highly motivated, achieve great benefits.

7.7 Quality costs

When an organization, usually a firm of quality management consultants, wants to capture the imagination of a potential client's managing director it will issue statements such as 'quality costs are 40% of turnover'. This looks good: if the company can improve quality it can make 40% more profit! Things, though, are rarely so simple and it doesn't really work like this. In any organization there will be costs associated with not achieving quality, called costs of non-conformance. Similarly, there will be costs of conformance, i.e. of meeting specifications. Quality costs are the sum of these two and you cannot affect the cost of non-conformance without affecting the cost of conformance. For example, operators may make 20% defective circuit boards because they do not know how to solder – clearly, a significant cost of non-conformance. However, this can only be reduced by spending more money on training, a cost of conformance.

Any organization will want to keep a tight financial control of its activities to ensure that budgets are not exceeded, that there is sufficient cash flow, etc. This should also apply to quality costs as they will clearly affect the financial viability of the company, and yet many companies do not bother. The main reason for this is that it can be difficult to decide what is a cost of conformance and what is non-conformance. It can be even harder, having decided what is what, to collect cost data, because of the limitations of many accounting systems. However, the determination of quality costs can provide indicators for monitoring progress in a quality system implementation and they can highlight areas to be targeted for improvement to reduce costs.

Two models for determining quality costs are defined by BS 6143: (1) the prevention, appraisal and failure model and (2) the process cost model. These are both described briefly below. It is worth remembering that the optimum quality cost is that associated with achieving quality; both not achieving quality or exceeding it can have severe cost implications for an organization.

7.7.1 Prevention, appraisal and failure (PAF) model

This is the model that is most widely known. Its basis is that all quality costs can be categorized as either prevention, appraisal or failure. The total quality cost is the sum of the three.

Prevention costs

These are the costs associated with preventing failure. They include the cost of training people so that they can achieve defined levels of workmanship and implementing statistical process control and quality systems. Thus, as more systems are put in place, prevention costs increase, as should quality.

Appraisal costs

These are the costs that come from trying to find out if goods meet the specifications set. In a manufacturing company appraisal costs will include the cost of inspection personnel and of equipment and overheads to maintain an inspection department. Appraisal costs should not vary significantly with the quality achieved. An inspection department costs the same amount if it detects 100% defects or 1%.

It can sometimes be very hard to distinguish between prevention and appraisal costs. Is a design review prevention or appraisal?

Failure costs

Failure costs fall into two categories. The first is internal failure cost and this should be easy to calculate. It is the cost associated with scrapping, reworking, and warranty. The second category, external failure cost, is the most difficult to quantify of all the quality costs. This is the cost associated with failure outside the company and would include the effects of customers not buying your product again, or telling others that your product is not worth buying.

The idea of the model is that by determining the costs in each of the categories the quality costs can be monitored and controlled. As prevention costs are increased, failure costs should fall. However, appraisal costs could theoretically keep on increasing without a fall in failure costs. If a machine produces junk then it does not matter if you do no inspection or 100%, junk is still produced. By monitoring costs in each of the categories it should be possible to make a correlation between the costs of a particular activity and its effect on quality. Thus it should be possible to optimize quality costs.

7.7.2 Process cost model

The process cost model has developed following the requirement in TQM to define processes. It has the advantage over the prevention, appraisal, and failure model in that it looks at total process costs and does not detract from the real issue, i.e. identifying opportunities to reduce costs, by forcing costs into categories.

The model requires that a process be defined in terms of its inputs, outputs, controls, and the resources that it uses. Within the process there will be four cost elements which can then be identified: people, equipment, materials and environment. The cost of operating the process to specification is the cost of conformance and any costs associated with inefficiency or waste in the process are the costs of non-conformance.

The costs of non-conformance clearly provide a target for improvement and eventual cost savings. However, the process might also be improved, leading to savings in the cost of conformance.

7.8 Summary

In this chapter we have considered why quality is so strategically important to so many companies today; their customers expect their requirements to be met. We have examined a number of techniques and management approaches that further the achievement of quality.

In Section 7.2 we looked at the international standard for quality systems, ISO 9000. The requirements of the standard were examined and we considered the way in which the standard is being used by organizations, as a specification for a quality system for which certification could be given. In Section 7.3 we have described the principles of total quality management and considered how this approach differs from that given by ISO 9000. We have presented a case study showing the importance that a major international company places on quality and TQM. Section 7.4 presented a number of tools and techniques that can be used as part of a quality initiative to control and monitor quality, and to improve quality. In the UK approximately 30 000 organizations have achieved certification to ISO 9000. We do not know the actual number using TQM, but it is significant. In Sections 7.5 and 7.6 we have described some of the factors that influence the approach to quality taken by organizations, and the issues that arise from a quality system implementation. Finally, in Section 7.7 we presented two methods of defining quality costs which can be used to both assess and control costs.

7.9 Revision questions

1. How can quality be defined for a product?

2. How does quality contribute to competitiveness?

3. What is the traditional approach to quality? What are the limitations of this approach?

4. How do the quality systems of TQM and ISO 9000 differ? How are they similar?

5. What is ISO 9000? How is it used by organizations?

6. Why are there three separate specifications for quality systems within the ISO 9000 series?

7. Describe the main principles of the TQM philosophy.

8. What is benchmarking? How is it different from certification? What value does it have for an organization?

9. What are the factors that influence organizations' approaches to quality?

10. What benefits can a manufacturing company expect to achieve by operating a quality management system?

11. How could an engineering design consultancy (five consultants) justify the expense of implementing a quality management system?

12. Determine what information flows are necessary between the personnel and quality functions.

13. How does the quality function impact on the product development function?

14. Describe the PAF model for quality costs. Using your own department, give examples of costs under each category.

15. What is the process cost model ? What advantages does it offer over the PAF model?

16. (i) Define the terms 'quality assurance' and 'quality control'.
    (ii) Describe the operation of a quality assurance department in a manufacturing company, indicating the main tasks that you would expect to be carried out by the department.

17. Why is document control important within an organization? Make a list of points. For each point describe an example problem that would occur if document control was not used. What sorts of documents need to be controlled?

18. What benefits does document control bring to people outside the organization?

19. Describe a method of document control.

20. What disadvantages might a document control system bring to an organization? Do these outweigh the advantages listed in question 18 above? Discuss.

Further reading

Bank, J. (1992), *The Essence of Total Quality Management*, Essence of Management Series, Prentice Hall International.

British Quality Foundation (1994), *The UK Quality Award: Guide to Self-Assessment.*

BS 6143: Guide to the economics of quality. Part 1: 1992 Process cost model. Part 2: 1990 Prevention, appraisal and failure model.

BS 7850: Total quality management. Part 1: 1992 Guide to management principles.

Caplan, R.H. (1982), *A Practical Approach to Quality Control*, 4th edition, Business Books.

Dale, B. G. (ed.) (1994), *Managing Quality*, 2nd edition, Prentice Hall International.

Deming, W. E. many works including: *Out of Crisis* (1986), Massachusetts Institute of Technology: Cambridge University Press (1988).

Feigenbaum, A. V., *Total Quality Control* (1983), McGraw-Hill, New York.

Hutchins, D. (1985), *Quality Circles Handbook*, Pitman.

ISO 8402: 1994 Quality vocabulary.

Oakland, J. S. (1993), *Total Quality Management*, 2nd edition, Butterworth-Heinemann.

Price, F. (1984), *Right First Time*, Gower.

Sinha, M.N. and Willborn, W.O. (1985), *The Management of Quality Assurance*, Wiley.

Project planning and management

Overview

Projects are the contemporary solution to resolving the conflicting needs of meeting requirements with limited resources and time constraints. All engineers are involved in projects at some stage of their career or undergraduate programme. In this chapter we look at projects, their definition, and techniques that can be used to plan them. Through the medium of a case study we also consider the way in which projects are controlled and executed in industry.

8.1 Introduction

The marketing department of Vammu Tins Ltd have identified a niche in the marketplace which they believe could be filled by a new product. The product they have identified is a self-evacuating tin that could store biscuits indefinitely. This product would be new to Vammu Tins as previously they have only produced conventional tins and boxes *and* it would require some considerable investment on the part of the company. We will look at how Vammu Tins can make a decision about whether or not to proceed.

First, the Board of Directors will need information so that they can decide if the project is feasible; they need a project proposal. This will have to identify what the aim of the project is, what it is going to cost the company, and what benefits will accrue; it will also have to show that the project is achievable.

A project specification should be used to identify the aim of the project. This is the document that sets the limits on the project and that can be used by others to evaluate the implications of the project. In the case of Vammu Tins the project specification would be drawn up by personnel from both the marketing and engineering departments. The engineers will be able to make an assessment of the broad activities that will have to be carried out in order to execute the project, such as designing, drawing, prototyping, and, finally, manufacture. They will also be able to make an assessment of the resources required for each of these activities and to give an estimate of the cost of the engineering aspects of the project. The marketing personnel will be able to identify the benefits that can be obtained from successful introduction of the new product. They will also identify any constraints, such as deadlines, as well as any marketing activities that have to be carried out and the costs associated with these. All this information can be put together in the form of an outline project plan which will show whether or not it is feasible to meet the deadlines with the resources available.

The specification, the cost–benefit analysis, the resource requirements, and the outline project plan form the project proposal. The Board can then use this to decide whether or not to proceed, knowing the implications of their decision for the company.

If the Board decide to go ahead then the project team can get to work. However, the deadlines set in the proposal have to be met, specific tasks identified and resources and work planned. In order to do this the project team has to carry out detailed project planning and resource scheduling. This is particularly important when, as in this case, a number of people will be involved since they all need to know what they are required to do and when.

The level of planning at the proposal stage depends upon the type of project to be considered, but, in general, if it is very speculative, a high risk for the company, or requires a lot of investment then more detail will be required.

Planning after the proposal has been accepted is much more detailed but the level of detail will vary with the time to the project completion. For example, if the Vammu Tins project is estimated to take six months it will be possible at the start of the project to schedule tasks for the first month on a weekly or daily basis. Beyond the first month the schedules may be given on a monthly basis. However, the project plan should be reviewed throughout the project and can be revised to include more and more detail. Therefore, after the first week the schedule for the next month is given on a daily basis but for the remaining four months and three weeks a monthly schedule is outlined. This review of the project plan and updating of schedules allows the project to be controlled so that either the targets set in the plan can be met or a change in the plan can be foreseen.

The alternative to planning would lead to a great deal of uncertainty for the company. The Board of Vammu Tins would not know if the new product is feasible and might decide against it, missing a major opportunity, or they might decide to design the product and find that it consumes more money than it could ever make. Worse still, they might realize that it is feasible and work on the design only to be beaten by the competition because the engineering and marketing departments each thought the other was working on a particular aspect of the project when neither of them were.

Projects are a major part of an engineer's job, whether a product development project, like the Vammu Tins example, or a more specialist one such as the installation of a new piece of machinery. Projects vary in size in terms of the time that they take, the resources that they require, and the impact that they have upon the business. However, whatever the subject matter of the project, it should be planned to ensure that the resources are used in the most effective way to achieve the desired result.

Most projects, such as the design of a new product, will have very well-defined goals. However, some are more speculative such as research into methods of manufacturing high-temperature superconductors for the electronics industry. An industry-based project will always require justification, showing that there will be some form of reward for the company. Often this will be in terms of financial reward. However, some projects are done because of more intangible benefits such as improving team-working skills by providing interpersonal skills training. The main reason for the project may be defined for you, from the corporate objectives. However, you will be constrained by resources, including time, and, of course, you will want to do the job well. Doing a good job starts with good planning. This chapter will help you to plan well.

The first step of project planning relates to what is to be done and why, and usually results in some form of project proposal. We will consider this step in Section 8.2. In Section 8.3 we will look at detailed planning and at two project planning techniques that can be used. In Section 8.4 we will describe how a project can be controlled using a project plan and we will also look at a case study of a project carried out by Edwards High Vacuum.

The learning objectives of this chapter are:

1. To understand how project management techniques are used
2. To consider some examples of project management as it relates to engineering
3. To see the constraints that can be imposed on projects and the effects that these can have
4. To learn how to plan a project using Gantt charts and critical path analysis
5. To understand how to allocate resources
6. To consider the problems of project control and how they might be resolved.

8.2 Defining the project

In this section we will examine how a project may be defined, using a project specification. We will also introduce ways to consider the implications of the project, and examine some of the things that will constrain what happens during a project.

8.2.1 Specifying the project

The project specification is a description of the project so that all interested parties know what is planned and what the outcome should be. In industry a project specification might form part of a proposal which would be used to sell the project to the people who control the resources. In an undergraduate programme the project specification would be used by the course tutors to ensure that the work planned is of an appropriate standard for the course and that it is achievable. It may also be used by external examiners and the engineering institutions as part of their assessment of the student.

The project specification is similar to a design specification in that it provides the terms of reference for the people employed on the project. From this initial specification the project team can analyze the project in more detail and they can prepare a project plan. Finally, a project proposal can be produced which can be used as a working document during the execution of the project.

A project specification should include:

1. The title of the project
2. The scope
3. The objectives
4. Any conditions under which the project is to be carried out
5. Priority in relation to other projects
6. Authority.

- *Title*: Giving the project a title ensures that everyone knows what the project is about and avoids confusion when there is more than one project.
- *Scope*: The scope defines the scale of the project. For example, a design modification project may be limited to the products supplied to one particular customer or a computer system implementation project may be limited to the purchasing department.
- *Objectives*: The objectives are what are to be achieved by successful completion of the project. Objective setting and use of corporate objectives are discussed in Chapters 11 and 2 respectively; the same rules apply to setting project objectives. The objectives must be clear and understood by all the people involved in the project and should also be consistent with company strategy. Project objectives should be realistic. In a design environment the objectives of the project may be fully defined in the design specification. Before finalizing objectives it is always worth discussing them more widely to get views and ideas from other people who may have interest and expertise in the field, to check congruency with other projects, and to ensure that your views are consistent with company policy.
- *Conditions*: The conditions under which the project is carried out need to be specified. This is particularly important when the limits of the project extend into other areas. For example, if a design project needs to have some manufacturing element, one of the conditions may be that daily production schedules must not be compromised by the project. Conditions will also include deadlines to be met, any special conditions relating to funding or personnel, and points at which decisions have to be made by other people.
- *Priority*: The level of priority the project should have within the organization has to be established and clarified in order to make sure that it

has the right level of support and commitment. In order to do this the project's compatibility with other projects in the organization needs to be assessed.

- *Authority*: The authorization for the project should be stated and the authority of people working on the project defined. For many projects one person will have sole responsibility for planning and executing the work, but for large projects there might be a whole team of people, and it is imperative that the authority for particular parts of the project, and the authority for making decisions at different stages, is defined at the start.

Figure 8.1 shows the project specification that the authors used for writing this book.

Title: *Management in Engineering* text book

Scope: This book is intended to provide an introduction to the management topics and techniques that an engineer will come across in the normal daily activities associated with working in a small or medium-sized electro-mechanical engineering company. It is intended that the book will provide a standard text for management courses that are run in the first and second years of many BEng courses and will also sell to qualified scientists and engineers already practising engineering.

The book is not intended to be exhaustive, and many other books will be referenced. It does not intend to cover the law as it affects management and engineers as this would dilute the material which it aims to cover, although it will mention the areas in which the law applies so that the reader can be guided towards further reading. Similarly, it does not cover the European Market as it is felt that this would be better dealt with by a subject specialist, and because of the amount of material, in its own course.

Objectives: The main objective is to produce a book, of approximately 450 pages, that meets the market requirements for an undergraduate mainstream text that is applicable to UK, European, American and Australian markets. Each chapter in the book will contain an overview of the topics to be discussed and will show how this relates to the other sections of the book. Each chapter will be able to stand alone, thus providing a reference text, or can be used as a whole, thus providing a core course text. The chapters and section will be illustrated with case studies and examples. There will also be typical exam questions and some worked exam questions, appropriate to first- and second-year degree.

Each section and chapter will be summarized, including a list of the main points made. Reference for each chapter will also be given, enabling the student to carry out further reading in order to develop each topic area further.

Conditions: The first draft of the book will be available to the publisher by the end of January 1992 in order to be available for the 1993/94 academic year. The book is to be written as a joint project with each author contributing 50% of the text. It will then be edited with each author having specific responsibility for a number of chapters.

Priority: The book will take priority over other research work but will take second priority to teaching commitments.

Authority: The copy will be shared out equally and individual chapters will also be shared out with each author having total authority for editing and amending copy in their chapters in preparation for the final draft.

Figure 8.1　Project specification

8.2.2 The implications of the project

The implications of the project extend to its benefits, costs, the resources required for its completion, and the risks associated with it. In order to consider these implications the tasks that have to be carried out must be defined. The level of detail with which the tasks are analyzed will depend upon what stage you have reached in planning. For instance, if an assessment is being done for a project proposal then the tasks might be very broadly defined, if a detailed work plan is being prepared then they will be much more specific.

Project costs
In order to cost the project we have to be able to identify the resources that the project will consume. The resources may take the form of money, people, machines and materials. For some tasks there will be no flexibility in the way in which the resources can be used, but for others there may be some flexibility. However, at this stage we are concerned with gross requirements and not schedules. For example, if laying a cable will take one person three days then the resource requirement is three person-days. It does not matter at this stage if the project will eventually use three people for one day.

In determining the cost of a project you should always consider the costs in terms of labour, materials and expense, as discussed in Chapter 4. The labour cost would be the cost for the time of the people who are required to work on the project. Both direct and indirect labour must be costed and the cost if people are required to work overtime, shifts and so on must also be counted. If new staff are required for a project then the costs escalate, as you have to account for the costs of recruitment and allow for the fact that initially they may not be as productive as you might expect. Material costs will include those associated with any new parts that have to be purchased specifically for the project. Expense will include the direct monetary costs associated with capital investment and the use of specialist services, as well as machinery and equipment costs that are calculated in terms of the time for which you need to use them. It is particularly important to identify this cost if it means that equipment is not available for manufacturing when it normally would be. Expense will also relate to the cost of scrapping equipment and parts that are made obsolete by the project. The costs of any project should be considered in terms of those that are one-off and those that are recurrent such as annual maintenance charges.

EXAMPLE

Consider the costs associated with implementing a new computer-aided design (CAD) system and conversion from a manual drawing system. The cost of the system will include hardware, software, training and documentation. You should include an allowance for consumables as there will have to be a first-off purchase, and there will be costs for installation and probably for delivery.

After assessing what is to be paid to the supplier for the system, consideration must be given to where and how it is going to be used, who is going to use it, and how the organization will move from the current situation to the planned one. Planning how and where the system is going to be used might mean that you have to consider converting office space. You might need new furniture and services, such as air-conditioning or stabilized power supplies. You will also have to consider the cost of administering the system, particularly if this means a new post or someone being given extra responsibility.

When you consider who is going to use the system you have to plan for training and perhaps for redefining duties and responsibilities. At a minimum, you will have to cost the time for which that training will take people out of the workplace, but it is also likely that the costs of travel and expenses during training will be significant. You may also have to cost the time for someone to prepare in-house documentation for using the system.

You will need labour to input data, such as drawing frames and parts libraries, or to transfer data from the manual system. Also, you will have to charge the time for the project manager and the person who is actually carrying out the installation of the system and writing the procedures for using it, and determining how the change from the manual system is to take place. Finally you should not forget the first-year running costs, which will include maintenance and consumables.

This all seems very detailed but it is important to consider all factors so that a decision can be made on the value of the project. The problems of not properly costing a project are numerous, not least of which is the fact that you might look silly (at best) or incompetent. The worst consequence of not having a correct costing is that of getting part-way through the project and the budget runs out. This may mean that you have to reach some sort of compromise on a scaled-down project, or even scrap everything that has been done. If the project cost is overestimated then the decision may be not to proceed because it does not appear to be economically viable.

Project benefits

If you are asking someone to 'buy' your project you need to give them a good reason to invest and to justify the project in terms of the benefits that it will bring. Generally, it is much easier to reach agreement on a project if the justification is done in terms of financial benefit, as the decision is then based on what you get back, in monetary terms, for what you spent, in monetary terms. However, there will be times when it will not be possible to give a financial equivalent as benefits can be both tangible and intangible. Tangible benefits are those that you can see clearly and which can be quantified, such as the amount of money you expect to be able to save from doing the project. Intangible benefits cannot be quantified in financial terms. Consider a value engineering project that will result in a product having fewer individual parts. This might lead to a cost saving in materials and labour, and these are the tangible benefits. However, a project which involves training to improve team working will have intangible benefits. It is very difficult to cost improvements in morale.

When calculating the cost and identifying the benefits of a project it is important to realize that the costs will mostly be one-off. However, the benefits will accrue over a long period of time.

EXAMPLE

Let us consider the case of the CAD system and look at how this might be justified:

1. The speed of producing drawings and parts lists will be increased. This will be a labour saving and therefore the cost saving can be estimated.
2. The speed of amending drawings will be increased. Again this is a saving in labour cost which can be calculated by estimating the number of amendments made each year.
3. Storage and retrieval of drawings will be more effective. This is more difficult to quantify in terms of cost but you may consider factors such as the time saving, the reduction in space requirements, or a cost saving by not having to make archive copies on microfiche.
4. It will be possible in the future to expand into a linked computer-aided manufacturing (CAM) system. This benefit may be very difficult to quantify in financial terms but that does not detract from its importance if computerized manufacturing is a company objective.
5. Use of the system should improve the quality of designs. This benefit may arise because you employ a higher calibre of staff to use the system or because they are able to use a more systematic approach to design by having recourse to a library of standard parts. It may be because the system offers 3-D graphics which allows the designer to consider the part in detail before it is manufactured. It is not likely that you would be able to quantify this benefit, but if it appears to be a real benefit it must be highlighted.
6. Finally, another intangible benefit is that having an up-to-date system would attract more, and better, recruits to the department.

This list is not exhaustive but it provides an insight into the sort of areas that really have to be thought about when you are putting forward a project.

Project risks

There will always be risks associated with a project, but it is the duty of anyone preparing a project proposal to make an assessment of these risks so that an informed decision may be made about pursuit of the project. Risks may follow from the possibility of not taking the project. For example, if a company does not develop its market-leading product it may be overtaken in the marketplace by a more highly developed or cheaper product from the competition. They may arise from taking on a project. For example, if the company puts all its research effort into one particular field it cannot simultaneously follow opportunities in other areas. Risks of failure of the project must also be assessed. For example, if the deadline for a new product is missed, will there still be a market for it? Another example would be if a chosen research path did not provide the desired result, what will happen to the company's next generation of products? In cases where projects are used to meet specific customer requirements, such as the Oxford Lasers project in Chapter 13, there will usually be contractual terms that mean a company will be penalized for failure. The most common of these is the penalty clause in which the supplier agrees to refund a percentage of the sales price for any delay in delivery. An assessment of risk provides an opportunity to develop alternative strategies before a crisis arises.

8.2.3 Constraints

Constraints are the factors that restrict what can be done in the project. Typical constraints might come from deadlines that have to be met for an external body. For example, in a design project there may be meetings with the customer. Other constraints that might be imposed could relate to the level of funding or other forms of resource available (for example, people, equipment, or a requirement to use certain suppliers).

Constraints are also imposed when you require something to be done that is outside your sphere of influence (for example, a director has to make a decision before the project can continue to the next stage). It is very important to identify the constraints because they do affect the success of the project.

EXAMPLE

Constraints on the CAD implementation project might include:

1. Not making anyone redundant
2. Not employing any new people
3. Must be compatible with existing company hardware
4. The software must be able to link with a CAM system
5. Financial restrictions
6. Deadlines for various stages of the implementation.

8.2.4 The project proposal

The project proposal is the document that brings together all the information about the project so that it can be accessed by other people. If it is your project then the proposal is your sales brochure. If accepted, it should also be your working document. The proposal must include the project specification with a clear statement of expected outputs, the costs, the justification, an assessment of the risks associated with the project, any constraints, and a project plan. It might also include some background to the project. It should be well written and appropriate to its audience, and there should be an executive summary so that anyone can see at a glance what you want to do, why, and how much it is going to cost. In the next section we will look at how you prepare the project plan.

8.3 Planning the project

In the previous section we considered how a project is defined and a project proposal prepared. Here we will examine how one determines what needs to be done in detail to plan the activities and resources in order to achieve the project objectives. We will look at two planning techniques: Gantt charts and critical path analysis (CPA).

In order to plan activities we have to determine what those activities are and how long each activity is going to take. We also have to use the information that we had collected previously defining deadlines, as these set the limits of the project plan.

8.3.1 Project activities

Activities

The activities are the specific tasks that have to be followed through in order to meet the project objectives. Determining what tasks have to be done can be very difficult and it is important to consult widely to ensure that all activities are considered. One way of trying to determine the tasks required is to work back from the planned project end point.

EXAMPLE

A product development project requires that you design a product and produce a prototype. If we look at the production of the piece of equipment we can see that this will involve a number of steps, as shown in Figure 8.2. From the full list of activities we have to establish the sequence in which they should be done and which activities are dependent on other activities.

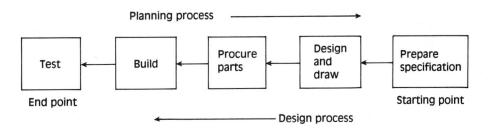

Figure 8.2 **Steps in prototype production**

EXAMPLE

The activities the authors identified for preparing this book were as follows:

1. Research and prepare copy for each chapter
2. Divide chapters and topic-writing responsibilities

3. Proofread each other's work and verify material
4. Assemble book and proofread
5. Prepare index and table of contents
6. Send to publisher.

Duration of activities

In order to determine the time it will take to complete the whole project and to plan the different activities you need to have an idea of how long each activity will take. The more realistic the estimates used for the plan, the more accurate it will be.

Estimating time can be very difficult initially, but it should get easier as you build up more expertise through the life of the project and more experience of other work of a similar nature. One way of estimating is to forecast on the basis of historical data, for instance if a similar task has been done previously. It is easier to estimate time for small pieces of work, and therefore another way of estimating is to break up the tasks into smaller and smaller pieces until it is possible to identify some particular activities and put a time to them.

It is impossible for anyone to be perfectly accurate at estimating time, but by concentrating on the things that affect the time required for each activity it should be possible to keep errors small. For example, in the preparation of this book we had to consider that we had other jobs to do and we had to take into account the time that we could really spend writing, allowing for teaching commitments, holidays and so on. Unless you are realistic and accept that other things still have to be done, you will end up always producing plans that can never be achieved. This is not only demoralizing personally but can also reduce the confidence of a supervisor in your ability to get things done. In contrast, care must be taken to make sure that times are not greatly overestimated, otherwise people will start to wonder what is being done and may assume either that you are lazy or that you are incompetent. When estimating activity times never underestimate the value of asking for advice from people who have been doing that type of work for some time. Chapter 11 provides further information on time management which will be useful when planning a project.

In the project proposal resources for the project will have been identified. These now have to be tagged to specific activities as they will affect the duration of the activities. For example, if is known that two people are available to carry out some activities then this should be acknowledged when the duration for those activities is being calculated. In Section 8.3.5 we will examine how projects can be scheduled on the basis of meeting particular resource requirements.

8.3.2 Milestones and targets

Milestones are natural key points in the project defining the end of one set of activities which must be completed before the next set can start. They divide large projects into a series of smaller ones.

Targets in a project are forced key points. They define certain tasks that have to be completed by a certain time. Targets must relate to the project objectives. They may be set because they are decision points, they are motivators, or they can be used to check progress so that remedial action can be taken before the project has advanced too far.

Milestones and targets are particularly important as motivators in long projects, as they provide goals that can be worked towards and achieved in a reasonable time. In an undergraduate project most of the project milestones would be set by the course tutor and they would relate to handing in reports, giving presentations, and having drawings available for the workshop. For this book we set some of our own milestones (for instance, when we would have copy ready to swap with each other, when we would have decided on the case studies we were going to use and so on). In addition, we had external targets to meet which were set by the publishers relating to preparing the first draft and responding to reviews.

Having determined the activities and milestones, the next step in the project management process is to plan, and, as mentioned previously, we will look at two planning techniques – Gantt charts and critical path analysis. Both of these techniques can be carried out on computer and there are a number of software packages available. However, as with any computer application the results will only be as good as the data entered and it will still be necessary to define times for activities and resources available.

8.3.3 Gantt charts

Gantt charts are named after Henry L. Gantt, who was a pioneer of scientific management working in the early part of the twentieth century. Gantt charts take the form of bar charts which provide a graphical picture of a schedule. On a Gantt chart you use the vertical axis to indicate the activities to be carried out and the horizontal axis indicates the time. An example of a Gantt chart is shown in Figure 8.3.

As you can see from the figure, it is very easy to see what needs to be done from the chart. For this reason, Gantt charts are widely used for planning in industry, particularly for production planning. They can be produced on a computer or written on a piece of paper but more commonly they take the form of a wall chart or planning board.

Gantt charts are limited in their use because it is not possible to show easily how activities depend upon one another. For example, if you are planning to manufacture a new product that requires special tooling you would have to show that the manufacturing activity was dependent upon the procurement of the tooling. You can do this by using tie-bars or arrows, as shown in Figure 8.4. However, this rapidly makes the chart confusing to look at when you have many dependent activities. It is possible to group tasks together vertically so that they

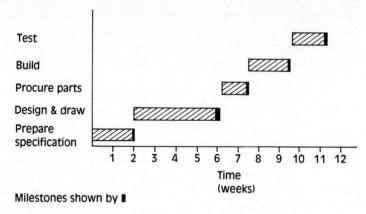

Figure 8.3   Gantt chart

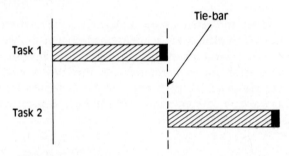

Figure 8.4   Tie-bars on Gantt charts

form blocks of similar tasks with a common output, thus reducing the complexity of the chart.

8.3.4 Critical path analysis

Critical path analysis is a more sophisticated planning technique that allows the representation of all the activities in the form of a diagram showing their sequence and any interdependence. It also allows analysis of each activity to determine when it should start, when it should be finished and how much room for manoeuvre there is if the project is to finish within a certain time.

We will look at a method of critical path analysis which involves drawing a network diagram of activities in which the activities are represented by arrows. We will consider the rules that apply for drawing the network and then look at how it is analyzed.

Critical path analysis does have disadvantages. It can be difficult to see at what stage the project is unless the network is drawn to scale, almost like a Gantt chart. It can also seem quite complex to people who are not trained in its use.

Drawing rules and conventions

A network diagram is made up of two elements: activities and events. An activity is a time-consuming task, and is represented by an arrow or line. An event is an instantaneous point and it may be the completion of one activity or the start of another. It is shown in the network as a circle or box.

A sequence of events is called a path and it is usual to number the events sequentially as they are viewed reading from left to right in the network. It is also conventional to draw the network with time increasing from left to right. A path of activities and events is shown in Figure 8.5.

The network is constructed by putting all the activities in a logical sequence, with no activity starting until all the activities upon which it depends have finished. The path in Figure 8.5 shows dependent activities, that is, where one activity has to be completed before the next can start. However, there may also be independent activities where two activities can go on at the same time, as shown in Figure 8.6.

The logic of a network should be checked to make sure that all activities have an appropriate start and end point. There should be no activities that do not lead to the end of the project, as shown in Figure 8.7. Similarly, paths should follow from the start to the end and they should not go round in circles, as shown in Figure 8.8.

It is vital to start and end a network with a single event since only a single state defines the start or end of a project. An example of a simplified network for assembling a printed circuit board is shown in Figure 8.9.

In this figure you can see that the activities are labelled, which makes the network easier to use if you can make the labels clear descriptions of what the

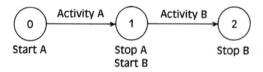

Figure 8.5 A path of activities and events

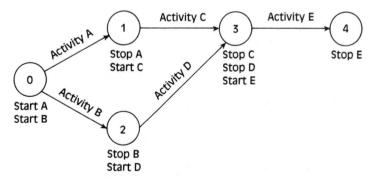

Figure 8.6 Network showing independent activities

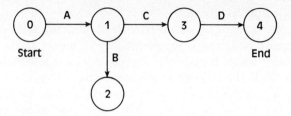

Figure 8.7   Incomplete network

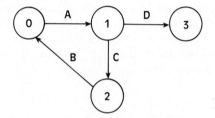

Figure 8.8   Illogical network

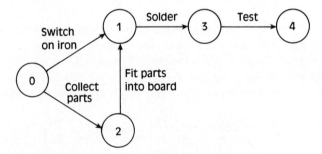

Figure 8.9   Simplified network for assembling a circuit board

activities are. The convention is to label above the activity arrow for horizontal arrows or to the right of the arrow for vertical arrows.

Activity times should be shown on the network below the activity label. You may also want to indicate how many resources are involved in each activity. This is also shown below the activity label as shown in Figure 8.10.

When you initially draw a network you may find that it takes several attempts to get it right. However, once finalized, it is always worth redrawing the network neatly. It will certainly be easier to follow and therefore easier to work with.

Summary of rules and conventions

1. Activities are drawn as arrows, events as circles or boxes.
2. Events are numbered sequentially, activities are labelled either above or to the right of the arrow.

Activity A
0 → 1
$d = 12$
$p = 1$

d: days
p: people

Figure 8.10　Labelling of activities

3. Time increases from left to right.
4. There should be no activities dangling and no loops.
5. Activity times and resource requirements can be shown below the activity label.
6. Draw the network neatly.

WORKED EXAMPLE

Draw the network for the following list of activities. The preceding dependent activities are given.

Activity	Predecessor
A	–
B	–
C	A
D	B
E	D
F	C,E
G	F

This gives the network shown in Figure 8.11. You see that the project starts with a single event, labelled 0, and ends with one event, number 6.

Dummy activities

Dummy activities are often used when drawing a network, although they have no time associated with them. They are represented by dotted lines and can be used

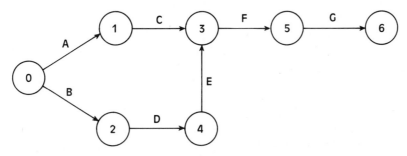

Figure 8.11　Solution to worked example

to ensure that the logic of the network is correct, to prevent activities having the same start and end points, or if their use makes the network easier to read.

WORKED EXAMPLE

Draw the network for the following list of activities. The preceding dependent activities are given.

Activity	Predecessor
A	–
B	A
C	A
D	B
E	D
F	B
G	C
H	B,G
J	H
K	E,F,J

This gives the network shown in Figure 8.12.

The dummy activity translates the end of activity B to event 5, indicating clearly that both B and G must be finished before H can start. To have shown B and G feeding into only one event would have been incorrect since D and F are dependent only upon B.

Analyzing the network

Analysis of the network will provide the overall time that it will take to complete the project assuming that you have all the resources that you need. You will also

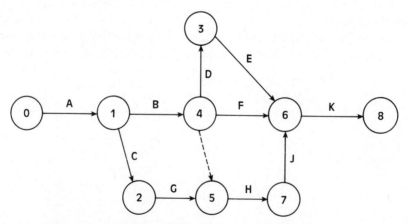

Figure 8.12   Solution to worked example showing dummy activities

be able to calculate the earliest start and latest finish dates for each activity. You can calculate the float on each path, and you will be able to identify the critical path.

When analyzing a network it is usual to start at time 0.

- *Earliest start date (ESD)*: The earliest start date can be indicated above each event, or it may be written within the event circle. Both methods are shown in Figure 8.13. The ESD for any activity is the earliest that it can start due to the activities that precede it having to be completed. The ESDs are calculated by working through from the beginning of the network by totalling the durations of all the previous activities. This is called doing a forward pass or forward scheduling. When two or more events have to be completed before an activity can start then the start date of the activity will be determined by the longest duration preceding the activity.

EXAMPLE

Activity	A	B	C	D
Predecessor	–	A	–	C,B
Duration (days):	1	2	5	2

The network, with earliest start dates, is shown in Figure 8.14. You can see that in order to start D both path AB and path C have to be completed. AB will take 3 days and C will take 5 days, therefore the earliest that D can start is day 5. When you reach the last event in the network the ESD is calculated for an activity subsequent to the project completion. Thus the ESD on the last event is the minimum overall project duration.

- *Latest finish date (LFD)*: The latest finish date can be indicated above each event, or it may be written within the event circle. Both methods are shown in Figure 8.15. The LFD is the latest time that any activity can finish if the project is to be completed in the earliest possible time. The LFDs are calculated by working

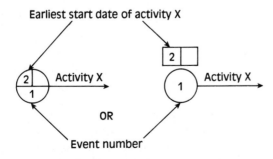

Figure 8.13   Earliest start dates in the network

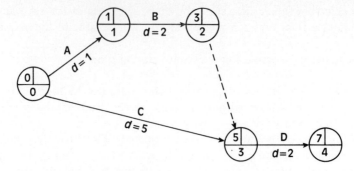

Figure 8.14 Calculating earliest start dates

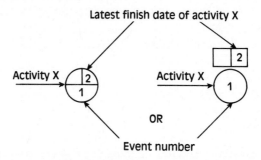

Figure 8.15 Latest finish dates in the network

from the end of the network, successively subtracting durations from the project completion date. This is called a backward pass or backward scheduling. When two or more activities precede one event the earliest date will determine the latest finish date for the activities.

EXAMPLE

Activity:	F	G	H	K
Predecessor:	E	E	G	F
Duration (days):	6	10	5	2

The above gives the network with latest start dates as shown in Figure 8.16. Note that the completion time for the project is 30 days. The LFD for the last activity will be the minimum overall project duration. Note that for both the first and the last events in the network the earliest start date should be the same as the latest finish date.

● *Float*: Float is the difference in time available for the series of activities in a path and the time required for the path. It is the amount of slack in the project. Knowing the float allows the project manager to schedule and optimize the use of resources while having full knowledge of the effect on the overall project duration.

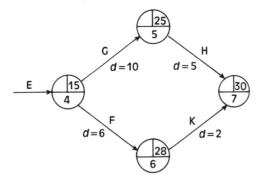

Figure 8.16 Calculating latest finish dates

Time available = Latest finish date for activity X (LFD) – earliest start date
for activity X (ESD)
Time required = Duration (d) of activity X
Float on activity X = LFD – ESD – d

Float is a characteristic of a path and not a single activity. For example if the float on a path is 5 days, if any activity takes 5 days longer than expected the float is reduced to 0 and none of the other activities can be allowed to overrun if the project is to be completed in the minimum time. Section 8.3.5 shows how float is used.

● *Critical path*: The critical path is the longest path through the network, and can usually be identified by the fact that the float on the path is zero. The critical path activities can have a float, although it will be the lowest in the network, if a project duration is imposed rather than if it has been calculated from the activity times. Consequently a critical path can also have negative float, which highlights the fact that too little time has been allowed for the project and therefore steps need to be taken to reduce activity times if the project is not to be delayed.

Activities on the critical path must be finished in the time given otherwise the finish date of the project will be affected. The only way in which you can bring a project forward is to reduce the duration of an activity on the critical path, such that a new critical path applies with a shorter overall duration. Determination of the new critical path is achieved by re-analysis of the network.

Summary of analysis
1. Calculate earliest start dates, using a forward pass.
2. Calculate latest finish dates, using a backward pass.
3. Calculate float on paths.
4. Identify critical path.

EXAMPLE

Draw and analyze the network for

Activity	Predecessor	Duration (days)
A	–	1
B	A	2
C	A	1
D	A	1
E	A	3
F	B	2
G	D	2
H	E	5
J	F	1
K	CFG	1
L	H	1
M	J	2

This gives the network shown in Figure 8.17.

The project overall completion date is day 10. The critical path is AEHL and the float on the other paths is as follows:

Path	Float (days)
BFJM	2
C	7
DG	5
K	4

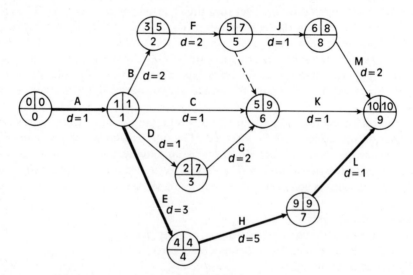

Figure 8.17   Solution to example showing full network analysis

This indicates that each of the four paths, shown above, can be scheduled within a time slot that is longer than the duration estimated originally. They can therefore be delayed to the extent of the float without affecting the overall project duration.

8.3.5 Resource allocation and levelling

We have seen that by calculating float it is possible to determine the time that can be used for a particular path of activities without affecting the overall project duration. It follows that if the changes cause the float on a path to be exceeded then this will affect the project duration. We have also seen that any changes on the critical path will affect the project duration. We can use this information to schedule all the resources that are required for the project. The scheduling of the resources is the most important aspect of project planning. For example, you may need to know exactly when certain monies are required in order to get them from the bank, or you may need to book a particular piece of equipment.

All projects are resource limited in some way. For instance, a project may be restricted by a certain number of people being available or by certain materials not being available until a particular date. In addition, there are always upper limits on the amount of money budgeted for a project. The project manager needs to know how these restrictions are going to affect the project.

EXAMPLE

Let us consider the following project:

Activity	Predecessor	Duration (days)	People required
A	–	2	2
B	–	4	2
C	–	5	2
D	A	10	4
E	B	1	2
F	C	3	1
G	D	4	2
H	EFG	3	5

The network is shown in Figure 8.18.

The project completion date is 19 days, assuming that all the resources are available as required. The critical path is ADGH and the float on the other paths is:

Path	Float (days)
BE	11
CF	8

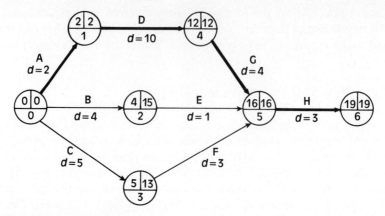

Figure 8.18   Solution to example

This shows that we have an extra 11 days within which we can schedule activities B and E, and an extra 8 days for activities C and F. Rescheduling up to these limits will not affect the overall project duration.

We can now draw a resource histogram which allows us to see when the resources are required and in what numbers (Figure 8.19). At this stage it is assumed that all the activities start at their earliest start date.

In real life it is more likely that you would have a fixed number of people available for the duration of the project and therefore you do not really want a situation where some of them will be idle one day and overworked the next: you will want to level the requirements. There are complex mathematical ways of doing this that are beyond the scope of this book so we will look at a simple graphical method.

First, we will assume that we do not want the project to go on any longer than necessary. In this case we do not want to reschedule any activities on the critical path unless absolutely necessary, so the first step is to redraw the resource histogram putting in the critical activities first, as in Figure 8.20.

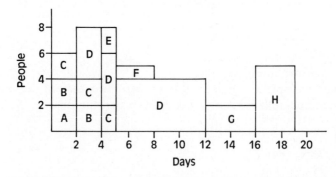

Figure 8.19   Resource histogram – all activities starting at their earliest start date

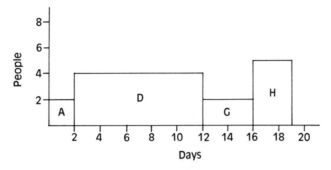

Figure 8.20 Resource histogram – critical activities only

Second, we need to look at where we have spare capacity. In this case if we consider that we need five people for a critical path activity (H) then we have spare capacity as follows:

From day	0–1	3 people
	2–11	1 person
	12–15	3 people

We can now look at the non-critical activities and see what resources are required and when. From the network we know the float available and the time slots that can be used for each activity. What we do not know from the information given is whether a particular activity, say C, has to be done by two people for five days, or if it is possible to use up some of the float extending the duration and using only one person for some of the time. This would obviously be known if it were a project with which you were involved.

In Figure 8.21 a solution is given which assumes that activity C can be done by one person for a longer period of time, and also that activity B can be split into two sections with a delay between the two sections. You will see that in this way it has been possible to reduce the maximum personnel required from eight to five,

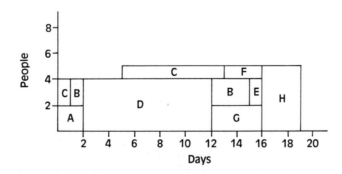

Figure 8.21 Resource histogram showing schedule to reduce personnel requirements

and the demand is constant at four people for the first five days and five for the remainder of the project.

This rather simple approach to resource allocation is equally applicable to other resources such as cash and materials. In this way we can schedule a project to make optimum use of resources available or to take account of resource constraints.

8.4 Project control

The project plan is a working document. If used correctly it will enable control of a set of activities and resources to achieve the desired end point.

In the same way that design work should be reviewed at regular intervals to ensure that the specification is being met, so project plans should be reviewed. Review dates should be incorporated into any plan and they should be appropriate to both the timescale and the complexity of the project. Reviews should be carried out by management in conjunction with the project team. The review should focus on progress made to date and look forward at the work that is to be done. It should highlight any potential problems so that action can be taken to prevent them, or so that the consequences can be acknowledged.

At the initial planning stage estimates of activity durations will have been made. As the project progresses it should be possible to refine the estimates based on experience. It will also be practical to incorporate more detail about the activities to be carried out in the near future.

The project review process should be formal, reflecting its importance for project management. In addition, the process should result in revision of the project plan so that it reflects the real situation.

Project management is no different from the management of any other engineering activity. It requires effective use of resources, particularly personnel. It is likely to involve financial control, which we have discussed in depth in Chapter 4. Most importantly, good project management depends upon strong leadership and team working. The team effort involves not only the project team but also all the people who will be affected by the project. The case described below highlights some of the issues relating to project management that have been discussed in this chapter. The case studies of Oxford Lasers (Chapter 13) also show these issues since the cases describe how a large design project was managed.

8.4.1 Case study – Edwards High Vacuum

Edwards High Vacuum is a division of BOC Ltd and manufactures a large range of vacuum equipment on three UK sites. Its site at Crawley houses the company headquarters and is also a manufacturing base for the company's range of diffusion pumps.

One of the pumps manufactured at Crawley is the DIFFSTAK MK2 diffusion pump. In October 1989 the factory at Crawley was organized by function and the manufacturing system had to cope with eight families of products, each with six variants, and production of between 100 and 200 per month for each variant. The DIFFSTAK MK2 pump was manufactured in batches to a monthly demand forecast indicated by the master production schedule (MPS). Buffer stock of finished pumps was held to cover any fluctuation in demand. The lead time for the pumps was three months.

Following a review by a team of consultants the management team at Edwards decided that the method of producing the DIFFSTAK MK2 could be changed to reduce lead time and they therefore established a project team to achieve this. The Project Manager chosen to lead the team was the Materials Manager. His team consisted of two professional consultants, one full-time and the other as an adviser, and two Edwards production engineers. The decision to include the consultants was made because they had previous experience of other production systems and in implementing modifications to existing systems.

The brief to the project team was that they should work towards a make-to-order (JIT) production system for DIFFSTAK MK2 pumps, starting with the fabrication and assembly processes. As this would inevitably entail significant changes on the shopfloor, communication with the workforce and the unions was of paramount importance. The project team realized that a major part of the project would be a method study to determine what methods were actually being used in the existing manufacturing system and that this data-gathering exercise would have to be undertaken before any decisions could be taken about the way forward. Edwards' General Manager therefore consulted with the union shop stewards and got their agreement to the exercise. Following this, the Project Manager made a presentation to the workforce, which included an introductory address by the General Manager. The workforce were also trained by the consultants in JIT techniques by means of practical exercise.

The first phase of the project, the method study, took two months. It was carried out by the two production engineers from the project team. During the method study they employed a part-time toolmaker and draughtsman so that any problems identified by the study could be quickly rectified. This showed the workforce that the project really was intended to have an impact. The method study confirmed that the suggested JIT system could be used successfully and a more detailed project plan was prepared and issued by the Project Manager. The preparation of the plan followed discussions with the project team. The plan covered the implementation of the JIT system and was presented in the form of a Gantt chart, allowing it to be understood quickly and easily.

Prior to the implementation phase the project team presented the proposed solution, including a new plant layout, to the senior managers and then to the workforce. The changes to the plant layout were scheduled by the plant engineer to take place during the Christmas holidays.

Project monitoring and control was achieved by daily meetings and monthly reviews. The daily meetings were held throughout the project and were to report on progress. They lasted usually for about 30 minutes and included the project team, any operators who wanted to go along, the Production Manager, and sometimes either the General Manager or the Works Manager. The daily meetings also served to engender team spirit and to foster good communications between the shopfloor and the project team. When the project entered the implementation phase the meeting also involved a manufacturing systems engineer and the plant engineer who had responsibility for plant layout. More formal project reviews were held monthly with a presentation being made to the senior site managers by the project team. In addition to the planned daily and monthly meetings *ad hoc* brainstorming meetings were held when problems were experienced.

The Production Manager took over ownership of the line from 2 January 1990 and the new production system, operating as a single unit flow line, started on time on 7 January 1990. The lead time for the DIFFSTAK MK2 pump was 10 days. The project team was disbanded after the line had been running for a week. As well as the reduction in lead time the project achieved a 30% saving in floor space and considerable reduction in work in progress and finished goods stock. Finally, continuous improvement groups were established to encourage worker participation in the continuing development of both the product and the production system.

8.5 Summary

In this chapter we have examined the role of projects in an industrial environment and the importance of planning and costing. We have considered the way in which a project can be fully defined in order to enable the costing and planning process to be carried out effectively. The role of the project proposal as a document defining the project, its implications, and the constraints imposed upon it were also described.

We have described two project planning techniques that are widely used: Gantt charts and critical path analysis. We saw that there were many advantages in using critical path analysis, particularly because of the way in which float can be calculated. Knowledge of the float on any path of activities allows optimum resource allocation and scheduling to be carried out. We showed a simple technique that can be used for this.

In Section 8.4 we gave a case study of a large industrial project and considered the way in which it had been planned and the techniques used for execution and control.

8.6 Revision questions

1. (a) For the information given below:

   (i) Draw an activity network.

(ii) Determine the earliest finish for the project assuming that it starts at day 0.
(iii) Determine the float on each path.
(iv) Determine the critical path.

Activity	Predecessor	Duration (days)
A	–	5
B	–	1
C	A	5
D	B	7
E	C	2
F	B	6
G	CDF	10
H	E	8
J	G	2

(b) (i) If activity A can be reduced by 2 days how does this affect the overall project duration?
(ii) If activity E is increased by 3 days how does this affect the overall project duration?

2. (a) Briefly explain the following terms:

Gantt chart
Network
Milestones

(b) Give two advantages of using Gantt charts rather than networks
(c) Draw and analyze the following network:

Activity	Predecessor	Duration (days)
A	–	2
B	–	1
C	–	3
D	A	1
E	B	3
F	C	4
G	D	2
H	E	1
J	BF	1
K	GHJ	3
L	J	2
M	L	4

Indicate the overall project duration, the latest finish and earliest start dates for each activity, and the critical path.

3. Draw and analyze the following network:

Activity	Predecessor	Duration (days)	Workers required
A	–	2	3
B	A	3	2
C	A	3	1
D	B	4	2
E	C	1	4
F	C	2	3
G	BE	1	1
H	D	1	1
J	F	2	2

(a) Draw a resource histogram showing the numbers of workers required each day, assuming that each activity starts at its earliest possible start date. What are the maximum number of workers required?

(b) Reschedule the activities so that the maximum number of workers does not exceed six at any time. Redraw the resource histogram and comment on the practical implications of this project if workers are to be hired specially.

4. A product is made up of five parts, W, V, X, Y, and Z. Parts W, V and X form a subassembly and parts Y and Z a subassembly. The two subassemblies are assembled to make the product. The supplier lead times and assembly times for each part are shown below. Show this information on a Gantt chart and determine the lead time for the product and the latest dates on which each part should be ordered if product is to be delivered at the end of week 6.

Part	Supplier lead time (days)	Assembly (days)	Assembly (days)
W	2		
V	10	2	
X	4		
			3
Y	3		
Z	4	2	

5. What information should be included in a project proposal? Give reasons for the inclusions.

6. Discuss the personnel management issues raised by the case study in Section 8.4.1.

7. Prepare an analysis of the communication techniques used in the case study in Section 8.4.1. What alternative methods could have been used? How did the communication methods adopted contribute to the project? What would have been the likely effects of the alternative methods?

Further reading

Lock, D. (ed.) (1987), *Project Management Handbook*, Gower.
Lockyer, K. G. (1984), *Critical Path Analysis and Other Project Network Techniques*, Pitman.

Part 3

THE MANAGEMENT OF ENGINEERS

9

Personnel management

Overview

Every engineer is responsible to someone, and so personnel is an issue for all. In addition, there are few engineers with no responsibility for personnel. Even in the early stages of their careers engineers may be called on to supervise technicians or fitters involved in the manufacture or test of equipment designed by them, and so the management aspects of personnel will be important.

In this chapter we will look at the issues that affect the management of personnel and consider their practical application.

9.1 Case study and introduction

Hilary Briggs is an excellent example of how an engineer can use her engineering talent and expertise, combined with effective management skills, to progress rapidly through a company.

Hilary joined the Rover Group after graduating in manufacturing engineering. Seven years later she is the Logistics Director for the Large Cars Business Unit at Cowley in the UK. The company is Britain's largest motor manufacturer and operates on several sites in the UK. It also services a number of international operations that manufacture under licence.

The Logistics Department covers production programming and material control for Cowley, KD operations, and Logistics development.

Production programming involves taking information from the commercial department, in the form of sales projections and orders, and translating this to production programmes and shopfloor schedules. Material control includes determining material requirements in order to meet the production programmes, transferring this information to suppliers, and then ensuring that the materials are available for the production line as required. This activity also includes quality assessment of suppliers and supply, in conjunction with the purchasing and manufacturing departments.

KD operations organizes the export of kits of parts to companies that carry out final assembly under licence, and provides associated technical support.

Logistics development has a long-term focus and is involved with looking at the way in which the different operations in the department are carried out. It seeks methods of improvement in the light of new technologies or greater understanding of the processes involved.

Hilary has been Logistics Director for eight months, since the department was formed; she previously worked as a Manufacturing Manager. The Logistics Department is obviously of great strategic importance to the company. It has a total of 365 employees and Hilary has five managers reporting directly to her. Hilary must provide the direction for the department and ensure that it meets company objectives.

Her main requirement is for effective personal and personnel management. Although she only has five people reporting directly to her, Hilary has to establish good communication links and ensure that the whole department is working to meet the company goals. She also needs to control the myriad activities that she is now called on to perform. Because this is a new venture for Hilary much of the information that she needs to do the job is held by the five managers. They also have considerable expertise in the logistics area and she needs to draw on this resource. As a first step, Hilary has established a fortnightly meeting with all the managers in which they can discuss the role, objectives and performance of the department. This meeting helps to ensure that everyone knows that they are part of a team, and is followed up by a weekly meeting with individuals where targets for particular activities are determined and agreed.

One of the major ideas that Hilary is currently developing is a training package to ensure that all the people in the department understand the processes used in the operation of the department, in particular the information flows in and out. This is particularly important in a large engineering plant where 'process' is often thought to refer to the production process only, and there are in fact considerable benefits to be gained from applying a production process type approach to information flows, which are critical to the functioning of a company. This will probably form part of a much wider discussion with the managers when Hilary carries out their appraisals at the end of the year.

Hilary is also part of another team, that of the other directors who are responsible for developing strategy for the business unit at Cowley. In this role she needs to be able to step back from the detail with which she is involved in the department and must be able to look in overview at the operation of the company. It is imperative that the Logistics Department fits in with the overall company objectives, that logistics implications are taken into account when planning strategy for other departments, and, that as a whole, the Large Cars Business Unit meets the Rover Group objectives.

At this stage in her career Hilary does not use the specific engineering knowledge that she gained from her degree course, although she does use the problem solving and more general aspects of her engineering training. However, the engineering background does allow her to talk the same language as the design department and her manufacturing experience ensures that she can empathize with, and understand the problems of, the manufacturing department which is serviced by the Logistics Department.

You can see from the above case study that personnel management is very important to Hilary's job. She needs to be able to motivate her staff to meet her departmental objectives, to have some method of determining their performance and then has to consider ways of developing and training people to modify or improve that performance. We saw from the case study that she will be involved in appraisal interviews and is developing training.

In the event that any of Hilary's staff leave the company or the department is planned to expand she would also be involved in the recruitment of new people, or she may delegate this.

Because Hilary has reached a high position in the company her personnel management load is considerably more onerous than would be the case for a more junior engineer. Nevertheless, the majority of engineers will be involved in personnel management at some level during their career. Therefore in this chapter we will look at some aspects of personnel management and consider some techniques that can be used by the manager. However, first we shall consider how people are deployed in organizations, looking at some of the organization structures that exist and the factors that influence the structure used by a company.

We will examine the methods by which you identify the people that are needed and how they are recruited. We will then consider some of the management issues relating to stimulating people to work effectively. We will examime the methods of developing people through training. Finally, we will consider job design and reward of staff.

The learning objectives for the chapter are:

1. To consider different ways that businesses can be structured and understand the implications of different organization structures
2. To understand the process of employing people and to consider some of the legislative and financial factors that affect the process
3. To consider the effects of motivation on productivity and to look at how motivation factors can be determined
4. To consider methods for developing people in organizations, including appraisal and training
5. To look at the role of job design and its effect on employee motivation.

9.2 Structure of organizations

In this section we will examine three ways in which organizations can be structured. We will then consider the factors that affect the structure adopted by the organization.

In order to provide a graphical representation of a company, organization charts are used and so we will start the section by looking at how these are drawn and what they show.

9.2.1 Organization charts

These charts provide a picture of the organization that can be readily understood by people internally and those external to it. The most common form of organization chart is the line chart, as shown by the manufacturing company chart in Figure 9.1. The amount of detail shown on the chart will vary with the audience for which it has been prepared. It may show an overview of the company's departments or the personnel deployment in a particular department or group.

Figure 9.1 illustrates the way in which the various functions within the company are addressed. For instance, it shows that the function of storekeeping falls within the remit of the production director. The chart also indicates the number of people working in each area.

The chart shows the company hierarchy, line management authority, routes for consultation and the span of control of each manager. For example, let us consider the chart in Figure 9.2.

The 'span of control' defines the number of people reporting to a particular person. Obviously, the span of control directly affects the task of each manager. However, it is important to recognize that the numbers of people involved are not the only factors that have to be considered when looking at the effective control of groups. One needs to consider the types of task involved, the ability of those carrying out the tasks, and the complexity of the task. For example, an Engineering Manager with six project engineers reporting to him or her might be overwhelmed, whereas a Production Supervisor responsible for twenty operators working on a production line may cope very easily.

Authority and reporting is shown by the vertical lines. In Figure 9.2 employee S2 reports to employee P3 who in turn reports to the 'manager'. If P2 wanted S2 to

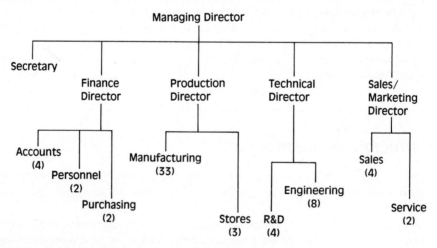

Figure 9.1   Organization chart for a manufacturing company

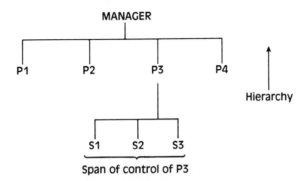

Figure 9.2 Organization chart showing personnel deployment

carry out a particular task then the correct route would be for P2 to consult with P3 and then for P3 to authorize the work. The horizontal lines indicate lines of consultation and the hierarchy indicates the levels of personnel in the organization.

9.2.2 Methods of company organization

There are three common methods of organizing companies: functional, divisional, and project structures.

Functional organization
This type of organization is the one most commonly used in smaller manufacturing companies. In this organization you group together all the people concerned with a particular function, giving separate functional departments, e.g. the personnel department. The company shown in Figure 9.1 is organized according to function.

The advantages of this type of organization are that it allows integration of people with similar expertise and knowledge. However, as a company grows, functional organizations can present barriers to communication with people in a different area because of the size of each area. As a company develops, it may be necessary to carry out subdivision of the areas.

Divisional organization
Many large companies organize themselves into divisions. In these cases each division will contain a number of functional areas. Consequently, some functions will be found in more than one division. There are a number of reasons for using this approach, two of which are listed below:

1. *Division by product or service*: When a company has a range of very different products or products addressing different markets it may make more sense to use separate divisions to allow more appropriate manufacturing and

marketing techniques. It would also allow more accurate costing of products by clearly differentiating the resources used for manufacturing each type of product. In addition, it may be difficult for a company to make a name for itself in a new market segment if it has a reputation in a different one. For example, if a company has an established reputation as a producer of high-volume low-cost goods it might be difficult to persuade potential customers that it can also make a range of high-quality, high-cost goods, even if this were for a different product range.

2. *Geographical division*: Geographical division makes organization and administration easier when divisions are far apart or when operating to different laws and regulations. It also eases communication and transport problems. Geographical division usually means that a company operates on different sites, and these may be in different counties or even in different countries. It is particularly common to find geographical division when the product must be close to the market (for example, perishable goods, or consultants who serve a particular locale and therefore need to be close to their clients).

Some companies mix both functional and divisional structure by retaining overall control of key functional areas such as personnel and finance. In this case each division would pay a contribution for the cost of these services.

Project organization

When a company organizes itself such that distinct groups of people address particular projects or pieces of work this is 'project organization'. This type of organization is commonly found in the construction and civil engineering industry. In the manufacturing industry project organization tends to be of a more temporary nature, with people being brought together to tackle specific jobs on secondment from their usual job in the function or division. However, this form of organization is used widely in product development (see Chapter 5).

Although we can now look at a company organization chart, as in Figure 9.1 and see how it arranges its functional areas we have not yet considered how people may be deployed within those areas. In Chapter 3 we looked at each functional area and the way in which they can be organized. However, there are some general rules that relate to the methods of organizing people, and we shall now consider these.

9.2.3 Deployment of personnel

Any organization must be concerned with fitting into its structure people of varying skills, abilities, and personalities so that they can be used to maximum effect within the organizational system. The organization must also recognize that when it buys the skills, energy, and abilities of people it must accept the individuality of people and their feelings of uniqueness and importance.

By dividing the workforce one can build a hierarchy in which people have other people reporting to them, and for whom they are responsible. The purpose of division is to enable people to manage the company more effectively and to give a sense of belonging to a discrete group.

Consider the structure shown in Figure 9.3. In this structure the Managing Director (MD) has a span of control of sixteen. With so many people to look after, the task of management becomes very difficult, if not impossible. The MD will not have sufficient time to carry out the planning of work and be able to spend the time necessary to motivate, organize, control, direct, and provide feedback to all sixteen people, let alone undertake all the strategic tasks that this senior role requires.

The other disadvantage with this type of structure is that as there is little (or no) hierarchy the possibilities for promotion and advancement are very low. This will inevitably be a major demotivating factor for many people.

A structure that alleviates these problems and provides more opportunity for advancement is shown in Figure 9.4. This structure also has the advantage that for the people involved in the lower levels it gives a much better-defined route to authority. In the previous structure it appears that the reporting route to the MD was very short and clear-cut. Unfortunately, in organizations that operate in this way the people involved tend to form their own hierarchy which, first, does not necessarily meet the needs of the organization, and second, causes confusion because it is not generally known.

Hierarchical structures can also be problematic and care must always be taken to ensure that people at the bottom of the organization are not left feeling remote

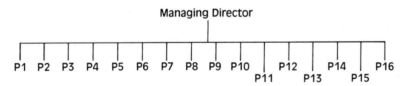

Figure 9.3　Single-level organization structure

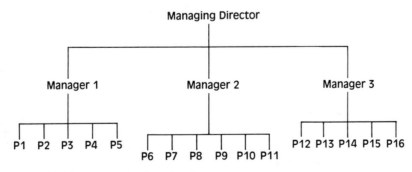

Figure 9.4　Multi-level organization structure

from the company management. One problem with imposing a very formal and hierarchical structure is that it can make communication very difficult for people at these lower levels. In particular, they can lose sight of the company's overall objectives. A consequence of this is that informal structures tend to be adopted by the workforce. This can be very good, as it gives the workers a sense of belonging because the communication links across the company improve. However, it can be quite disadvantageous for the company. An informal structure implies that the management have no control of it. It can lead to an excessive burden being placed upon the people who comply with the requirements of this arrangement and are used as advisers or mentors in place of the managers who should really have the responsibility for this.

The structure shown in Figure 9.4 is the commonly used line structure. In this structure everyone has a well-defined manager and there is clear definition of the routes of authority and communication. The advantage of this type of structure is that everyone knows where they stand, in terms of responsibility, authority, and opportunities for advancement. The disadvantage is that this type of structure requires consultation across the organization to be done formally at fairly high levels, with the consequent loss of benefits this might have had.

Another type of structure is 'staff structure', also known as a 'matrix organization'. In this type of structure there are no line managers – each manager has a specialist functional responsibility and a pool of staff which can be drawn upon for a specific project. The problem with this type of structure is that people at lower levels in the organization, and some at higher levels, find that working for more than one boss is both inefficient and demotivating. It also requires people in the organization to be fairly adept at time management and diplomacy. For the people at higher levels it can be very frustrating not having total control of a particular resource.

The most common organizational structure is a compromise of line and staff, with the level of compromise being appropriate to the organization involved. An example is shown in Figure 9.5.

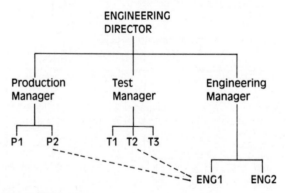

Figure 9.5   Line and staff structure

In the structure shown in Figure 9.5 the production personnel, P1 and P2, have a line manager, the Production Manager, as do the test personnel and the engineers. However, the engineers also have some technical responsibility, which means that they work directly with the people in the other functional areas when required. For example, the engineer ENG1 has technical responsibility for a particular product which is to be manufactured by P2 and tested by T2. P2 will be authorized by the Production Manager to work on that product. P2 will also receive line management from the Production Manager. So the Production Manager will produce work schedules, be responsible for approving holidays and overtime, and have responsibility for any other line management tasks. However, during the period of working on the product P2 will work closely with ENG1, perhaps receiving detailed work schedules or discussing manufacture problems.

This line and staff structure can be very effective. It ensures that the managers of each functional area are able to manage their areas and they know what is happening in terms of work schedules. They are in control. However, for the technical input there is a quick and easy route directly to the engineers. The only real disadvantage of this type of structure is for the engineer who may have considerable technical responsibility and responsibility for meeting project deadlines. However, the engineer has no direct authority over the resources needed to do the job. This type of organization is very common for companies operating 'project offices', and the project engineers need very good interpersonal skills in order to meet their objectives.

9.2.4 Factors that affect company organization

We have considered some of the organization structures that are widely used. We will now look at the factors that affect the way in which a company is organized, and the people deployed within it. These are:

1. Product and manufacturing system
2. Functions and expertise
3. Size
4. Responsibility and authority
5. Centralization and decentralization
6. Communication.

Product and manufacturing system
The type of product and the way in which it is manufactured will have the greatest effect on the organization of the business. In particular, it will dictate the emphasis put on areas such as research, sales, purchasing, and product development. However, the manufacturing system will also specify the way in which people are deployed within the manufacturing area. For example, it will define whether they work in groups on a complete product or individually on discrete parts.

The state of product development of the company's products will also affect the organization. If the company has a stable product range and plans very little development then it lends itself to the functional organization. However, if the product range is changing a lot and the company requires flexibility to be able to react to that change, then a matrix organization may be more appropriate.

Functions and expertise

Depending upon the nature of the business in which the company is involved, there will be various functions that have to be carried out (for example, manufacturing, testing, exporting). The company must have expertise in these functional areas and its organization will reflect the distribution of that expertise. In small companies in particular this can lead to some unexpected combinations of functions due to limited personnel with appropriate expertise. For example, personnel and finance may be combined because the company accountant has some previous personnel experience.

Size

The size of the company will have great influence on the way in which it is organized because of the need to establish a level where the number of people being managed is low enough to allow effective management and delegation but sufficiently high to make sure that there are enough people 'doing' rather than 'organizing'. If there are too many managers the company will have a large financial burden that has to be charged against profits, and which is not directly contributing to those profits. If there are too few managers it is likely that they are spending the majority of their time scheduling or carrying out day-to-day management activities. The company suffers because no-one has the time to look forward and consider company development.

Responsibility and authority

The amount of responsibility and authority that an organization gives to its managers will affect the structure because of the amount of resource you can commit to any particular area, in particular how many people they are prepared to commit to one area and whether line management, functional, or technical responsibility can be given to a particular person.

Centralization and decentralization

Whether or not an organization operates in a centralized organization or decentralized manner is largely dependent upon the style of the Chief Executive or Managing Director. Both types of organization have benefits which should be considered. Centralization should produce strong leadership and direction, and defined strategies. However, decentralization supports the workforce, therefore morale should be higher, communications improved and decision making brought closer to the situation it is addressing.

Communication

Good communication is essential in any company and the lines of communication, both essential and desired, should be considered as an inappropriate structure can easily lead to poor communication to the detriment of the company as a whole. A hierarchical structure, for example, can inhibit communication between people at very different levels. This may be unacceptable in a company where there is a lot of technical knowledge at the 'top' which has to be accessible by people at the 'bottom'.

9.3 Employing people

In this section we will examine the legal and contractual issues associated with employing people.

We will start by considering the recruitment and selection processes used to attract potential employees to an organization and to select the person most likely to meet the needs of the organization. In order to recruit likely candidates there needs to be a clear definition of the job to be filled. We will therefore examine the role and the preparation of job descriptions.

Following selection, a formal job offer is made. We will look at the way in which this is done and examine a contract of employment. In addition to contract law, employment practice is regulated by a considerable amount of legislation in all countries. In this section we will look specifically at some of the legislation that applies to recruitment and selection in the UK.

Following the offer of a job, a new employee will join the organization. Alternatively, an existing employee will take on a new role within the organization through an internal appointment. The process of introducing employees to their new job is called induction. We will examine this process and the effects that it can have on both employer and employee. We will also look at the costs associated with employing new staff.

Occasionally it may be necessary to terminate someone's employment and so we will also consider the different ways in which this can take place and what must happen if it does.

9.3.1 Recruitment

Recruitment is the process of attracting potential employees to the company so that the selection process can be carried out. It follows from a need being identified following loss of staff or from manpower planning.

Manpower planning

Manpower planning matches the company's requirements for people and skills with those that are available. It highlights deficiencies and surpluses in advance of their making an impact on the business. There are two levels of manpower planning, the highest level being the overall company requirements to meet corporate objectives. These would be identified in the business plan and would not normally be very detailed. However, at a lower level the line managers in each department would have to plan in detail their manpower requirements in order to meet department objectives. For example, the business plan might say that in order to meet expansion plans the manufacturing area will have to grow by three people over the year. However, the manufacturing manager will have to specify what those people are to do and exactly when they are required. This is the information that feeds into the recruitment process.

Manpower planning must accompany any planning activity, whether for mass manufacturing or for a specific project. Its absence can lead to unrealistic targets being set. For more information on planning and planning techniques see Chapter 8.

Whether recruitment is to take place as a result of an identified need in the manpower plan or in order to replace someone who has moved on, it is always advisable to carry out a job analysis to ensure that the vacancy exists, that the vacancy really does need to be filled, and that the requirements for the jobholder have not changed. It is very easy to replace a person who is leaving without assessing what job they were really doing and determining what needs to be done. A job analysis should result in a clear definition of the job in the form of a job description. In addition, the manager responsible for the job will have to consider the type of person that is needed to do the job well. A personnel specification should be developed to provide this definition.

Recruitment policies

Formal recruitment policies should ensure that recruitment is carried out in a similar way across the organization and so that everyone is treated equally. Recruitment policies may include rules relating to where jobs are advertised, the geographical location of employees, the rules relating to internal recruitment, expenses paid for interviews, and the way in which interviews are carried out. Policies should define how job descriptions and personnel specifications are prepared, the way in which information about jobs is communicated, and the methods by which information about potential candidates is gathered.

We will now look at the preparation of job descriptions and personnel specifications, advertising, and the use of application forms.

Job descriptions

A job description documents the main tasks and objectives associated with a job. Many employers now provide job descriptions for their employees although there is no legal requirement for them to do this in the UK. However, some employers positively dislike job descriptions because they are felt to be symbolic of a bureaucratic system and are seen as removing management flexibility by defining responsibilities. This can be a problem both in new and well-established companies, and it is therefore essential that any system of implementing and using job descriptions should not place a large administrative burden on the company. Similarly, there needs to be an appropriate level of flexibility built into the job description that does not detract from the purpose of the job description, which is to define a job.

The job description is equally useful for both the employer and the employee. For the employer it forces important issues to be considered, addressing the areas relating to why the job exists and what is to be achieved by the jobholder. In addition, it forces analysis of the needs of the jobholder in terms of resources to do the job described.

The job description should give a broad indication of the responsibilities of the job and indicate how it links with other jobs in the company; in particular, how it contributes to meeting the company's goals. In preparing this statement the employer is forced to think about why the job exists and if it is really necessary for the prosperity of the company. It also allows the employer the opportunity to consider whether or not filling this job is the most effective way of contributing to the company goals.

Objectives should be included in the job description allowing the employer to state what is most important for the jobholder to achieve. These are especially important for reviewing performance and progress (see Section 9.5).

The job description indicates tasks that are to be carried out by the jobholder. In defining these tasks the employer is able to confirm that they need to be done and is given an opportunity to review whether or not there are other tasks that also have to be done. If a job description is being prepared following someone leaving their post then consideration should be given to all the tasks that were carried out by that person against what was stated in the job description. Job descriptions define jobs and should be separate from personalities.

The employer needs to consider what level of authority the jobholder should carry in relation to the resources that can be committed, such as the authority to commit financial resources, to recruit staff, and to dismiss or suspend staff. This is an important issue, for it also ensures that when considering someone for the post the employer considers their ability to carry this authority effectively. Similarly, consideration needs to be given to the level of responsibility required to do the job effectively. The employer should define the resources that are available to the jobholder such as the budget for the job, the number of staff, and the equipment, plant and buildings.

For employees the job description is important because it allows them to know what is required, what authority they have, and what standards are expected.

The form of a job description

There is no defined form for a job description but a hypothetical one would normally include the following:

1. Job title and grade
2. To whom you report

3. What authority you have
4. Definition of those for whom you are responsible
5. Your main objectives
6. Key responsibilities and tasks
7. Reporting methods and requirements.

A clause may also be incorporated which provides the employer with a degree of flexibility when asking the job-holder to carry out specific tasks. An example of a job description is given in Figure 9.6.

Personnel specifications

The preparation of personnel specifications follows from consideration of the job. Many small companies do not prepare formal personnel specifications but, if carrying out recruitment, must have views on the type of person that is being sought. The formalization of these views is essential if a number of people are involved in the selection process, or if the person doing the selection will not be in direct contact with the jobholder.

When preparing a personnel specification there are a number of things that can be considered:

1. *Qualifications*: The level of qualification essential for the job, and that which is desirable, should be defined. In order to do this there will have to be a decision made about what is expected from the jobholder and how the job is likely to develop. It is important that neither too many nor too few

---

Job Description: Technician Engineer

The technician engineer in the Engineering Department will provide engineering support to other members of the department. The key tasks of the person are:

- To provide engineering technician support to members of the Engineering Department.
- To test, assemble and develop components and products, as instructed, in accordance with the requirements of the quality management system.
- To assist in the production of documentation for components and products including work instructions, bills of material, and operating manuals.
- Any other appropriate tasks required of the Engineering Department.

Job title:	Technician Engineer
Grade:	3
Responsible to:	Head of Engineering Department
Responsibile for:	Engineering Support Technician
Purchase authority:	None
Contract authority:	None
Reporting:	You will be required to produce a written progress report for the Engineering Department progress meetings.

Figure 9.6 Job description for a Technician Engineer

qualifications are required, as both of these can lead to future problems. An underqualified person will have difficulty with the tasks and will not be able to do them as well as required and may become demotivated by lack of achievement. An overqualified person will become bored and either not do the job satisfactorily or in a rather perfunctory manner, or will use the job as a fill-in occupation until a better job can be found.

2. *Skills*: The type of personal skills required might include defining someone who is a good team worker, someone with leadership qualities, someone who is self-motivated or who will be led, or someone who is a good communicator. In addition, the technical skills required for the job will have to be considered.

3. *Experience*: Consideration should be given to requirements for previous experience, whether appropriate experience is required and if so whether it was in a similar company, or in a similar job, or at a similar level. It is important to balance this carefully in the light of what is currently offered and what the prospects are. For example, if the job requires some specialized knowledge and there is little training provision it will be necessary to bring in someone who already has that knowledge. A balance of skills and experience should also be achieved by a consideration of how the experience of the jobholder will complement that which is already available in-house.

4. *Attributes*: If there are any physical attributes that are essential in order to be able to do the job then these must be clarified. This will require care to make sure that there is no inadvertent discrimination against any particular group by the imposition of unnecessary restrictions.

Consideration should also be given to other requirements such as an ability for geographical mobility either nationally or internationally, or to have a driving licence. A person's ideological beliefs may even be important.

Advertising

The job description and personnel specification provide the blueprint for the post to be filled and the person required. Advertising is the process of communicating the vacancy to potential applicants.

When advertising, the job and the working conditions should be expressed in a way that makes them attractive to the right sort of people. It is usual to include some detail of the job and the terms and conditions of employment. At the same time, there is a need to limit the number of potential applicants to those that are most likely to be able to do the job. It is, however, important to avoid the opposite extreme. An example of appropriate advertising for the post of an Engineering Manager is given in Figure 9.7. The advertisement includes details about the job, limitations on applicants, the application procedure, something about the company, and the closing date for applications. It should attract the attention of potential applicants. The placement of an advertisement should be appropriate to the audience that is to be attracted. This might indicate use of national rather

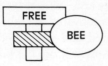

ENGINEERING MANAGER

Competitive Salary

FreeBee Ltd is part of a multinational group and one of Europe's leading manufacturers of rotary engines. The company operates in the domestic applicance market and its activities are primarily directed at the UK and Europe.

Applications are invited to join FreeBee as Engineering Manager to develop, establish and oversee all aspects of the design of new products.

Your enthusiasm, superior technical skills, and flexibility to work in a multi-disciplinary team on an aggressive schedule will make you a successful candidate. You will have at least 5 years' experience of managing your own team and are likely to have Chartered Engineer status.

If you would like to play a key role in a fast-growing, innovative and friendly organization, apply now. Please write with CV to:

> Dr D. Dodd
> PO Box 123
> London

Closing date for applications: 23 June 1993

Figure 9.7 Advertisement for an Engineering Manager

than local press, or of certain trade or professional journals. In view of these issues many companies resort to the use of advertising agencies.

In large organizations the chances of being involved in preparing and placing job advertisements are slight. This is more likely to be an activity undertaken by the personnel department. However, this is not the case in small companies and often line managers will be involved.

Application forms

Application forms are an effective way of ensuring that you get the right information from applicants and that it is in an easy-to-analyze form. Depending upon the type of person to be employed, a very simple form that asks for facts might be used. Alternatively, there could be a requirement for a longer, more complex form that requires the applicant to argue their case for employment. Some companies ask for a *curriculum vitae* rather than send an application form, but this can result in vital information being omitted and in inconsistencies in the way in which the information is presented.

Typically, the organization will want to know some basic information such as:

1. The job the person is applying for
2. Where the person found out about the job (this gives feedback on the effectiveness of different recruitment methods)
3. Name and contact address
4. Qualifications and experience
5. Health and any disabilities
6. Age, if appropriate
7. Interests outside work.

This gives some idea about the type of person you are dealing with as well as providing information on which to base informal conversation at the interview.

In the more detailed application form you might also ask open-ended questions such as 'What can you bring to the job that others may not be able to?' 'Where do you see yourself in five years' time?' 'Why have you applied for this job?' The answers to these types of questions can reveal a lot about the candidates and will help when preparing a shortlist. However, as with all forms, they can be used by candidates who will put in only information that shows them in a good light.

9.3.2 Selection processes

Having attracted applicants for the job, the next stage is to select the candidate that is most suitable for the job and the company.

The first round of the selection process is shortlisting. This is the process of reviewing the application forms against your job specification and selecting the candidates that are likely to meet your specification. When using a recruitment agency they will normally do the shortlisting.

Following shortlisting, the company should have a list of the people that are likely to meet their requirements and are worth further investigation. This is done by interview and some companies also use tests. References are used to provide assurance that someone meets the job requirements. Interviews, tests and references are discussed below.

Interviews
Selection includes at least one interview, and many companies use two, one of which may involve the candidate making a presentation. The aim of the interview is to determine which candidate from the shortlist best meets the requirements of the job. It is also to present the job and the company to the candidates, and, as such, is very much a two-way process because the organization will require good candidates and will be competing with other companies for their services.

The interview should allow the company to determine what the candidate is like, whether or not they could do the job, and whether they will fit in with the existing team. This can only be achieved if candidates talk frankly about themselves and their abilities. Candour is important, since the company will want both to employ and retain the best candidate. It is important that promises are not made if they cannot or will not be kept. Effective interviewing is an essential part of all managers' jobs and is discussed in detail in Chapter 12.

Tests

Tests are becoming increasingly common in the selection of graduates and others at junior and middle-manager level. There exist many types of test and fashion seems to play as large a part in the selection of tests as does scientific basis. Tests range from simple intelligence quotient (IQ) and reasoning tests to residential outward-bound courses which test candidates' ability to work with others, identifies leadership qualities, examines behaviour under stressful conditions, and tests physical stamina. Handwriting tests (graphology) are common in some countries, such as France, and are becoming more widely used in the UK.

References

References from previous employers, college lecturers, or people who have known the candidate for a significant period of time are usually required. The value of references is not high since it is unlikely that anyone would name a referee if they thought something bad would be said. However, references do give the opportunity to get a different perspective of a candidate and are usually taken up after a job offer is made 'subject to satisfactory references'. Anyone who cannot find someone to say something positive about them should be treated with caution (for instance, they may have left their previous employment under dubious circumstances). For this reason, the presence of references is often more significant than the content.

9.3.3 Making a job offer

After the process of selection, interviewing and testing the company must make a decision about which, if any, of the candidates it wants to employ and make that person an offer. Many companies allow a period of time between carrying out interviews and making an offer. This can be useful, as it allows time to consider each of the candidates carefully and fully. However, it can also be disadvantageous in times of high employment or if the company is recruiting in a skills shortage area because while the decision is being made a good candidate may have received other offers. In addition, it should not be forgotten that selection is a two-way process and the candidate that is chosen may not choose the company, so it is always worth considering the fall-back position. If an offer is made on the same day as the interview it allows an offer to a different candidate if the first refuses, thus avoiding a whole new round of selection.

When a person is employed they are contracted to work in return for remuneration, and the terms and conditions under which they work will be defined in a contract of employment. The contract of employment as used in UK companies is described below, as are the legislative requirements defined in English law.

Contract of employment

A contract of employment exists from the moment the two parties reach agreement on the basic terms; the parties to the contract being the employer and the employee. There is no requirement for the contract to be in writing, although this is usual. However, under English law a written statement of the main terms and conditions of work must be made available to each employee within 13 weeks of them starting their employment.

Under Contract law there are four main points that have to be satisfied for the contract to be legally binding. This applies as equally to the contract of employment as to a contract to purchase a house. The four points are as follows:

1. The parties must have a legal capacity to contract. Those who do not have this capacity include minors, the mentally ill, and people who are obviously drunk at the time of making the contract.
2. The parties must have intended it to be legally binding. This is automatically assumed by English courts.
3. There must be a clear offer that has been accepted. The offer does not, in law, have to be made in writing but it is sensible to ask for a written offer, particularly if promises have been made at interview. Acceptance of the offer must be positive, that is, the acceptor must do something to confirm acceptance and it must be unqualified. Any qualification is a counter-offer and would have to be accepted by the other party. For example, if you receive a job offer with four weeks' holiday per year and your answer is an unqualified 'yes' then this is a positive acceptance. If, however, you reply 'yes, but only if the holiday is increased to five weeks' this is a qualified acceptance and requires further acceptance by the company. In practice, going to and fro like this is quite common particularly when salary is being negotiated.
4. There must be an element of consideration. Consideration is what you give in return for what is offered. In the employment contract this is the work that you promise to do in return for the remuneration package offered.

In addition to these four main points the objects of the contract must be legal. A contract to do something illegal, committing a crime, is not, not surprisingly, legal.

The existence of a contract gives rise to both explicit and implied terms of employment. The explicit terms should be stated in the contract as described below. The implied terms are that an employer is expected to pay wages, take reasonable care of the employee, indemnify the employee for expenses and liabilities incurred in the course of their employment, to treat the employee with courtesy, and to provide a safe place to work. In return the employee is expected to do the job agreed, to take reasonable care in the performance of duties, to obey reasonable instructions, to act in good faith towards the employer, and to refrain from impeding the employer's business.

Main terms and conditions

An employer is required to notify the employee in writing of the main terms and conditions of work within 13 weeks of the start of the employment. The terms and conditions that must be addressed are as follows:

1. Names of the employer and employee. These are the parties to the contract.
2. The date of starting employment.
3. Whether any previous employment counts towards the employee's period of continuous employment. This may be important as the period of continuous service is used to calculate sickness, maternity and paternity leave, and pension payments. It is also used by some companies as a way of allocating bonuses or determining eligibility for fringe benefits such as share options.
4. The date at which the contract expires if it is for a fixed term.
5. The rate of pay. This should state the normal rate of pay and is usually given as either a weekly or annual sum. It should also be clear when overtime payments will be made, if at all, and at what rate overtime will be paid. The majority of 'professional' jobs are salaried and overtime is not normally paid. However, a salaried person is often expected to work beyond 'normal' hours.
6. The intervals of pay. This defines how often payment will be made and will indicate whether it is in advance or in arrears.
7. The normal hours to be worked. These will be the hours of work that will be paid at the standard rate. It is usual to indicate any fixed breaks such as lunch-time.
8. Holidays, sick pay and pension arrangements. This will define the number of days that can be taken as holiday and should specify if there is a requirement to take any holiday at particular times, for example during a factory shut-down. The rate of holiday pay should also be given and the method of payment. The terms and conditions for sick pay entitlement, methods of notifying sick leave, and method of payment should all be defined. The pension arrangements should be included. Full details of the employee's contributions to the pension scheme and any made by the employer should be given, but it is usual to refer to a separate document containing all the other pension details. Some companies will contribute to a personal pension. If this is the case it should be indicated.
9. Terms of notice. Notice is required to bring the contract to an end. If notice periods are not included in the contract then the statutory minima apply as laid down in the Employment Protection (Consolidation) Act 1978.

10. Grievance and disciplinary procedures. A statement about the disciplinary and grievance procedures operating in the company, or information about where these details may be found, must be included.

There may be other rules relating to the job, such as those concerning safety. These should be referred to in the terms and conditions indicating where full details can be found. An example of a contract of employment is given in Figure 9.8.

Remuneration
Remuneration is the term used to describe the package of pay and other benefits that you receive from your employer in return for work. Usually the greatest part of any remuneration package will be the salary. However, the value of other benefits such as pension, company car, and medical insurance can be considerable. As part of the remuneration these benefits are generally taxable. Remuneration is discussed in detail in Section 9.7.2.

9.3.4 Legal aspects of recruitment and selection

Employment practice is regulated by legislation in all countries. In this section we will consider English law as it applies to recruitment and selection. However, it should provide a guide to the types of legislation that will exist in countries which do not use English law.

During the recruitment and selection processes care must be taken to avoid discrimination on the basis of gender, colour, disability or ethnic minority. Discrimination is not only held to be socially unacceptable, it is also illegal in many countries. In the UK the Equal Opportunities Commission and the Commission for Racial Equality oversee anti-discrimination legislation. There are three main pieces of legislation: the Race Relations Act 1976, the Sex Discrimination Act 1975, and the Equal Pay Act 1970. These are described briefly below.

Race Relations Act 1976
This Act is enforced by the Commission for Racial Equality which points out breaches and can apply penalties for future breaches. A breach of the Act is not a criminal offence, but anyone who feels that they have been discriminated against can claim damages at an industrial tribunal. Alternatively, the Commission can issue a non-discrimination notice ordering a business to stop discrimination during the 5-year period of the notice. An employer is liable if its employees discriminate unlawfully.

Employers can discriminate on the basis of race where there is genuine occupational qualification. This covers the following circumstances:

1. Authenticity of entertainment. For example, you can advertise for a black man to play Othello.

Project Engineer

Terms and Conditions of Employment

1. **Employer**: ....

2. **Employee**: ....

3. Employment Period
The contract is a fixed term of two years commencing .... It is not renewable. However, in the last six months of the period .... (the Company) will indicate whether or not a permanent position is to be offered.

4. Job Details
The job description is attached. The job title is Project Engineer (PE). The PE will report to

....

Project monitoring will be carried out on a fortnightly basis with minuted meetings between the PE and the Project Manager with progress being reported at the quarterly Engineering Projects Management Committee Meetings. In addition, the PE's personal, professional, and career development will be assessed at six-monthly performance reviews with supervisors.

5. Hours of Employment
Normal hours are 37.5 hours per week excluding lunch breaks. The hours of work are 8.45–5.15 Monday to Friday.

6. Notification of Intention to Cease Employment
(a) Employee: One month's notice in writing.
(b) Employer: One month's notice in writing.

7. Holidays
(a) In addition to bank holidays, 20 working days by prior arrangement with the immediate supervisor, with each holiday limited to two weeks unless agreed by immediate supervisor with the approval of two directors.
(b) In the event of termination of employment you are entitled to accrued holidays or, with management approval, pay in lieu at the current rate.

8. In Case of Absence From Work
If you are unable to work for any reason you must notify your immediate supervisor as soon as possible or phone in to pass on a message.

9. Pay and Pension
You will be paid £...... p.a. This will be paid in arrears, in twelve monthly instalments on the last Thursday of each month.
The position is non-pensionable.

Figure 9.8   Contract of employment

10. **Disciplinary Procedure**
(a) A written notification will be given if disciplinary action is necessary for any reason.
(b) If two written notifications for disciplinary action are issued then dismissal may result.

11. **Grievance Procedure**
(a) If there is dissatisfaction with any disciplinary decision or a wish to seek redress of any grievance relating to the employment the employee may apply in writing to the immediate supervisor.
(b) The employee will be advised of any action taken as a result of the above within 10 working days.

Figure 9.8 (continued)

2.   Authenticity of modelling. You can select people of a certain race if you are trying, for example, to illustrate certain cultural distinctions.
3.   Authenticity of food and drink establishments. This only applies to premises where the food and drink is consumed, normally, on the premises. For example, you can advertise for Thai people to work in a Thai restaurant.
4.   For work with a particular ethnic group. A particular ethnic group may be defined if someone is required to work with people in such a group. For example, an employer may advertise for a Bangladeshi person to work as a counsellor to Bangladeshi groups in the community.

Sex Discrimination Act 1975
This Act is overseen by the Equal Opportunities Commission. In a similar way to the Race Relations Act you can lawfully discriminate in some circumstances:

1.   To achieve authenticity of entertainment or modelling
2.   For jobs in single-sex establishments
3.   For jobs providing personal welfare and education
4.   For jobs where there are legal restrictions on the employment of one sex
5.   For employment in a private household
6.   For employment carried out wholly, or mainly, outside the UK.

With both sex and race relations indirect discrimination is unlawful, i.e. imposing a condition or requirement which, although applied to both sexes or all racial groups, is such that a considerably smaller proportion of one sex or racial group can comply (for example, stating a requirement for a senior engineer to grow a beard).

Positive discrimination can be unlawful, but is sometimes used in the form of special training schemes and initiatives for under-represented members of one sex or race in order to encourage more people from these groups into the workplace. In the UK it is also used to promote the employment of disabled people.

Equal Pay Act 1970
This Act provides for an implied term in the contract of every female employee,

known as the 'equality clause', which entitles her to similar rates of pay, holiday and benefits as those for male employees in the same organization where she can prove that one of the following applies:

1. She and the male are employed to do like or broadly similar work.
2. She and the male are doing work rated as equivalent by a job evaluation scheme.
3. She and the male are doing work of equal value to the organization. Whether or not this is the case will ultimately be determined by an independent expert appointed by an industrial tribunal.

The third clause was introduced in 1984 as a result of rulings in the European Court of Justice. However, although this is a potentially very useful clause it is limited because it applies only where comparisons can be made between men and women in the same employment. If the workforce is all one sex no claim can be made.

Many companies now implement equal opportunities monitoring in the hope of identifying problem areas and resolving them.

9.3.5 The induction process

Induction is the process of introducing new employees to the company and to their jobs. It should not be restricted to people who have just joined the company, it may be as appropriate for people who have transferred from one department to another.

The alternative to having an induction programme is to 'throw people in at the deep end'. This idea, which still has its devotees, is rather outdated and is clearly impolite and arrogant, giving the impression that the new employee is not worth the effort of common courtesy. In addition, many people are very good at their jobs when given the opportunity to settle in but not at coping with the stress of being left alone in an alien environment. This method discriminates against them and can alienate them further.

The induction programme, which must be a planned activity, should allow people the time to settle in while enabling them to gather the information that will make them most effective in their jobs. Thus, any induction programme should include the following as a minimum:

1. Introductions and meetings with people with whom the new employee will have to interact
2. A tour of the facilities where the employee will be working
3. An introduction to the company, including some history and an overview of its operating systems, including organization charts
4. An introduction to the department in which the employee will be working, particularly introducing the systems with which the employee will have to

become familiar, such as the quality system and the design system, as well as any processes that the employee will have to use

5. An introduction to the company's products
6. An explanation of the safety requirements and procedures operating within the company, including the issue of safety equipment and a tour of emergency exits and places where first-aid and safety equipment are kept.
7. A discussion of the objectives of the job and the setting of initial tasks.

Obviously, an induction programme will take time and will cost money but its benefits will accrue from increasing the speed with which the employee will be able to settle into the job, thus reducing the time when their effectiveness will be limited. It can also be used to encourage team working if responsibilities for different parts of the induction are shared among the team, and it should also encourage a transfer of loyalty from the previous employer by making the new employee feel 'at home'.

9.3.6 The cost of employing new staff

The cost of recruitment, selection and employment can be very high and should always be borne in mind when making decisions about employing new people. Even internal recruitment, when an existing employee changes to another job within the organization, can have a significant cost.

The costs of recruitment and selection, as with any type of costing, can be broken down into labour, materials and expense. If we consider the labour costs first, these will include the time of people from the personnel department who will be involved with the administration of the process, including sending out application forms and arranging interviews. They will also include the time devoted to preparing job specifications, shortlisting, and preparing and carrying out interviews. Consider that many interviews for graduate appointments take at least an hour, will require at least an extra 20 minutes for preparation and review, and will be conducted by more than one person. If two people are interviewing it will cost the company of the order of two days of time to interview five people and this is likely to be middle-manager time.

The materials cost is not likely to be significant and would probably be limited to stationery. However, the total expense could be very high if you are paying a recruitment agency, which is generally paid on the basis of a percentage of the new employee's salary. If this is not the case there will still be the costs of interviewees' expenses, postage, telephone and advertisements.

Costs will be enhanced considerably if the company uses any form of testing requiring the time of a consultant, use of premises outside the company, or is recruiting from overseas.

When the employee is in post the cost of employment is not limited to the salary paid but includes an amount for national insurance and, what can be quite a significant amount, for benefits and perks such as pension, car, health insurance, and share allocations. In addition, there are the costs of induction and training.

Because of the significance of these costs it is important to make sure that a job needs doing before you recruit for it, that you select the most appropriate person for the job, and that you retain that person by providing a good working environment and room for personal development. In Sections 9.4 and 9.5 we will look at some aspects of employing people that relate to their wellbeing while working.

Unfortunately, however, it is sometimes necessary to terminate a person's employment, and we will now look at this aspect of personnel management.

9.3.7 Termination of employment

An employer can terminate an employee's employment by giving them notice and then dismissing them. An employee can also give notice in order to terminate the contract of employment legally. In addition, employment may be terminated because an employee has reached the agreed pensionable age or a fixed-term contract has come to an end.

We will now look briefly at notice and then consider some of the reasons that can be used for dismissing someone.

Notice
Terms of notice should be defined in the contract of employment. However, if they are not then there are statutory minima that apply, defined in the Employment Protection (Consolidation) Act 1978. If notice is given by an employer to an employee and that employee has been working more than 4 weeks but less than 2 years the statutory minimum notice period is 1 week. Each additional year worked adds one additional week to a maximum of 12 weeks after 12 years' service.

The minimum notice from employee to employer, defined in the Act, is one week regardless of length of service. However, the typical notice period for salaried staff defined in a contract of employment is a month from either employee or employer, although the higher your position in the organization, the more likely it is that this period will be increased. Notice can be written or oral and pay in lieu of notice can be agreed.

Dismissal
An employee can be dismissed for a fair reason. The reasons laid down in the Employment Protection (Consolidation) Act 1978 are conduct, capability, redundancy, legal prohibition, and any other substantial reason. These are each discussed below.

(1) *Conduct*: Conduct relates to the way in which employees go about their duties and to what is accepted with regard to custom and practice. It covers acts of gross misconduct where the action of the employee is such that the contractual relationship is destroyed. There is no legal definition of gross misconduct but it is usually seen to include acts such as theft, violence, drunkenness, or other 'unacceptable behaviour'. Gross misconduct can lead to dismissal without notice.

(2) *Capability, including health*: This would arise if an employee was no longer able to perform the duties required at an appropriate standard. For example, if a warehouseman with a bad back was unable to lift boxes, he could be fairly dismissed if this had previously been one of the tasks that he was required to do.

(3) *Redundancy*: Redundancy occurs when the whole or main reason for dismissal is that the employer's need for an employee to do work of a particular kind in the place of employment has either diminished or ceased. Examples of when redundancy can arise include a reduction in the workforce because of a fall in sales, a move to a new site some distance away, or a change in work requirements due to implementation of new technology.

Redundancy is deemed to be fair dismissal but selection of candidates for redundancy must also be seen to be fair. This is normally accepted as 'last in, first out', i.e. the newest recruits to the company are made redundant first. This is not always practical because the company may be looking at particular skills or areas of operation. Nevertheless, the onus is always on the company to be seen to be fair. It is unlikely that people being made redundant will think it fair, and therefore they must always be dealt with carefully and sympathetically.

When a company decides that redundancy must take place, any recognized trade union in the company must be informed, at least 90 days beforehand if more than 100 employees are involved and 30 days beforehand for 10–99 employees. At the same time, the Secretary of State for Employment must be notified in writing. In addition, employees who are being made redundant must be given reasonable time off to look for other work.

Many companies will negotiate redundancy terms with either a trade union or staff organization, but any terms must be equal to or better than those laid down by law. Minimum redundancy pay is only paid to those below the age of retirement. The minimum rates are calculated on the basis of age, the number of years of service with that employer, and the wage received just prior to redundancy. There are maxima set on the number of years of service that can be counted, and on the weekly wage that will be used for the purpose of the calculation. There is a state redundancy fund which pays minimum redundancy pay if a company becomes insolvent.

Rather than making people redundant a company may opt to offer alternative employment in which case an employee can only claim redundancy if:

- The alternative offer is inferior, such as being offered a lower level of pay or benefits.
- The new job is unsuitable to the employee by way of skill or training. It is therefore not possible to offer someone alternative work for which they

are unqualified and then avoid giving redundancy pay if they decline to accept it.

- The new job is an unreasonable distance from home.

(4) *Legal prohibition*: An employee may be dismissed if they are are no longer able to carry out their duties because they are not legally allowed to undertake certain tasks. For example, a lorry driver could be fairly dismissed if an offence had been committed which resulted in the driver's driving licence being revoked.

(5) *Any other substantial reason*: This is likely to arise if there are prohibited activities in the workplace that have been previously enforced but have been carried out regardless. For example, if there was a company rule that decreed no-one should consume alcohol at lunchtimes and you had been doing so then you could be fairly dismissed, as long as this rule had previously been upheld or warnings that it was now going to be enforced had been given, and you had been warned that this behaviour was likely to lead to dismissal.

In almost all cases, for dismissal to be fair a series of warnings should be given, the exception being an act of gross misconduct. An employer is required to give the reason for dismissal in writing, if requested, if the employee has worked for more than 6 months.

Cases where employees believe that they have been unfairly dismissed must be taken to an industrial tribunal within 3 months of the dismissal. To claim unfair dismissal the employee must have worked continuously for at least 2 years. The employee has to show that the dismissal was normal, which means that either the contract was terminated by giving notice or a fixed-term contract was not renewed, or that the dismissal was constructive. Constructive dismissal is where the employee is put into such a situation that they no longer feel they are able to carry on with their employment and so resign. In all cases the employer is required to show that dismissal was for a fair reason. Remedies for unfair dismissal are reinstatement, re-engagement or monetary compensation.

9.4 Motivation and leadership

There are two main areas to consider when managing personnel if they are to work effectively. The first is motivation, and we will look at how motivation affects people and how motivating factors can be assessed. The second aspect is leadership and the way in which people can be managed in order to achieve objectives. We will look at the area of leadership and consider some of the styles of leadership that can be adopted by managers. We will also review the factors that affect the style of leadership adopted.

9.4.1 Motivation

When you start to plan your career and your first job, you will be concerned with many things, for example what type of work, salary, and training are being offered. Many factors affect why people take jobs and the weightings applied to each factor will vary with each individual. When you take a job the way in which you approach your work, whether you are enthusiastic or indifferent, or whether you take on extra responsibility, will also depend on a number of factors which relate to you, the job itself, the organization and so on. It is important to know which factors affect you and the relative importance of each so that you can make the best decisions and end up with a happy and satisfying career.

Everyone understands the problems of a lack of motivation, when people do not feel like doing a certain job so keep putting it to the bottom of the pile (in the hope that one day it might just go away), or it's done but not very well so at least it's out of the way. Apart from this, if your motivation to work is weak you may not worry if you get to work late or if you miss deadlines. Motivation can therefore have a profound effect on productivity and the quality of your work. The implications for a company of having a poorly motivated workforce are therefore great, and so managers have to understand what motivates their subordinates and provide an environment in which they will be motivated and will produce good-quality work at an acceptable rate.

Psychology has much to offer in the way of motivational theories, and fortunately there are plenty of people who have done a lot of work in the field of motivation that an engineer/manager can draw upon in trying to determine what might motivate themselves and their subordinates. Some of the motivation theories that have been produced are summarized below.

By considering the motivation theories you will have some ideas that will help you to plan your initial strategy. They are not rules and they do not tell you how to do the job; people with experience may look at them and say that they are just common sense.

Maslow and Herzberg

The two motivation theories that are most commonly referred to are those of Professor A. H. Maslow and Professor Frederick Herzberg. There are many excellent texts that discuss these theories in detail and some are given in the Further reading at the end of this chapter.

In summary, Maslow and Herzberg define specific things they found to motivate people based on experiments carried out in the 1950s and 1960s. Maslow's results are presented as a 'hierarchy of needs' in which needs have to be met at different levels so that people can be motivated by the next set of needs at a higher level. The hierarchy, which is graphically represented as a triangle, starts with the most basic physical needs and works up to the pinnacle of self-realization needs. The work of Herzberg led him to define two areas that affect people at work and he called these hygiene factors and motivators. The hygiene factors correspond to the more basic needs defined by Maslow and relate to working conditions. Herzberg concluded that if the hygiene factors were not adequate then morale and productivity would be low, but improvement of these factors beyond a satisfactory level would not lead to an increase in productivity. This could only be achieved by improving the motivating factors such as the level of responsibility of the job and the type of work done.

A review of these two theories gives you a list of things that you should consider when trying to determine what gets people to work and what gets them to work better. Some of these are salary, job security, the tasks involved, the people and working relationships, and the working conditions.

Salary was found not to be a motivator by Herzberg but a hygiene factor. If salary was at a suitable level then it would enable someone to be motivated by other factors but would not in itself motivate. While it is true that many people are not motivated by salary, a large number do seem to be, particularly with pressures in society to obtain material possessions and a high standard of living.

Job security is very important to many people because people generally like to know what they might be doing next month and next year. The pressure of society encourages many people to live beyond their means or take out large loans, so the security of a job is of paramount importance. In times of high unemployment the need for job security will override considerations of salary, hours to be worked, working conditions and so on.

The job that you are required to do may be challenging and will stimulate you, it may challenge you a little, or it will not challenge you at all. If you are not interested in the job you are not going to be motivated by it, but you may like the status or responsibility associated with it, or even find the perks invaluable.

The people that you work with can affect your motivation. If you have to work closely with someone who is well motivated and enthusiastic this will affect you and will increase your motivation. Conversely, if you are working with someone who is dissatisfied with their work or depressed then this will demotivate you. It is for this reason that some companies prefer to give pay in lieu of notice if someone resigns because their presence could spread bad feeling and reduce productivity.

The working conditions relate to the environment in which you have to work and the physical comfort associated with your place of work. In a foundry where the environment may be unpleasant people may find work uncomfortable and will minimize the amount of time spent in that atmosphere. Therefore thought should be given to trying to remove some of the pollution. Similarly, office chairs that do not provide adequate back support can make someone feel so uncomfortable that they do not stay at their desk for very long. An alternative to considering 'needs' has been put forward by John W. Hunt. In his book *Managing People at Work* Hunt proposes that the most effective way in which to predict people's behaviour is to consider their personal goals. From this you can motivate them by creating an environment in which their goals can be satisfied while those of the organization are also met. Hunt puts individuals' goals into the eight categories of comfort, structure, relationships, recognition and status, power, autonomy, creativity, and growth.

Whereas the theories of Herzberg and Maslow are presented as constant, in that the same needs have the same priority throughout life, Hunt puts forward the view that people's goals evolve as they do because the goals and the factors that influence them are based on people's backgrounds. Consider that at different stages in your career your motivation will be affected by not only the place in which you work but also the external environment and, very importantly, your home-life. For instance, engineers after graduation may have as a priority getting into a company which offers a proven training scheme to management and engineering certification. When this is done they may be more interested in salary if they want to buy a house or geographical location if they have a partner to consider. Later they may be more keen on a job which provides flexible working so that they can spend time with their children.

Motivation in practice

A problem with having to deal with motivation is that there are no correct answers. Many engineers who have a very rationalistic approach to work seem to have difficulty with this. There are no universal solutions. What motivates one person will not motivate another, therefore the engineer who manages has to be prepared to compromise and develop strategies that form the best overall solution for a number of people, including themselves.

Managers have to try to build up a picture of what their subordinates are really like and then establish what it is that drives them. If you are asked what motivates you it is unlikely that you will be able to find a satisfactory answer in a short time. Similarly, if a manager asks a subordinate this question it is unlikely that a reliable answer would be supplied unless the subordinate had been given plenty of time to think about it first. Because motivation is a complex issue, affected by many different things, it is difficult to quantify. Yet a manager needs to have this information. Whether the 'needs' of Maslow, the hygiene factors and motivators of Herzberg, or the 'goals' of Hunt are appropriate depends upon the manager's view, but they only provide pointers which allow the psychology of subordinates to be considered.

So how can this information be collected? At the selection stage when employing people there is no reason why they should not be asked the question. However, many people may give the answer they think the interviewer wants to hear. A more accurate method is to look closely at the way the candidate acts when not constrained by others, in particular what sort of outside interests the candidate has, what initiatives the candidate has instituted, and what the candidate feels is important in the job you are offering.

When a person is in post it is possible to look at what that person really does at work and their attitude towards it to indicate what they find most stimulating. The appraisal interview (see Section 9.5) provides an opportunity for pursuing this type of investigation.

Motivation can be improved, or the opportunity for improvement provided, only when a subordinate's motivating factors are known. Job descriptions and appraisal interviews can both help in improving motivation as they can be used to set goals, clarify what is expected, and explore any expectations the employee might reasonably have. It is widely felt that in manufacturing particularly, for operatives that have rather tedious jobs, that anything that can improve the job will improve motivation. A technique which is becoming more widely used is that of quality circles where people are encouraged to take an interest in the wider aspects of their jobs, in particular considering ways of improving working conditions and techniques to improve product quality. These are described in Chapter 7.

9.4.2 Leadership

Leadership is the way in which managers cause people to meet the objectives of the organization. Managers have the authority to require their subordinates to do what they are told, but this does not always prove to be the most reliable method. The aim should be to achieve authority through respect so that people do what is needed because they themselves appreciate the need. Establishing a leadership style is difficult, particularly as many people have no experience of leading when they first start work. However, it is a management skill which can be developed like any other.

Leadership is intrinsically linked with motivation; people have to be motivated through sound leadership in order to meet the company's objectives. Unfortunately, as with motivation, there are no correct answers defining how to lead a particular group. As with all management skills, the effective solution will be one that takes into consideration the situation being dealt with, the way the manager feels, and likes to operate, and the subordinates. This means that as a leader the manager cannot formulate a plan of action to cover all circumstances. Every new situation must be assessed and appropriate action taken 'thinking on your feet'.

There is, however, guidance as to the types of leadership styles that managers use. In the same way that the motivation theories provide a framework for developing your own ideas, the leadership styles broaden your view as to what is possible without laying down what is right.

Leadership styles

It is important that, whatever leadership style is adopted, it should be consistently used when dealing with specific people or situations, or it should be made clear that the style is being changed for a specific reason. Continuous changes in style will mean that subordinates never know what to expect, and this can be very unsettling, and consequently demotivating, for them.

The Continuum of Leadership Behaviour, developed by Tannenbaum and Schmidt, is useful for considering the different styles of leadership that can be adopted and how these relate to the style expected by subordinates. The Continuum defines a natural progression from extremes of the authoritative, allowing the subordinates no role in the decision making, to a democratic style where the decision making is delegated to the subordinates.

Two other contrasting styles are those of the task-centred leader and the employee-centred leader. The task-centred leader is concerned predominantly with the task to be accomplished and sees subordinates as tools that are used to get the job done. The employee-centred leader takes the converse view and is mainly concerned with the welfare and wellbeing of subordinates, with the view that if the subordinates are cared for then the task will be achieved through their commitment.

A view put forward by Douglas McGregor is that there are two ways of viewing our approach to work – theory X and theory Y. According to McGregor, many 'old-fashioned' managers are strong believers of theory X. This proposes that people dislike work and must therefore be manipulated and forced into it, and that people prefer to be told what to do. The opposite theory Y proposes that most managers have underestimated the worker and that work is as natural as play if the conditions are favourable. Control imposed from above and the threat of punishment are not the only ways for getting workers to work satisfactorily. People will exercise self-direction and self-control in the service of objectives to which they are committed. Theory Y also promotes the view that people will actively look for responsibility and that a large proportion of people have the ability to use their imagination, ingenuity, and creativity to solve problems.

Factors that affect leadership style
The ideas described above, the authoritarian, the democrat, the task-centred leader, the employee-centred leader, theory X, and theory Y, all define potential extremes of leadership behaviour. We will now look at what determines the behaviour adopted by a manager. As mentioned earlier, there is no correct way to lead. The most effective method will take into account you, your subordinates, and the situation which obtains.

Your behaviour as a manager will be influenced greatly by your personality. You will also perceive any leadership problems in a unique way based on your background knowledge and experience. Among the factors which will affect you as a leader are your value system and the confidence you have in your subordinates. Your value system relates to what is important to you, such as whether you feel that individuals should share in making the decisions that affect them.

Your values might mean that you are convinced that if someone is paid to do a job then they should be able to make all the decisions associated with that job. Your value system will affect where you would be seen by your subordinates to be operating between the extremes of autocrat and democrat. However, your behaviour will be further influenced by the relative importance that you attach to organizational efficiency, personal growth of subordinates, and company profits.

Your confidence in your subordinates will be affected by your knowledge of your subordinates' previous experience and performance. As well as evidence from previous records, people do tend to vary in the amount of trust they concede to others. Therefore there will inevitably be some people you could trust with one job but not with another. When given a task, as a manager, you should ask yourself 'Who is the best qualified to deal with this problem?' Unfortunately, many managers have more faith in their own abilities than those of their subordinates and consequently are not good at delegating.

Before deciding how to lead a group of people you must consider the factors that affect the behaviour of the people in the group and the goals that are motivating them as individuals and as team members. You need to judge whether or not the individuals work well together and whether they are all interested in the task. You have to identify whether the team has the necessary knowledge and experience to deal with the task, and whether they are able to share knowledge and make decisions. Finally, a leader must consider whether the team members feel a need for independence or if they still want to be guided.

The situation will affect the leadership styles that you can choose because of the way in which the company expects you to perform and the culture that permeates the company. You will also be affected by the constraints that are imposed on you by the task to be undertaken, such as time constraints, whether or not the task will be repeated, or whether there is a requirement for secrecy or confidentiality.

As you can see, there are no rules about the kind of leadership style that must be used, or which indicate what style will be most effective. All situations must be assessed separately and you must consider your own feelings and those of your subordinates. The ability to apply an effective leadership takes practice and thought, particularly in evaluating the effects after the event.

9.5 Appraisal of employees

Appraisal takes place all the time, with managers making assessments of their subordinates' work and considering their progress in the company. Informal appraisal like this cannot be stopped because judgements have to be made before a piece of work is allocated to someone or a decision is made about whom to promote. However, one problem with informal appraisal is that it does not actively encourage the participation of the person being appraised, and it is left to the manager to decide when, and if, the judgements that are made are relayed to that person. Additionally, the process is dependent upon the objectivity of the manager's judgement.

Formal appraisal is used by many companies to overcome these problems, although it should be used as a complement to, not instead of, informal appraisal. In this section we will look at formal appraisal schemes, considering their aims, operation, and the costs involved. We will also describe the advantages and disadvantages of such a system for both employer and employee.

9.5.1 The aims of an appraisal scheme

Appraisal is the feedback loop used to assess the performance of employees. It usually takes the form of a documented discussion between manager and subordinate with the manager carrying out an assessment of the individual. An important factor of appraisal is that both parties should agree on the record of the discussion. Many companies use appraisal systems for some, if not all, of their employees. Appraisal is a judgement of an employee's performance in their job and therefore it should be done in a formal way at a time that is arranged by both parties. It should also allow a comparison of performance over time that is meaningful and therefore the time intervals should be appropriate to the employees' objectives. It is also a forum for discussion of the appraiser's approach to the subordinate and should allow a free debate without fear of repercussions.

Appraisal can be defined as an objective assessment of employee performance against objectives, and it often involves:

1. A review of actual performance over the period covered by the appraisal, and a discussion of improvements required in the performance if appropriate
2. A statement of the present skills and abilities of the appraisee
3. A statement of the objectives and goals for the next period, including criteria by which performance will be measured
4. Identification of the best way to develop from the existing level of skill and experience to one likely to prove more desirable in view of the objectives set.

From the point of view of the person being appraised there are four questions that should be answered during appraisal. Open discussion of these questions can help people be more effective in their work and their level of motivation should improve. The questions are as follows.

1. *What should I be doing?* Everyone needs to know what they are supposed to be doing, and what responsibility and authority they have. A job description will provide an answer to this question but a job description will change as the company's circumstances and needs change and as the jobholder develops.
2. *How am I getting on?* The balance between what the company expects and what is being provided has to be clarified. An honest answer to this question allows assessment of achievements and will enable the appraisee to effectively direct their efforts and modify career plans if necessary.
3. *What will my next career step be?* The appraisee will want to know what the possibilities are for growth within the job, or for advancement into other jobs in the company. Again this allows individuals to manage their careers.
4. *How can I achieve my goals?* Finally, the appraiser should identify what help is available from the manager and the company so that goals can be achieved. This is the area of training and development which is dealt with in Section 9.5.

For people who do not have the opportunity to have appraisals at work self-appraisal may be carried out. This is described in Chapter 11.

9.5.2 Formal appraisal schemes

There should be a clearly defined approach to appraisal, and in order to ensure fairness it must be monitored throughout the company. This consistency is generally achieved by everyone using the same appraisal form to record the discussion. Monitoring of the forms is then done centrally. The organization of the appraisal system is usually undertaken by the personnel department although this requires careful management due to the problem of maintaining confidentiality. However, in order to be effective, appraisals must be carried out locally between manager and subordinate, and anyone performing an appraisal should be trained in both counselling and appraisal interviewing.

Appraisals should be made on a regular basis; annual or six-monthly appraisals are common. Both appraiser and appraisee should have sufficient notice to prepare properly for the appraisal and to consider the issues that have to be raised. However, the appraisal system should never be used as an excuse to put off dealing with a problem at the time that it occurs. This will only allow the problem to grow, thus worsening the situation. What the appraisal can do is focus on the reasons why such problems occur and allow a strategy to be developed for long-term prevention.

The main part of the appraisal process is the appraisal interview, which is documented using an appraisal form.

9.5.3 The appraisal form

An example of an appraisal form is shown in Figure 9.9; most companies develop their own forms so that they are appropriate to the type of workforce being appraised. The purpose of the appraisal form is to provide an agenda for the subsequent interview. In particular, it allows the prospective appraisee to consider the topics that will be raised and to plan what is to be achieved by the interview. The appraisal form also provides a record of the interview, since it should be filled in during the interview and then signed by both appraisee and appraiser as a record of what took place, indicating what decisions or promises were made and what objectives were set. There are advantages to using standard appraisal forms because they ensure that people are familiar with the topics addressed and it helps review and collation of data following appraisal. However, one has to counter this with some method to ensure that the appraisal form is not in itself seen as overly bureaucratic or indeed inappropriate to the situation. The form will ensure that the discussion covers all aspects of work but must be used in conjunction with good interview practice such as asking open-ended questions (see Chapter 12).

It is, of course, important that the appraisal form is not too complex or seen to detract from the most important part of the appraisal process which is the discussion between manager and subordinate.

9.5.4 The appraisal interview

The appraisal process will include at least one interview carried out between the appraisee and their first-line manager, the appraiser. The appraisal interview should have a defined agenda, whether based on an appraisal form or not, and this should concentrate on reviewing actual performance and discussion of required improvements and of future prospects. It must be a two-way discussion if it is to be effective, with the appraisee contributing significantly to the discussion. As stated previously, a record of the interview should be agreed and kept for reference throughout the period for review at the next interview.

The success of the appraisal interview relies upon the appraiser's interpersonal and interviewing skills to put the appraisee at ease, encourage discussion and ensure that all pertinent information is reviewed. Interview skills are discussed further in Chapter 12.

A basic agenda for an appraisal interview would include:

1. Review of performance
2. Discussion of improvements
3. Discussion of potential
4. Objective setting.

The review of performance since the last appraisal interview should review the targets set previously and include a discussion of any problems, such as the reasons for targets not having been met.

There should be a discussion of any required improvements or changes in the appraisee's work which have been identified since the last appraisal.

Discussing the potential of an employee is an important part of the appraisal process; it is equally important to establish whether or not the appraisee is interested in further career development. This is usually achieved by assessing the appraisee against dimensions considered important by the organization. In discussing potential the appraiser should examine what might be possible for the appraisee if given appropriate training or if allowed to develop certain themes. It is possible to use this discussion as a way of motivating people whose current performance appears unsatisfactory. The discussion should consider training and development. Training and development, as described in Section 9.6, are appropriate for those people who are willing to improve themselves and will be motivated by the training provision. However, substandard performance in those who are not motivated is unlikely to be rectified by training.

PERFORMANCE APPRAISAL

SECTION 1   TO BE COMPLETED BY THE INDIVIDUAL BEING APPRAISED

NAME...

JOB TITLE .... START DATE ....

PERIOD COVERED BY THIS APPRAISAL.. TO ....

TRAINING UNDERTAKEN DURING THE PERIOD OF THIS APPRAISAL

SECTION 2   TO BE COMPLETED BY THE APPRAISEE, PRIOR TO THE INTERVIEW. LIST THE MAIN DUTIES OF YOUR PRESENT JOB AND ANY OBJECTIVES WHICH WERE AGREED FOR THIS PERIOD.

....

....

....

....

SECTION 3   TO BE COMPLETED BY THE APPRAISER

ASPECTS OF PERFORMANCE:

MARK EACH ASPECT ACCORDING TO THE FOLLOWING SCALE. PLEASE ALSO COMMENT OR STATE 'NOT APPLICABLE'

Outstanding	A
Performance above requirements	B
Performance meets the requirements of the job	C
Performance not up to requirements	D
Unsatisfactory	E

a) Written communication   A B C D E

Comment

....

b) Oral communication   A B C D E

Comment

....

c) Technical knowledge and ability   A B C D E

Comment

....

d) Application of latest technical knowledge   A B C D E

Comment

....

e) Acceptance of responsibility   A B C D E

Comment

....

f) Ability to get on with others   A B C D E

Comment

....

g) Ability to produce constructive ideas   A B C D E

Comment

....

Figure 9.9   Appraisal form

h) Reliability under pressure   A B C D E
Comment
....

i) Timekeeping and attendance   A B C D E
Comment
....

j) Planning skills   A B C D E
Comment
....

k) Ability to meet targets   A B C D E
Comment
....

l) Analytical skills   A B C D E
Comment
....

m) Ability to foresee and avoid problems   A B C D E
Comment
....

SECTION 4 RECORD OF INTERVIEW TO BE COMPLETED BY APPRAISER
This should be an accurate record of the main points of the interview, drawing attention to
strengths and weaknesses and including a statement of any disagreements between the
interviewer and interviewee. State whether any of the identified weaknesses were
discussed during the period of this appraisal (i.e. before the interview).
....
....
....
....
....
....
....

SECTION 5 RECOMMENDATIONS
Give details of any recommendations that have arisen from the interview.
....
....
....
....
....
....
....
....

SECTION 6 AGREED RECORD OF THE INTERVIEW
a) Appraiser's Agreement
Signed .... Date
b) Appraisee's Agreement
Signed .... Date

Figure 9.9 (continued)

When carrying out this type of discussion it is important that the appraiser takes care not to promise what cannot be readily delivered, and also to be honest about the appraisee's potential. Everyone has their performance ceiling. Dishonesty and promises that are not kept can lead not only to demoralization of the appraisee but also to a lack of confidence in the appraiser's ability to manage effectively. Any action that is planned for the future should be carefully checked with all the other parties involved before any commitment is made, even if this means deferring part of the appraisal interview.

Setting objectives and targets for the period to the next appraisal is the most important aspect of appraisal from the management point of view. Objectives allow performance to be measured as well as providing the input so that plans for work can be prepared. An objective is a prediction about the state of affairs that will exist at the end of an activity. The more precise this prediction, the clearer you can be about what you have to do in order to make it happen. In the workplace objectives ensure that all resources are effectively employed in order to meet the company's goals. Objectives should be agreed by appraiser and appraisee for the next appraisal period. This can be very motivational, since the objectives will provide goals for the individual, and good objectives allow the opportunity for the employee to determine how the objectives will be met. Objectives can also be very demotivating if they are unrealistic or adequate resources are not provided. Objective setting is discussed further in Chapter 11.

9.5.5 Two-interview appraisals

Some appraisal systems also use a second interview where the appraisee has a discussion with a person at a higher level in the organization than their own manager, usually the manager's manager. This interview is not intended to duplicate the initial interview but to cover other aspects of the employee's job, in particular a discussion of the wider company objectives.

The use of the second appraiser helps to reduce favouritism or prejudice. This person will appraise many more people than an individual manager and will therefore be able to provide an overview of standards set and the way in which people are dealt with. It will also ensure that personality conflicts do not affect the objectives or standards set. This is important if employees feel that they are being unfairly treated because they do not get on with their manager.

The second appraiser ensures that a uniform approach is taken across an organization and allows the appraisee to air views more widely. This appraiser should not overrule decisions made in the initial interview and should always avoid criticizing the manager in order to show solidarity with the appraisee.

The operation of a two-tier appraisal system can absorb a significant expense and a company should be committed to the appraisal system and be aware of the costs that it will incur before adopting such an approach.

9.5.6 The implications of an appraisal system

The implementation of an appraisal scheme needs financial as well as personnel planning. There are costs involved with training the appraisers and administering the scheme, but these can be small compared to the time that is lost to other business activities while appraisals are taking place. Although one should not limit the interview time, so that it becomes more of form-filling exercise than a discussion, it should be remembered that each hour of an appraisal interview costs at least two person-hours of time. Therefore there must be some advantages to appraisal for the employer. These advantages are described below but, as you will see, they are largely intangible and it is difficult for a company planning to introduce an appraisal system to produce a numerical cost justification.

Advantages of appraisal for employers as follows:

1. It allows an objective assessment of an individual's performance to be made.
2. It provides a database of information concerning the personnel, their skills, and abilities, on which the company can draw. From this, it can lead to better and more effective use of staff by means of transfers, training and planned projects.
3. It can identify difficulties and potential problems so that they can be dealt with before they become major problems.
4. It can improve the performance of personnel through the use of objective setting and increased motivation.
5. It is an ideal tool for use within a management system that relies on objective setting to control and motivate the workforce.

The disadvantages to employers, apart from cost, relate to the way in which the appraisal is carried out. If it is used as a means of criticizing rather than developing personnel it can have an adverse effect on performance. The other problems that might arise are that the system may be seen by the employees as a control device rather than as a development procedure, and that any interview of this type will be affected by the manager/subordinate relationship.

Appraisal schemes are less appropriate, and are less widely used, for employees whose work situation rarely changes, such as machine operators.

9.5.7 Linking appraisal to pay review

Many companies link the appraisal interview to pay review. This can be very cost effective for the company as it means that only one process has to be administered. There are two other advantages for a company operating this system. It can provide a fair basis on which to divide the salary budget by awarding merit points for performance and giving salary increases on the basis of points achieved. In this way those who contribute the most get the best reward. In addition, it can increase employee motivation to reach company objectives.

However, there are disadvantages to linking appraisal and pay, and these are often deemed to outweigh the advantages. First, it will increase the amount of defensive behaviour and reluctance to admit faults on the part of the appraisee, and some people will not listen until they hear news of their new salary level. In addition, it can cause people to use the appraisal as an opportunity to justify why their salary should increase, and this can lead to conflict between the appraiser and appraisee, damaging the personal relationship and undermining the purpose of the appraisal.

9.6 Training and development

Well-managed departments have clear objectives. In order to meet these objectives a manager will require a battery of skills from the department's personnel. Since these objectives change with time, and because personnel come and go, there will frequently be times when the manager does not have the required skills and abilities within the department. The prudent manager has a constant eye for potential mismatches between the skills required and resources available, and aims to correct such a mismatch once it is identified.

There is a spectrum of response that the manager may use. At one extreme, the manager may do nothing and leave the individuals to cope on their own. At the other, new personnel are recruited who have the required skills. In between these two extremes lies the middle ground of development and training. In this approach existing personnel are encouraged to develop and acquire the desired skills.

It is important to to distinguish between development and training. Training presupposes that the desired skill is already within the capacity of the individual and that they only need to be shown how to do it. Development, however, involves preparation for tasks or behaviours that are currently beyond the individual's range of responses. A training objective might be 'to become competent in the operation of the production control computer system by the end of next week', while a development objective might be 'to acquire skills necessary to run the group of three sales engineers responsible for initial customer visits and after-sales commissioning within the next year'.

Landy and Trumbo have defined training as:

> Planned activities on the part of an organization to increase job knowledge and skills, or to modify the attitudes and social behaviour of its members in ways consistent with the goals of the organization and requirements of the job.

There are two prerequisites for successful training: intellectual capability and the desire to learn. One cannot train a mathematically inept individual to produce creative mathematical ideas, nor can one force knowledge into an unwilling pupil.

It should be realized that there are limits to how much a person may change as a result of training. Training cannot affect basic psychological attributes. An introvert will never be made an extrovert simply through training. However, an individual might go through such a change in their own time and as a result of their own personality development, but such a process cannot be fundamentally altered by training. An individual's value system is equally deeply rooted and not amenable to modification by training. Attempted modifications to such attributes comprise brainwashing and the results are not only unethical but unreliable when ultimately tested. Brainwashing techniques include such things as 'reinforcement', where a particular idea is restated in different ways on hundreds of occasions, or the so-called 'love bombing' where acceptance of the ideal being presented results in convincing affectionate behaviour from the 'trainers'. Sleep-deprivation is often used by filling the day and night with 'important' tasks that support the idea being conveyed. Also, the absorption of material by repetitive regurgitation such as chants or the incantation of secret verses can be effective. After prolonged exposure to such techniques the individual becomes less able to apply rational criticisms to the material presented and even the most resilient individuals adopt it. Brainwashing works by replacing the normal supply of stimulus from the world with a biased and one-sided view. The more completely the biased presentation replaces the real world, the more complete the brainwashing becomes. Some extremist religious sects use brainwashing techniques to engender beliefs that are inconsistent with reality. Individuals who return to normal society soon recover their rationality and reject the brainwash. It is not therefore surprising to find such sects reluctant to expose their members to normal society which tests their brainwashing and usually overturns it. While the isolation, and with it the illusion, persists loyalty, and usually money, are commanded by the sect.

9.6.1 Conducting a training programme

There are three phases to conducting a training programme:

1. *Identify the needs*: Using the organization's objectives and a summary of the given individual's abilities as starting points, generate a list of training needs for each individual in the form of objectives.
2. *Select and apply the appropriate training method*: There are many techniques with which people may be trained. Each is suited to particular circumstances. The most appropriate technique should be selected and applied. A number of techniques are described in Section 9.5.2.
3. *Monitor the effectiveness*: At this stage the manager assesses the response of the group to the training. This should be measured both by how the individuals are performing and by discussing it with the trainers. The manager will wish to know whether any further training is required, whether the individuals are now able to produce the required skills, and whether

value for money has been realized. There are implications for the monitoring system here. For instance, how will one measure whether value for money has been realized, and what criteria will the organization use to assess the 'before' and 'after' cases? Usually these questions should be answered in terms of the original problem which highlighted the need for training. Often the benefits of training take some time to be realized and are difficult to quantify. How does one quantify and measure an improvement in personal effectiveness, for example? The appraisal forms an excellent opportunity for the individual and the manager to examine the success of training, and it is common to have a review of training received as a standing item on the agenda of an appraisal.

9.6.2 Methods of training and developing personnel

There are many ways in which individual performance may be enhanced and they all offer scope for either development or training to take place. The way in which the material is presented, the content of the material and the abilities of the individuals attending indicate whether development or training is in progress. The most effective methods for enhancing performance are as follows:

1. On-the-job experience
2. Coaching
3. Roleplay
4. Study
5. Games, simulations and case studies
6. Internal training courses.

On-the-job experience
This type of training occurs all through the working day. The trainer is anyone else involved in the daily running of the department but is usually more experienced or more senior. It is training that results from the daily human contact of colleagues at work, but is a very effective teaching method often chosen quite deliberately to be the main learning method. For example, an apprentice potter is assigned to assist a master craftsman, or an engineer has a desk in the project office in which many other engineers of different levels are at work. Simply by being there, job knowledge is absorbed.

Training of this sort is extremely relevant and cheap. It affords ways for individuals to develop in their own environment. The training introduces the trainee to the whole of the job and hence offers a complete training system. The process does not disrupt the flow of work. Unfortunately, it is often necessary to choose trainers by position, rather than by their ability to teach. The need to perform the training in a working department makes providing a good learning experience difficult. If the trainers are coached to be effective teachers this very cheap training method is enormously successful.

As part of the training, individuals may be given particular projects within the normal tasks of the department. A collection of such projects is often a good way to introduce the individual to the operation of the department. Individual projects can also be used as training vehicles for larger, more risky projects that may follow.

Sometimes trainees need to visit and see a special process or situation. When dealing with the unfamiliar one can always be advised of how things are, but there is never a good substitute for seeing things yourself. A sales engineer joining a company may be shown the various departments, special processes, or operations of the organization. This is done even though the job does not directly involve such knowledge since benefits may be derived in terms of appreciation of what is involved and the ability to have knowledgeable conversations with customers. Simply telling the engineer is unlikely to provide the amount of appreciation and recall that even a brief visit gives. Visits and demonstrations can be arranged as part of on-the-job training.

Some types of on-the-job training can be very structured. For example, it is common for large organizations to have well-organized and lengthy training courses for the apprentices. Similarly, graduate engineers usually go through a very structured training programme to achieve professional status. For the individual concerned the quality of provision of such training can be a decisive factor when accepting or rejecting a job offer.

Coaching

This is usually used to instil particular skills, often of a physical nature. For example, coaching is often used for sporting or operating skills. It involves a coach who will take the students through the learning process and on to independent reproduction of the desired skill. Coaching would be used to teach an aircraft technician to perform an emergency shut-down of a jet engine at full power while standing beside the engine.

Coaching often involves teaching individuals to go against behaviours they currently hold. In general, the more deeply the skill or behaviour being coached conflicts with normal behaviour, the more coaching is required and the closer the relationship between coach and student needs to be. For example, a trainer will have to command a great deal of respect if required to persuade trainees to jump out of an aeroplane and be saved by a parachute they have packed!

Coaching is very effective at instilling skills and is of particular relevance when personal safety or credibility are threatened. It can seem expensive but the cost of the coaching should be compared with the cost of errors made by the untrained. When this is done the suitability of expensive coaching is often revealed. It is one of the few ways that behaviour in unusual or undesirable situations may be made automatic. In all cases of coaching the importance of excellence on the part of the coach cannot be overstated. Coaching is a skill in its own right and there is no point using inappropriate individuals. Two special cases of coaching are worth noting: mentors and shadowing.

Mentor training is simply one-to-one coaching. It requires an experienced coach who follows the trainee almost everywhere during the period of training. The method is particularly appropriate where the material to be taught is hard for the trainee to learn. The technique is very dependent on the mentor. It is clearly intrinsically expensive. For example, a trainee marketing executive might be mentored by a qualified training consultant who takes notes as the day progresses and discusses these with the trainee at the end of each day. The consultant will also deliver short bursts of training at times when important examples have just occurred. The mentor system is able to deliver one of the most specific and appropriate training systems of all. When given by a competent mentor it is a very effective training method but it has a price tag to match.

Shadowing still involves one person learning from another but does not use the close teaching relationship of the mentor system. Shadowing means learning the job content of another by following them through a typical period of work, perhaps a day or two, or even a month. The shadow does not interfere with the person being shadowed but observes all that goes on and asks questions when required to clarify confusion. It is appropriate for gaining understanding of job content but not particularly good for acquisition of skills, on-the-job training is used for this. It is especially appropriate to train individuals where there are many other departments about which understanding is required.

Roleplay

In roleplay the trainees use their ability to see things from another's point of view and learn from the experience of playing another person. The trainee takes on a role which is provided and a scenario is played out. Since this is done away from real life, mistakes do not matter and learning can be made less risky. For example, a graduate engineer in a company might be sent on a business skills training course and roleplays a telephone conversation with a difficult customer. The role play is videoed and the small group watch afterwards and debrief the process. Another group member then tackles a slightly different situation using the skills learnt. After the experience the individuals may modify their behaviour and improve their effectiveness.

The process usually involves two or three people roleplaying in accordance with instructions. One or more members of the group are appointed to record what happens and try to see the situation from all sides. A video camera may also be used. The technique is particularly appropriate to interpersonal skills training since it revolves around feelings and how to deal with them. It is not appropriate to try to use the technique for factual tuition. One of the limitations of the technique is that not everyone is good at seeing things from another person's point of view and so may find the process difficult. It is also worth noting that when difficult scenarios are presented they may unknowingly upset trainees by dwelling on sensitive areas. An example might be someone recently widowed being sent on an appraisal training course and having to roleplay someone whose work has suddenly deteriorated because of a bereavement.

Roleplay offers the only real way to study and develop personal skills in a safe environment. The trainees themselves, however, can limit the effectiveness of the technique. Any individual who cannot act at all, or who finds it difficult to assume the role of another person in front of others, is not likely to learn much because they simply cannot make the technique work. Interestingly, it is often exactly this sort of person who has the most to gain. The lack of this skill can itself be the source of the poor interpersonal skills the training is seeking to rectify. Consequently, those who have the most to benefit from the method can be the least able to do so.

The low cost, wide availability and instant playback attributes of video cameras and recorders mean that they are extensively used for roleplay purposes. Seeing yourself on video is usually both surprising and informative, although for some people it can also be embarrassing. The trainees first perform the scenario and are filmed by the trainer, and then view themselves. A trainer goes through the video making observations on specific relevant points and so debriefs the replay. Seeing yourself on video allows an individual access to a record of performance that cannot be questioned. Other means of providing feedback depend on the observations of others and therefore lack the objectivity of the camera. People are often amazed by the sound of their own voice, which is very different when heard as others hear it. People become aware of many annoying idiosyncrasies or habits and some can be demoralized by the experience. For this reason, care should be taken with debriefing and it needs to be handled sensitively if it is to facilitate learning. Feedback should never be used in such a way that individuals are embarrassed by their performance. Videos can be particularly effectively used in 'before' and 'after' scenarios to make clear the improvements that the training has brought.

Study

Study is especially appropriate to the personal acquisition of factual knowledge. It is not, however efficient for the acquisition of practical skills. When you need to learn about the history of France you reach for a textbook, if you want to juggle you ask a juggler to coach you.

During study the individual often acquires the material alone, through books or correspondence courses, and is in complete control of the learning activity. If the study is in the interests of the organization the commitment may be supported by study leave or use of company resources. The study might take the form of formal higher education courses or less structured courses at other institutions. Occasionally the trainee will be given sabbatical leave to complete the course.

In some kinds of study, as in a formal higher education course, it should be remembered that an individual gains not only the factual body of knowledge from the study but also the methodology of thought associated with it; this indirect asset is hard to value. For example, many computer companies recruit extensively among classics graduates. This is clearly not because of their computing skills but because their systems of thought and ability to analyze options are useful to computing organizations.

Formal study within a structured syllabus gives high-quality knowledge. Formal qualifications are portable and an organization may find a good portion of its trainees marketing their skills elsewhere. However, low cost, training outside of working hours, and wide applicability combine to make the option attractive. Private study might be used as a route for a marketing manager to gain a Masters of Business Administration or an organization might offer a distance-learning course in German to an export clerk.

Studying for a formal qualification while working demands a high level of motivation on the part of the student. The pressures caused by formal study on top of a full-time job are not to be taken lightly by either side. Such a life style leaves little recreation time for the student and the long study hours can reduce the student's on-site performance, especially at exam time.

Games, simulations and case studies

These are usually used within other training material such as books or courses to illustrate particular points. Training games usually take the form of competition between teams who have some aim and must employ the tactics being taught in order to win. Often the games are based on imaginary companies engaged in competition and are similar to popular board games.

In simulations, the trainees are presented with situations in which they must devise strategies to solve the problem or achieve the aim. The simulation allows the trainees to practise in a safe environment away from the dangers of real life until they are confident of success. The simulators can vary greatly. Flight simulators are examples of how practical skills can be acquired in a situation so lifelike that the trainees are coached as if they really are in an aeroplane.

Case studies provide examples or illustrations of the issue being taught and are sometimes used in a purely illustrative way. Often they may achieve their instruction through involved and detailed analysis of a lengthy case history. Case studies provide the trainees with access to material just as it was faced by those involved at the time and so brings them much nearer to the real world.

Internal training courses

It is quite common for organizations to run their own training courses. The larger the organization, the more likely it is to make economic sense.

Internal training courses may be tailored to the individual organization and can provide a good vehicle for learning about the company's culture. On the other hand, they take a considerable amount of organizing and there is the danger that they will become outdated and stale compared with external consultants, whose constant exposure to the marketplace keeps them up to date.

Organization-specific training usually takes the employee off-site for a period of time, usually between one and ten days. Often the courses are residential and they are typified by long hours of intense study and learning. A great team spirit can develop on such courses and the absence of external factors leads to a sustained and intense atmosphere. Sometimes television, newspapers and incoming telephone calls are banned. The learning process is very rapid in such situations. Provided the courses are well constructed and relevant, they are very effective. Often such courses, especially factual ones, are supported by good documentation which can become an invaluable reference.

The trainee quickly gains highly relevant experience. The opportunity to develop a good working relationship is provided. It brings the disadvantage that the trainee is not available in his or her own department for normal duties and that colleagues may be resentful of the training and consequent career advantage gained. In addition, such courses are often expensive, especially when high-quality hotels or conference centres are used, and fatigue can interfere with the learning process.

We have now looked at six methods of training and development. The education of a professional person is something that continues throughout life. A qualified engineer can expect to experience most, if not all, of the above techniques during an engineering career. By understanding the different techniques and having a knowledge of them it is possible, not only to select the best method but also to get the most from it. Such improvements in training efficiency are clearly of interest to successful engineers.

9.7 Job design and payment systems

Employment is a relationship between an organization and an individual. The individual provides services and in return receives two things: first, payment of some form, and second, job satisfaction. Money is clearly an important part of the relationship but it is not all of it. No-one is truly happy while engaged in work they find uninteresting. The simple fact that not all vocations receive the same remuneration yet people still compete for jobs even in the poorly paid ones, shows that people attach great importance to the nature of the job as well as to the remuneration associated with it.

As engineers we may be called upon to assist or even take responsibility for the creation of new jobs. This might come about through a need to expand a department in which we work or because we have too much work ourselves and need assistance in the form of a junior. If we are to do this important job well and attract committed applicants for selection we will have to understand something of how a job may be designed to be motivating and how to arrive at a figure for the remuneration that the job should carry. In some organizations, particularly larger ones, there may be a department with just this responsibility; alternatively, it may be the responsibility of the individuals concerned. Either way, it is important that as professional engineers we have an understanding of what the process involves. We will now look at two subjects which address these issues. 'Job design' deals with engendering job satisfaction and 'payment systems' with the provision of fair pay or remuneration.

9.7.1 Job design

Job design aims to make a given job required by the organization motivating to the employee while retaining the benefits it brings to the organization. In the first half of this century attitudes towards employees were very authoritarian. Employees were seen as obstinate and lazy, people who had to be forced to work. As time went on, it was realized that this attitude was quite wrong and that people are highly motivated if given responsibility for their work.

Modern job design takes note of these factors and the needs of the organization are seen in conjunction with the needs of the individual. Designing a job well means finding a happy marriage between the needs of the organization and the motivation of the individual so that both benefit from the committed attitude of the other. This task may involve the inclusion of obvious motivational factors such as developing responsibility. It may also include the avoidance of demotivating factors. No-one likes to be given inappropriate tools or insufficient facilities for a job and job design is a means of ensuring that this does not occur.

There are many aspects of job design and below we will consider four particularly important ones: (1) ergonomics, (2) work study, (3) consistency, and (4) job specification. The first two, ergonomics and work study, are large subjects in their own right. A trip to the engineering section of any reputable library will soon uncover many texts on these subjects. In this section we are considering job design as a part of personnel management. We have therefore referenced ergonomics and work study since these subjects impinge greatly on the design of jobs. However, the sections should be considered only as a brief introduction to large subjects that might need further investigation if much job design is in hand. The second two aspects of job design, consistency and job specification, also have a great impact upon job design. These subjects, however, are very dependent upon the nature of the job in question and the circumstances of the organization. For this reason, the important issues are raised but these may only be addressed by considering them in the context of the organization.

Ergonomics

Ergonomics is the study of the interface between humans and machines. The aim is to make this interface an efficient one and allow the machine to be operated at speeds limited by the operator, rather than the interface. If we are designing jobs indirectly, that is, by designing equipment that will be used while doing a job or designing an environment in which a job will be carried out, a knowledge of how best to design the interface between humans and machines is desirable.

The design of control consoles or instrument panels for aircraft may draw heavily on the concepts of ergonomics. For any given situation the ergonomic approach is first, to understand the way in which the operator acquires information and gives instructions or moves controls, second, to characterize this exchange, and third, to design an interface that permits optimal exchange.

Different situations have different needs. A workstation for an assembly worker may require speed to be maximized. Alternatively, accuracy may be paramount as with the control room of a power station, or a balance of both may be important as in the case of a jet fighter control panel.

To be successful in such a task the ergonomist must have a thorough understanding of the human body's capabilities. Data relating to humans is called anthropometric data. Books on this can readily be obtained and it is usual to display the various data in percentiles. For example, a diagram showing the typical reach of the human arm will have the dimensions for the mean shown and will also have a table to show how the dimensions change for each percentile of the population.

The environment also affects the performance of the interface and so ergonomics is also concerned with optimal levels of lighting, shift lengths, working temperatures and so on. Again, many standard reference tables of such data have been prepared.

Work study

Work study means exactly what it says – studying work. The aim is to produce the optimal methods of performing a given job. This involves consideration of the work and the machines or other resources involved. There are clear benefits for the organization in doing this well but it is also motivational for the worker. It is certainly satisfying to know that one has the best method and tools for the job in hand.

The study can take place on any scale. For example, in the assembly of a watch the work study engineer will consider things such as the distance of hand travel in picking up the components and the optimal balance of the tools used. The engineer might weigh up whether two tools should be used for a particular task or if efficiency would be improved by the use of one slightly heavier tool which could do both tasks. Alternatively, in a shipyard the engineer might consider the distance hull plates have to be carried, how many hot rivet stations are required, or how many times a particular point has to be visited and so on. These allow the preparation of optimal work schedules.

Work study has many areas which are often considered separately. Method study, motion study, and time measurement are three important components. Frank B. Gilbreth did much work in these areas and is often attributed with being the founder of work study as a discipline in its own right. His wife was also a considerable force in work study and they jointly ran Gilbreth Inc. between 1910 and 1924. The company was hired to apply their scientific examination of the best way to complete specified jobs by many prestigious industrial organizations. While conducting research into work study Gilbreth even gave his name (in reverse, that is) to a symbol he invented. The 'Therblig' is a small symbol used to represent a specific body movement when writing down a particular task as a series of component motions. Once proficient in the 'language' of Therbligs one can write down complex operations with a view to manipulating them and finding improved ways of accomplishing the task.

The unique thing Gilbreth brought to his subject was his unwavering dedication to identifying the absolutely optimal method of performing even the simplest task. He was very clear and objective about his work and his attitudes extended into his private life. A novel, *Cheaper by the Dozen*, written by one of his twelve children, gives a humorous insight into the family life of the Gilbreths and does much to illustrate Gilbreth's own obsession with his subject.

Consistency

We have already made various observations on motivation in Section 9.4.1. These need to be considered during the process of job design and in this section we shall not repeat the ideas but rather will make some observations on their application to job design.

Although jobs need to be individually designed, a job designer must also consider consistency across the organization. One should not, for example, combat low pay in a company by introducing a few jobs with initial inappropriate remuneration. Attempting to rectify the situation in such a way will, in reality, have a demotivating effect on those whose pay is left behind. Individuals will naturally look at the fortunes of colleagues as well as themselves when judging issues like fair salaries or recognition. It is therefore vital to ensure that the design of all the jobs is consistent.

Consistency must also be maintained over a period of time. As skills develop an individual must be able to move on to new tasks which assist in the quest for self-fulfilment. During this process the job designer must avoid the pitfall of failing to acknowledge advancing abilities and responsibilities. For instance, there is no point in increasing the motivational aspects of responsibility and authority without raising remuneration to a level consistent with the new job.

Job specification

To specify a job completely one needs to include all the aspects that impinge upon the job. Such a specification is of use both to the designers of jobs and also to applicants who are interested to know as exhaustively as possible the nature of the job. Section 9.2 contains much detail about the process of employing people. Times change, and as circumstances move on, the job description and job scenario that used to apply may also have to change. Jobs do change gently and it is important to monitor the changes to be certain that the job actually being conducted, as opposed to the original description which may be out of date, is still of optimum design from the employment relationship point of view. The following four dimensions may be used to define any given job:

1. *Job descriptions and contract of employment*: These two documents have previously been discussed in Section 9.3, and are referenced in this list since

they are clearly documents that specify what a job entails. Of all the items included within a complete job specification, the job description is often the only one a job applicant is clear about before accepting a job.

2. *Standard codes of practice for the organization*: Many organizations have their own guidelines and codes of practice that affect jobs within them. Some organizations will have particular safety codes that have to be adhered to at all times. Others may have company meetings or regular departmental awaydays. Such activities can shape the way communications occur and can affect jobs in all sorts of ways. For instance, in some companies the normal practice is to answer a ringing telephone within three rings, regardless of what you are doing. Such a practice is clearly harder for some people to endure than others, and therefore has a bearing on the nature of the job.

3. *Law*: Organizations that provide more than the legal minimum on issues such as redundancy benefit or pension arrangements will attract loyalty in a way that those who begrudge their employees even the legal minimum will not. The law also makes a requirement to employ people in non-discriminatory ways and it is illegal (except for certain special cases) to stipulate, for example, the sex of an employee in an advertisement. Section 9.3.4 explains such issues. The whole issue of employment law is complex and beyond the scope of this engineering management textbook but a search of any library will soon produce a list of titles on this topic. However, it is clear from this paragraph that the law has the potential to affect the specification of a given job. It is therefore the responsibility of a job designer to be familiar with the law as it applies to the particular industry and to accommodate it when designing jobs.

4. *Organizational culture*: The culture of an organization can have a profound effect on a job. The culture tends not to operate in writing and is much more a socially controlled issue. For example, in the contract of employment for many professional jobs one will usually find a normal working week of less than 40 hours specified. In some organizations, however, it is common to find employees working far longer than this stipulated minimum. This extended working week is soon taken as normal, and those who work less are seen as uncommitted and are less likely to receive promotion. Unless such factors are accounted for and included in a job specification the resulting picture of the job is inaccurate.

9.7.2 Payment systems

Remuneration is the name given to the package of benefits that an employee gains from their work. It does not normally consist only of pay and the amount of other benefits included depends on the nature of the job in question. Such additional benefits may add up to as much as 20% or 30% of the basic wage and form an important contribution to the employee's standard of living. For example, the employee might receive:

Productivity bonuses	Share issues
Personal insurance	Pensions
Health care cover	Christmas hampers
Paid holiday entitlement	Book allowances
Subsidized meals	Reduced prices on company goods
Subsidized social club	Free sponsorship tickets
Company car	Expense account
Company petrol	Sports facilities
Crêche facilities	

This list is by no means exhaustive but clearly shows that remuneration means more than a basic wage. The organization will usually have its own policy concerning the allocation of such benefits. For instance, it is common for all employees to have health care cover but only a few to have a company car. In most cases the organization assigns a monetary value to each element of the remuneration package and the wage is calculated by subtracting the value of the other benefits from the monetary value assigned to the job. Of course, different individuals may rate the value of the benefits differently. For example, a car will be valued much more highly by someone who has a long way to drive to work than by a local employee who rarely leaves town.

The central aim of a payment system is to define the way in which the employees of an organization will be remunerated. There are two extreme views that may be taken with regard to such a system.

In the first, everyone receives the same wage. This rather ideological system has been tried but never seems to survive. Such systems are soon undermined by the inevitable movement of personnel. The senior people leave the organization to earn more elsewhere and the resultant skill imbalance ruins the organization.

The other approach is to say that everyone should receive remuneration in proportion to their contribution to the organization's success. This proposal is normally accepted as the most equitable method and many organizations freely state it as their aim to remunerate employees in this way.

The central question is how to rate the jobs within an organization. For example, how does one decide the relative importance of a draughtsperson and a managing director? Described below are three ways in which this may be achieved.

External comparison
A comparison is made with other jobs available in other organizations. This is often more difficult to do than it seems at first since companies are, in general, reluctant to disclose details of their own payment systems. Salary surveys are

regularly performed by independent journals and the engineering institutions, and these can help in providing raw data. However, they may not consider the value of benefits other than salary.

Job ranking

For a payment system to be fair the relative value of each job must be known. A list of the jobs within the organization is compiled which reflects the importance that the organization places upon each one. Since the jobs are in order it follows that the remunerations should also be in order and a glance down the column of salaries will easily reveal any anomalies.

Job assessment

A comparison is made of the content of all the jobs and their specifications within the company against the people who fill the posts. This is done to ensure that there are no inconsistencies within the organization. People of a certain standard should have similar levels of competence and remuneration across the organization. Nothing is more likely to cause trouble than having two people doing the same work and receiving different levels of remuneration. To ensure that employees are fairly rewarded, personnel assessments may be conducted. In this process personnel are assessed by their qualifications and skills and given a ranking relative to others. The more highly qualified and skilled the employee, the higher the remuneration should be.

Once these investigations are complete, the organization has a list of the jobs in order of importance and various benchmarks of external comparison against which to judge the level of pay. There is both relative and absolute information, and so a remuneration value for each employee can be defined. Two commonly used systems for choosing the final numerical value of a salary are described below.

Graded systems

The jobs are divided into 'grades' and within each grade there are various points. Each grade corresponds to a given level of seniority and there is a considerable difference in the job description of people between grades. To each point a salary is assigned. Everyone in the organization has a salary somewhere on the chart. An example is shown in Table 9.1.

Changes in salary are then brought about by three separate means. The first is through an annual increase of the number associated with every point. In this way the salaries of the company may be moved as a whole to allow for general economic effects such as inflation or recession. Second, an individual may be moved up the points ladder within the grade in response to good work. This may be done following an appraisal or simply after a given length of service, often each year. Third, the salary may be increased through promotion, which involves moving on to the next grade up and into a more senior job altogether.

Table 9.1 Salary structure for Tic-Toc Co. Ltd

Point	Grade 1	Grade 2	Grade 3	Grade 4	Grade 5
1	10 500	14 500	17 900	21 780	25 780
2	9 300	13 100	16 100	19 964	24 124
3	8 700	12 400	15 200	19 056	23 296
4	8 100	11 700	14 300	18 148	22 468
5	7 500	11 000	13 400	17 240	21 640
6	6 900	10 300	12 500	16 332	20 812
7	6 300	9 600	11 600	15 424	19 984
8	5 700	8 900	10 700	14 516	19 156
9	5 100	8 200	9 800	13 608	18 328
10	4 500	7 500	8 900	12 700	17 500
Increment	600	700	900	908	828

Free position systems

In companies using free position systems individuals are permitted to take on any salary level the organization sees fit. The advantage is that it permits people with quite similar jobs to be paid very different salaries if their performance merits it, since there is no scale against which everyone compares themselves. The disadvantages are that employees can feel that the organization is paying the least it can get away with in each case and the more secret approach leads to increased salary speculation. Payment systems must be fair and be seen to be fair. The system does allow considerable flexibility, and provided the employees feel that it is indeed the greater contributors that receive the greater rewards the system will endure. The system makes extra demands on the appraisal system if this is used to assess performance and therefore salary increases.

9.8 Summary

We started this chapter by looking at the case of Hilary Briggs, a Logistics Director in the Rover Group. We saw that in a few years since graduating Hilary now has a very considerable personnel management task. She has five employees reporting directly to her but manages a department of 365, and she has to motivate her staff, appraise their performance, and manage development and training for them. In addition, as with all companies, the staff have to be recruited and paid.

In this chapter we have considered each of these aspects of personnel management and at the ways in which businesses can be organized. We saw that the greatest influences over the method of organization are the type of product and the manufacturing system. We have also seen that in order to employ someone there are a number of legislative and economic factors that have to be taken into account, and we saw that there was a process that had to be followed of recruitment, selection and induction. We then considered how you might encourage your staff to work to their best and some of the ideas put forward by people such as Maslow and Hunt and how you can use these to determine motivation factors for individuals. Appraisal systems were reviewed in Section 9.5 and we saw that the process of appraisal, while costly, can have great benefit for both employer and employee. One of the main reasons for appraisal is to identify potential and determine training and development needs of employees. In Section 9.5 we discussed the training methods that could be used to meet these needs. Finally in Section 9.7 we looked again at some factors that affect motivation, but from the viewpoint of job design and methods of remunerating people for their work.

Personnel management is obviously an enormous topic and many people specialize in particular aspects of it, such as motivation or recruitment. As engineers you are unlikely to have to follow a specialized route in this area, but you will undoubtedly have a need for a general understanding of the topics. This text has provided an overview of those topics and in the Further reading we have directed you to some more detailed texts.

9.8 Revision questions

1. How can companies using a free position system for determining the level of pay of its managers ensure that it still complies with the Equal Pay Act 1970?

2. (a) List the main features that you would expect to see in a contract of employment.
   (b) What is the legal position with regard to the preparation and issue of job descriptions?
   (c) What role do job descriptions have in terms of personnel management?

3. A friend at work, also a graduate engineer, has asked for your advice on how she might resolve what she feels is a clear case of sex discrimination. You both work in an office with 15 other graduate engineers and she is the only woman.

   The company that you work for publishes no information on pay scales, and all pay negotiations being done on an individual basis. Generally no-one discusses their pay, but two weeks ago your friend overheard a colleague talking to someone about pay in the pub and she had the impression that this person was being paid more than she was. She feels that this is particularly unfair because (1) she has been in the company longer than the other person and (2) she feels that she has a much more responsible position than he has, as she works on a much more expensive project.

   During the last two weeks she has thought of nothing else and is now also convinced that she has been discriminated against in other ways. For example, she was not given the supervisory job that she had expected. She is now determined to walk into the manager's office and hand in her resignation, having made her views on the situation clear.

You had noticed that she had seemed particularly upset today and when you asked what was happening she told you the story and finally asked 'What would you do?' What advice would you give taking into account both legal and personnel implications?

4. As a personnel manager consider the legal and personnel aspects of the following two cases. How would you deal with them?
   (a) An engineering apprentice has a history of using bad language at supervisors. He was verbally reprimanded for this three months ago and until this week there had been no recurrence. Yesterday he was asked by his supervisor to do a particular job and responded with a stream of abuse which culminated in his walking out of the factory. He has not returned.
   (b) Due to the introduction of new machining centres your company wants to make a fifth of the people in the machine shop redundant. There are currently 50 people working there.

5. What is the role of a job description for (i) employer and (ii) employee?

6. Why should personnel specifications be prepared when planning recruitment?

7. Outline the process of finding and taking on a new employee.

8. What costs are involved in recruitment and selection?

9. What options are open to a company that fails to recruit a suitable candidate to a key managerial post?

10. What is the purpose of the contract of employment?

11. For what reasons can an employee be legally dismissed?

12. Under what circumstances can a company lawfully discriminate when recruiting?

13. What do you understand by the term 'motivation'?

14. What motivates you now, and how do you think this will change over the next five years?

15. How can a manager find out what motivates his or her subordinates?

16. How does leadership affect motivation?

17. How can a leadership style be developed?

18. What benefit does the manager gain by having a knowledge of the 'motivation theories'?

19. What are the benefits of an appraisal scheme for an employee?

20. What costs are involved in operating an appraisal scheme and how can these be justified?

21. What is the purpose of an appraisal form?

22. What are the requirements for carrying out an effective appraisal interview?

23. Why is brainwashing not appropriate to training in the workplace?

24. Difficulties can arise when trying to monitor the effectiveness of training. What are these, and how can they be overcome?

25. Management training makes much use of roleplay. What benefits can be achieved from this training technique rather than study?

26. On-the-job training is widely used in industry. What are the requirements for this to be successful?

27. What advantages can a company gain from using on-the-job training rather than any other method?

28. What are the aims of job design?

29. How can a knowledge of job design benefit a design engineer?

30. If a company has an unfair allocation of responsibilities between its managers how could it determine a fairer apportionment?

31. What are the advantages for (i) the company and (ii) the employee of using a graded payment scheme rather than a free position one?

32. What are the effects of using a single graded payment scheme in a company that operates on a number of sites nationally?

33. What factors affect the way a company is organized? How might the emphasis on each factor change as a company grows?

References and further reading

Bentley, T. J. (1991), *The Business of Training*, McGraw-Hill (training series). This book discusses the role of training and its impact on business. It looks at the need for training and the way in which training strategy can be developed, and training designed to meet those needs.

Gilbreth, F. B. and Gilbreth, Carey, E. (1949), *Cheaper By The Dozen*, Pan Books Ltd.

Handy, Charles B. (1985), *Understanding Organizations*, 3rd edition, Penguin Business.

Herzberg, F., Mausner, B. and Snyderman, B. B. (1959), *The Motivation to Work*, John Wiley.

Hunt, J. W. (1986), *Managing People at Work – a manager's guide to behaviour in organizations*, McGraw-Hill. This is an excellent book which is easy to read. It fully explains Hunt's theories on the use of goals as motivators in such a way that their practical application can be understood even by people without industrial experience.

Landy, F. J. and Trumbo, D. A. (1976), *Psychology of Work Behaviour*, Dorsey Press, page 222.

Maslow, A. (1954), *Motivation and Personality*, Harper & Row.

McGregor, D. (1960), *The Human Side of Enterprise*, McGraw-Hill.

Sidney, E. (ed.) (1988), *Managing Recruitment*, Gower. This book is a compilation of articles from different contributors covering all aspects of recruitment and selection. In particular, it discusses techniques for recruitment of special categories of people such as graduates.

Tannenbaum, R. and Schmidt, W. (March/April 1958), 'How to choose a leadership pattern', *Harvard Business Review*.

Torrington, D. and Chapman, J. (1983), *Personnel Management*, Prentice Hall International. This is a comprehensive text covering all aspects of personnel management. It is recommended by the Institute of Personnel Management and is an invaluable reference.

10

Team working and creativity

Overview

In the first chapter we introduced engineering management and explained the skills that the engineering profession demands of its members. One of these skills is the ability to work in teams. Successful teams greatly extend the abilities of the team members and allow the team to achieve far more together than they can singly. The ability to work successfully together in teams is a prerequisite of creative engineering. We do not work in isolation, we are all in teams of some sort. In this chapter we shall take a detailed look at the important science of team working.

10.1 Introduction

The history of science and engineering is filled with great advances made by brilliant individuals. We all know of Newton's laws of motion, Stephenson's *Rocket*, or Brunel's civil engineering achievements. These pioneers seem always to be from the past. Why is this? It clearly is not because progress has slowed down. In fact quite the reverse is true, progress has never been faster than it is today. The reason for an absence of such individual pioneers is simple. It is no longer possible for individuals to accomplish the vast intellectual challenge that is associated with even a small advance today. People cannot do it alone, but they can do it in teams. Teams of scientists unravel the structure of matter. Teams of engineers design and manufacture airliners and teams of professionals bring together the many skills that are necessary to introduce new products like compact discs to our markets. As engineers, we are clearly interested in studying this productive cooperation and in this chapter we will examine its operation.

The chapter is divided into three main sections. In Section 10.2 we shall examine how and why teams work. We will start by looking at the problems and advantages that team working brings. We shall then investigate team working from a theoretical point of view and introduce a particular theory together with the experimental evidence that supports it. Section 10.2.3 will examine team working from a practical point of view and make clear some of the limitations that apply to team theory in practice.

In Section 10.3 we will introduce the subject of group dynamics which describes the actions and results of the interpersonal forces that exist within any collaborating group of people. We will then consider the needs of a group. In Section 10.3.1 we will see how various human behaviours meet these needs and so make humans natural team workers. Such behaviours are the dimensions of group dynamics.

In Section 10.4 we will introduce the most important techniques for managing a process that is often thought of as being unmanageable. Teams are often formed to solve problems and so need to be creative. Some teams are much more successful at this than others and in this section we examine what actions may be taken to improve the effectiveness of the creative process. The section starts with 'Planning Innovation' which introduces some techniques for planning tasks whose progress depends on the unknown outcomes of future work. The next area considered is 'Problem Solving'. This section includes material on brainstorming and lateral thinking. Finally, in 'Decision Making' we investigate and explain how good decisions are made, and end with a look at how the team issues of Sections 10.2 and 10.3 apply to decision making.

None of the issues above can operate effectively in isolation. Many of the management and psychological principles in other sections of this book are therefore relevant to this chapter. Section 11.3 is of particular importance in ensuring that the team is directing its efforts to the actual task and not wandering off the point. All aspects of Chapter 9 apply to the management of, and participation in, teams as well as to individuals interacting with others to solve problems. Chapter 4 and, in particular, Section 4.5 are relevant to team work where financial autonomy has been delegated to the team along with task-oriented objectives. Chapter 8 is again important for teams that are autonomous and are therefore responsible for achieving their objectives without assistance. Finally, Chapter 12 applies to team working just as to all management issues.

It might seem from the above that practically everything that applies to the management of an organization is relevant to a team and, in general, this is true. It is not surprising, however, when one considers that a team is an organization within an organization.

The learning objectives of this chapter are:

1. To examine the benefits that team working brings
2. To review some theories of team composition and to see how these theories may apply in practice
3. To understand group dynamics
4. To present some contemporary problem-solving ideas
5. To understand how the creative process may be planned
6. To understand the various aspects of decision making
7. To explain several decision-making techniques.

10.2 Team working

For us, team working is a life-long habit. We enter this world in the team of our family. We are schooled in the team of our classmates. We choose our own team of friends and we join teams for sport and recreation. Some teams last a long time, others do not. Team composition is often beyond our control. In employment, we choose a team with a common interest but have little choice over our team mates. Most adults have been members of many teams and know well the fortunes and dangers of life in teams. In becoming parents we rejoin our first team in a new role and pave the way for another generation to live their lives in a world, filled with teams.

10.2.1 Holistic teams

The adjective 'holistic' describes an entity whose whole is greater than the sum of its parts. We are all familiar with the concept applied, for instance, to colonies of termite ants. These colonies are among the oldest of all life forms. By good management, the colonies are capable of far exceeding the puny abilities of the individuals. This example from nature illustrates vividly the benefits that holistic teams can bring.

In life, we are used to seeing teams of people whose group abilities amount to more than the sum of their individual ones. When we observe this, however, we must be careful to distinguish between the simple numerical advantage that team working brings and true holism. The ability to lift heavier objects or perform more operations simply by having more people is an example of the former. The group's acquisition of new abilities, that the individuals could not have provided alone, is the latter. A simple holistic example might be a song-writing duo in which one member writes the lyrics and the other the words. Separately they are nothing, but together they are everything. Another example comes from a case study of engineering students.

Case study – holistic teams

Some final year undergraduates had to present a design to their colleagues after a term's work of preparation. One team had to design a rubber band powered buggy to negotiate an obstacle course defined at the start of the term. One of team members describes the work:

> Our team worked really well together. In the end we came second and it wasn't really very difficult. It wasn't even that we were particularly good at design or anything, we just seemed to find it easy. Joe was better at ideas than the rest so he did most of the actual design, I suppose. But then Angela did loads of reading and chasing things up; I hated that bit of the project so I was really pleased when she volunteered to do it. Dave was good too, he isn't

really good at anything in particular but he did make sure we all had our say and knew what we were all going to do. He gets on with everyone. In a way he was probably the boss but it didn't really feel like that and he never got bad-tempered or anything like that. Then there was me, I like pulling things together, I made up the report from the information the others gave me. I had to chase them up over a few loose ends and get them to explain what some of it meant but that served to tie up all the loose ends. We really enjoyed it, our marks were high and the buggy worked. It was great!

Of course, not all teams are so productive. Most of us can think of examples when we have been in teams that just didn't work together. Perhaps a group of friends who went off on holiday and things didn't go too smoothly; perhaps a group at work charged with some special task which never quite got done; or perhaps a meeting or committee in which many people gathered, much was said, little came of it and nothing was done. Most people can think of such cases. There are many reasons why a meeting can be ineffective. Section 12.6 explains the mechanics of meetings, but even when everything has been properly prepared, group meetings can still go wrong if the team does not have the capacity for holistic behaviour. It is easy to look at the above case study and see how it could have been different, how the members might have fitted less well together and how if any one of the tasks performed had been done badly, or there had not been someone to do it well, the whole thing would have gone wrong.

The potential benefits of being able to put good teams together are enormous. Managers need to know how to bring about the benefits of holistic teams and how to avoid forming ineffective teams. We will now examine team composition in theory and then make some observations on it in practice.

10.2.2 Optimizing team composition – theory

The ultimate question for a theorist in team work is 'how can one form the best possible team?' Since good teams must have good people in them, one might try to make a good team by choosing someone good at each of the roles the team needs. For instance, to form a winning football team, simply select a star player for each position. No-one with any experience of team games would actually try such an approach since it takes no account of how the team members would operate together. Potentially great football teams composed of star players at each position occasionally do perform well but they are by no means a guarantee of success.

A good team theory must therefore describe not only how to select individuals but also how to select groups that will work effectively together. This process is called 'team balancing'.

Team theory
There is one team theory which has the characteristics we are seeking. It has been experimentally proven and it does provide predictive results.

The celebrated theory is the work of R. Meredith Belbin and is described in his writings. What follows below is a brief summary of Belbin's work, together with some direct quotations from it. His description of the eight team roles he identified is reproduced, together with the questionnaire he produced to identify an individual's team roles. Also reproduced is his description of certain disastrous combinations of roles within a team. All the material described and quoted comes from R. Meredith Belbin, *Management Teams: Why they succeed or fail* (William Heinemann Ltd, London, 1981: reproduced by kind permission of the publishers).

The theory was developed over a period of about nine years and was predominantly researched at the Administrative Staff College, Henley, England, by the Industrial Training Research Unit from Cambridge. Additional experiments were conducted in Australia.

Belbin's research was centred around the performance of teams at a particular management game, 'Teamopoly'. The teams played the game competitively against each other. The game, loosely based on 'Monopoly', produced winners and losers after many rounds of play. The game was developed to eliminate the usual weaknesses of luck and chance from having a major effect. The game not only placed the teams under stress but also afforded the teams the opportunity to use imagination and skill in winning. Above all, the game was played by teams whose collective decision selected their strategy for the next round.

Belbin and his research colleagues used proven psychological classifications to generate their team roles. They started with the dimensions of introvert versus extrovert and anxiety versus stability. This immediately produced four team roles, although by the end of the work the researchers had identified eight 'pure' team roles.

Belbin's team model features a questionnaire-based analysis of team role and these psychometric tests are used to produce a numerical rating of each individual. The experimental verification of the theory is the most impressive aspect of the work. After many years of developing the theory, Belbin and his researchers were able to predict, with indisputable statistical correlation, the likely outcomes of given team compositions. They were able to do this from the results of the psychometric tests alone and did not even have to meet the team members.

Interestingly, the researchers found it easier to predict 'losing', rather than 'winning' teams. This led to the preparation of a list of disastrous combinations which is quoted below. Often these problems stem from a missing role or from two members that clash.

An important empirical fact supporting the value of the theory is that it has already found its way into the practice of many organizations who use it for all sorts of tasks, from making sensible choices about the composition of the board of directors through to general recruitment. The fact that such organizations can justify using the theory in these ways is an important indicator of the theory's value. The eight team roles developed by Belbin and his colleagues are as follows:

- *Company Worker* (*CW*): As a team role, specifies turning concepts and plans into practical working procedures; and carrying out agreed plans systematically and efficiently.
- *Chairman* (*CH*): As a team role, specifies controlling the way in which a team moves towards the group objectives by making the best use of team resources; recognizing where the team's strengths and weaknesses lie; and ensuring that the best use is made of each team member's potential.
- *Shaper* (*SH*): As a team role, specifies shaping the way team effort is applied; directing attention generally to the setting of objectives and priorities; and seeking to impose some shape or pattern on group discussion and on the outcome of group activities.
- *Plant* (*PL*): As a team role, specifies advancing new ideas and strategies with special attention to major issues; and looking for possible breaks in approach to the problem with which the group is confronted.
- *Resource Investigator* (*RI*): As a team role, specifies exploring and reporting back on ideas, developments and resources outside the group; creating external contacts that may be useful to the team and conducting any subsequent negotiations.
- *Monitor-Evaluator* (*ME*): As a team role, specifies analyzing problems; and evaluating ideas and suggestions so that the team is better placed to take decisions.
- *Team Worker* (*TW*): As a team role, specifies supporting members in their strengths (e.g. building on their suggestions); underpinning members in their shortcomings; improving communications between members and fostering team spirit generally.
- *Completer Finisher* (*CF*): As a team role, specifies ensuring that the team is protected as far as possible from mistakes of both commission and omission; actively searching for aspects of work which need a more than usual degree of attention; and maintaining a sense of urgency within the team.

Belbin's Self-Perception Inventory

DIRECTIONS: For each section distribute a total of ten points among the sentences which you think best describe your behaviour. These points may be distributed among several sentences: in extreme cases they might be spread among all the sentences or ten points may be given to a single sentence. Write the point distributions down beside the questions.

SECTION I. What I believe I can contribute to a team:

(a) I think I can quickly see and take advantage of new opportunities.
(b) I can work well with a wide range of people.

(c) Producing ideas is one of my natural assets.
(d) My ability rests in being able to draw out people whenever I detect they have something of value to contribute to group objectives.
(e) My capacity to follow through has much to do with my personal effectiveness.
(f) I am ready to face temporary unpopularity if it leads to worthwhile results in the end.
(g) I am quick to sense what is likely to work in situations with which I am familiar.
(h) I can offer a reasoned case for alternative course of action without introducing bias or prejudice.

SECTION II. If I have a possible shortcoming in team work, it could be that:

(a) I am not at ease unless meetings are well structured and controlled and generally well conducted.
(b) I am inclined to be too generous towards others who have a valid viewpoint that has not been given a proper airing.
(c) I have a tendency to talk a lot once the group gets on to new ideas.
(d) My objective outlook makes it difficult for me to join in readily and enthusiastically with colleagues.
(e) I am sometimes seen as forceful and authoritarian if there is a need to get something done.
(f) I find it difficult to lead from the front, perhaps because I am over-responsive to group atmosphere.
(g) I am apt to get caught up in ideas that occur to me and so lose track of what is happening.
(h) My colleagues tend to see me as worrying unnecessarily over detail and the possibility that things may go wrong.

SECTION III. When involved with other people:

(a) I have an aptitude for influencing people without pressurizing them.
(b) My general vigilance prevents careless mistakes and omissions being made.
(c) I am ready to press for action to make sure that the meeting does not waste time or lose sight of the main objective.
(d) I can be counted on to contribute something original.
(e) I am always ready to back a good suggestion in the common interest.
(f) I am keen to look for the latest in new ideas and developments.
(g) I believe my capacity for cool judgement is appreciated by others.
(h) I can be relied upon to see that all essential work is organized.

SECTION IV. My characteristic approach to group work is that:

(a) I have a quiet interest in getting to know colleagues better.
(b) I am not reluctant to challenge the views of others or to hold a minority view myself.

(c) I can usually find a line of argument to refute unsound propositions.
(d) I think I have a talent for making things work once a plan has to be put into operation.
(e) I have a tendency to avoid the obvious and to come out with the unexpected.
(f) I bring a touch of perfectionism to any team job I undertake.
(g) I am ready to make use of contacts outside the group itself.
(h) While I am interested in all views I have no hesitation in making up my mind once a decision has to be made.

SECTION V. I gain satisfaction in a job because:

(a) I enjoy analyzing situations and weighing up all the possible choices.
(b) I am interested in finding practical solutions to problems.
(c) I like to feel I am fostering good working relationships.
(d) I can have a strong influence on decisions.
(e) I can meet people who may have something new to offer.
(f) I can get people to agree on a necessary course of action.
(g) I feel in my element where I can give a task my full attention.
(h) I like to find a field that stretches my imagination.

SECTION VI. If I am suddenly given a difficult task with limited time and unfamiliar people:

(a) I would feel like retiring to a corner to devise a way out of the impasse before developing a line.
(b) I would be ready to work with the person who showed the most positive approach, however difficult he or she might be.
(c) I would find some way of reducing the size of the task by establishing what different individuals might best contribute.
(d) My natural sense of urgency would help to ensure that we did not fall behind schedule.
(e) I believe I would keep cool and maintain my capacity to think straight.
(f) I would retain steadiness of purpose in spite of the pressures.
(g) I would be prepared to take a positive lead if I felt the group was making no progress.
(h) I would open up discussions with a view to stimulating new thoughts and getting something moving.

SECTION VII. With reference to the problems to which I am subject in working in groups:

(a) I am apt to show my impatience with those who are obstructing progress.
(b) Others may criticize me for being too analytical and insufficiently intuitive.
(c) My desire to ensure that work is properly done can hold up proceedings.

(d) I tend to get bored rather easily and rely on one or two stimulating members to spark me off.

(e) I find it difficult to get started unless the goals are clear.

(f) I am sometimes poor at explaining and clarifying complex points that occur to me.

(g) I am conscious of demanding from others the things I cannot do myself.

(h) I hesitate to get my points across when I run up against real opposition.

Enter the points assigned to each question into the table below and then add up the vertical columns to give a score for each team role.

SECTION	CW	CH	SH	PL	RI	ME	TW	CF
I.	g___	d___	f___	c___	a___	h___	b___	e___
II.	a___	b___	e___	g___	c___	d___	f___	h___
III.	h___	a___	c___	d___	f___	g___	e___	b___
IV.	d___	h___	b___	e___	g___	c___	a___	f___
V.	b___	f___	d___	h___	e___	a___	c___	g___
VI.	f___	c___	g___	a___	h___	e___	b___	d___
VII.	e___	g___	a___	f___	d___	b___	h___	c___
TOTAL	___	___	___	___	___	___	___	___

INTERPRETATION OF TOTAL SCORES

The highest score on the team role will indicate how best the respondent can make his or her mark in a management or project team. The next highest scores can denote back-up team roles towards which the individual should shift, if for some reason there is less group need for a primary role.

The two lowest scores in a team role imply possible areas of weakness. But rather than attempting to reform in this area the manager may be better advised to seek a colleague with complementary strengths.

A successful team is a balanced team, one in which all roles are present, and an unbalanced team will be a losing team. The process of using the analysis of the individual's performance to choose a team in which all the required roles are present is called 'team balancing'. This result is contrary to many first opinions of how to produce a good team. One might suppose that the best way to form a team is to pick an expert in each field required, even doubling up on experts in the most important fields. Belbin and his researchers did construct such teams, much to the annoyance of the other competing teams who thought these teams certain to win, but to the researchers' surprise they nearly always fared badly. So spectacular were their failures that they earned the description 'Apollo syndrome'. These highly intelligent teams were noted for their infighting, each member seeming to gain more satisfaction from using their intelligence to defeat the proposals of other team members in a sort of 'fight to the death' approach. Before long, intellectual supremacy became their unacknowledged aim and the objectives of the group fell by the wayside. Belbin and his colleagues produced a list of the dangerous group combinations as follows:

- A Chairman along with two dominant Shapers both above average in mental ability. (The CH will almost certainly fail to get the job of chairman.)
- A Plant together with another PL, more dominant but less creative, and no good candidate to take the chair. (The plant will be inhibited and will probably make no creative contribution at all.)
- A Monitor-Evaluator with no PL and surrounded by team workers and Company Workers of highly stable disposition and good mental ability. (The team is likely to generate a climate of solid orderly working and not foresee any need to evaluate alternative strategies or ideas.)
- A Company Worker in a team of CWs with no PL and no Resource Investigator. (The company will lack direction and the organizers will not have much to organize.)
- A Team Worker working with TWs, CWs and Completers but no Resource Investigator, Plant, Shaper or Chairman. (A happy conscientious group will be over-anxious to reach agreement so that the presence of another TW merely adds to the euphoria.)
- A Shaper working with another Shaper, highly dominant and of low mental ability, a 'superplant', anxious and recessive, plus two or more Company Workers. (The SH will find that any display of drive and energy is likely to increase provocation and aggravation and disturb further an already unbalanced team.)
- A Resource Investigator with other RIs and PLs but no TWs, CFs, MEs or CH. (A formula for a talking shop in which no one listens, follows up any of the points, or makes any decisions about what to do.)
- A Completer Finisher with ME and CWs but no RIs, PL or SH. (The CF, if he or she intervenes at all, will probably only cause an already slow-moving group to become bogged down in detail.)

10.2.3 Optimizing team composition – practice

So far, we have met and can now apply an experimentally substantiated theory of team balancing. The question we must now ask is why this theory is not used all the time and teams always balanced in this way? There are two basic reasons. First, it is not always appropriate, and second, a number of practical reasons often make it difficult. We will now look at each of these points.

It is not always appropriate to use the team-balancing approach. The well-balanced team is a formidable weapon with an enormous capacity for creative work. It is only appropriate to create such a team if there is a task available to challenge its abilities. A balanced team will not be fulfilled with tasks beneath its abilities. It may well be an expensive group, providing another incentive to give it complex problems with potentially valuable solutions. In general, the technique will work well for selecting teams to face rapid change, competition, a need for innovation, and for action. The converse is also true. There is little point in using the model for line-management applications or where relatively repetitive work is in progress. On the one hand, team balancing is extremely appropriate when establishing a new group or project team to undertake ambitious and challenging work. On the other, it is not appropriate for repetitive work or work without a need for innovation performed by groups with loose working affiliation and who do not face rapid change.

It is often difficult to apply team balancing for logistical reasons. The opportunity to recruit the majority of a team at once does not often arise. It is common for existing employees to take precedence and so limit flexibility in team balancing. Even when recruitment is taking place there are the two problems of not necessarily having applicants of an appropriate profile and of having those whose team profile is acceptable but who cannot be selected for other reasons, perhaps being too weak technically. However, even when these problems apply, team balancing still has benefits to offer. It is certainly better to make some effort at balancing a team than making none. Perhaps the disastrous team combinations can be avoided even if the ideal ones cannot be achieved. Also, a team that applies the model to itself can use the analysis to assist performance without necessarily changing its composition. If a weakness is identified it can be compensated for; either the person who is best at the missing role can take it on or the group can have a constant eye for the problems brought by the missing role and act before the weakness lets them down. If a team cannot be balanced by changing its composition at once it may be possible to change it slowly over a period of time, perhaps after several rounds of recruitment. Having a policy on the sort of person to be recruited next time the opportunity arises is clearly better than taking each case on an *ad hoc* basis.

In these ways, many of the benefits of team balancing can be realized in situations even when it is impossible to form a new team.

10.3 Group dynamics

The word 'dynamics' is familiar to engineers, and is used in mechanics to describe the effects of forces in motion. The term 'Group Dynamics' is well chosen, though it is often misunderstood. It means the forces in motion within a group. Of course, here the dynamics refers to personal forces instead of mechanical ones. Group dynamics is therefore the study of personal forces within groups. Much work on group dynamics has been conducted on groups in the widest sense of the word, such as families or gangs. In this book we are studying management, and our interest in the subject comes from the fact that engineers normally work in groups to achieve their aims.

The most basic question that one can start with is 'Why do groups ever form in the first place?' The answer to this is very straightforward. A group together can achieve very much more than the individuals can alone. In Section 10.2.1 we saw how this can be true for teams of people in their work. In group dynamics we take the examination much deeper and realize that even from a zoological point of view this benefit of group work still applies. Wolves hunt in packs, whales live in schools, most animals exhibit group behaviour of some sort. The difference with humans is that there is a lot more of it.

This programmed team working instinct is there to make each species more successful. Inherent in the genetic coding is much more than a desire to be in a group. Many special behaviours, such as the acceptance of authority or the need to conform, are aimed especially at life in groups. We are not just conditioned to operate in groups, we are actually designed to do so.

Group dynamics is the study of personal forces in motion within groups of people. It deals with the behaviour of individuals and their interactions with others. These are very complicated things and making accurate models of them is not easy. Understanding always precedes rational control. Therefore as engineers, if we wish to benefit from rational suggestions to improve team weaknesses, or to engender those factors that assist our teams, we shall have to start by understanding group dynamics.

In the next two sections we shall introduce group dynamics first, by looking at the needs of the group. We have seen that the basic motivation for group working comes from the much-increased capacity for activity that it brings. This benefit does not just happen: many problems must be solved in order to reap the benefits. We will then see how various group behaviours exist to meet these needs and overcome the problems they bring.

10.3.1 The needs of the group

In this section we shall consider the needs of the group. These are things that the group require in order to provide the benefits of group membership to the society of individuals that constitute it. We are studying group dynamics in this section and so our examination of group needs does not include material aspects such as shelter or finance. We are, however, interested in things that the group must have before the interplay of personal forces that is group dynamics can start to operate. There is no definitive list of group needs but the following six are quoted in many studies in one form or another.

The need for a task
The whole purpose of forming a group is to achieve some task. The nature of the task can have a great impact upon the life and times of the group. If the group

task is continued prosperity, as might be the case with a team of company directors, the perpetual nature of the task shapes the group. They take a long-term view, are a strongly bound group and there is much scope for powerful political infighting. By contrast, if the task is short, such as a ten-session, one-hour-per-week training course the opposite applies.

The need for the group to have a focus is very powerful. It is responsible for creating substitute tasks where no real ones exist, as is the case with the imagined enemy of the 'boys' gang' or the instinctive recourse to competitive behaviour between groups engaged in similar tasks. The task is also responsible for influencing the type of person attracted to the group. Technologically advanced companies engaged in rapidly changing markets attract one sort of person while traditional craft industries, practising skills noted for their constancy rather than their transience, attract another.

More than anything else the task colours the group. It pre-selects group members and favours those with certain values or aims. Often even a particular background and intelligence are favoured. When you ask people about themselves they tell what they do, not what groups they belong to, such is the importance of the task.

The need for creativity

We now have a team that has something to do. Depending on the nature of the task, there will also be a need for some level of problem-solving skills. The team has been formed to provide greater benefits to the members than they could provide for themselves, therefore a mechanism is required for ensuring that good ideas are plentiful. This in turn requires that each individual is able to put forward ideas and that the best among them are selected. The concepts explained in Section 10.4 will be of use to the team here. The important thing is that the group are able to generate a solid bank of good ideas when necessary.

There are two ways in which the group can improve the likelihood of producing the ideas it needs. The first is to ensure that the group has members with the skills to produce the ideas it needs. These might be intellectual skills and may be provided by the team role of 'Plant' that we met in Section 10.2.2 which allow the group to produce the ideas for themselves. Alternatively, the ideas may be found outside the group and a member to perform the 'Resource Investigator' team role is required.

The second way that the group can improve their chances of creative output is to have an environment that is conducive to good work from the 'Plant' or 'Resource Investigator'. Some humans exhibit a process called facilitation. This describes the improvement in work quality and creative thought that can occur when an individual is surrounded by other productive minds. Some individuals almost have to be surrounded by others before they can produce much in the way of creative ideas. The question of what constitutes a conducive environment is a matter for the group to decide for themselves. Aspects that are frequently associated with the creative process are enough time to think, freedom from petty pressures, a lack of deadlines or strict plans, and the opportunity to work on one thing at a time. These needs are almost exactly opposite to the overall management needs of the group, where a constant eye for progress against plans, pressure to achieve and constantly working on more than one thing at once are good ways to pursue the group's task.

The group therefore has a need to be creative, either by having creative people or by using the creativity of others. In either case, an appropriate climate is required before group members can perform these tasks to the best of their abilities.

The need to resolve disputes

A group engaged in using new ideas to solve a task will very often have to choose between alternatives. One member of the group may favour one course, another the opposite. If the issue is a very important one, such as how much profit to retain in order to fund next year's work, or how to design the main components of the next new product, one can expect the debate to be intense. However, the group can follow only one path and a decision has to be made.

It is clearly the aim of the group to ensure that the most rational decision is made. There are many ways in which the rationality may be assisted and Section 10.4.3 describes good decision-making practice. From the group dynamics point of view, however, rather than from the purely logical one, the important issue is that personal motives and interpersonal contests do not allow the group decision to drift into the irrational. Plenty of decisions taken by groups are actually the product only of one member. While that member's judgement is good, the situation may be tenable but it clearly exposes the group to an undesirable risk.

Where the decision is dependent upon the special skills of one team member it makes sense to question the decision and, if found acceptable under examination by the rest, to adopt it. This might mean delegating the design of various components to certain experienced engineers. Often delegation is achieved through the group leader who, although ultimately responsible for the decision, delegates the authority to choose to one of the group.

The need to survive

For a group to be successful, it must survive for a period of time that is consistent with the task in hand. The structure of the group must be dynamic. If a member leaves, a replacement must be produced and the completion of the task interrupted as little as possible.

In primitive societies the contest for group positions is driven by the ambitions of individuals and this ensures that there are a steady stream of contenders pushing upwards. Usually they are selected against some ritual or custom that aims to assess their suitability in some way. In modern organizations the process is formalized into promotion and recruitment. Without the improved status and achievement that such a hierarchy brings there would be no fierce contest for the top roles and the task would not have champions to ensure that it is completed.

The need for optimal resource deployment

To be at its most effective, a group must have a means of ensuring that its resources are deployed in an optimal way. In the case of special skills or equipment, optimal deployment is usually obvious. In other situations it is less so. For instance, a junior engineer might be given a task which is at the upper limit of that individual's abilities – such a task would almost certainly be more competently accomplished by a more senior engineer. However, in doing the task the junior engineer has the opportunity to develop and so increase the resources of the group. In such a situation, comparison between the advantages of employing a junior engineer and improving the skill base of the group against the advantages of the senior engineer's speed and quality must be made. Such comparisons are constant in the optimal deployment of group resources.

The need for unity

After a course of action has been chosen for the group there is no benefit in individual members directing efforts in other directions. This reduces the resources that can be employed to achieve the task. The group needs unity if it is to survive, and this comes from either rejecting those who do not support it or convincing members of the sensibility of the chosen course.

If previously sceptical individuals are to be converted to the chosen route, respect for the group decision-making process and a commitment to the team are vital. A united stand maximizes the group's chances of survival and achievement. It makes it less prone to defeat from rival groups and provides assurance to the group members of their security. It should be remembered that having made a good decision is the most important thing. Even a strongly united team heading in the wrong direction will eventually fail.

The need for unity applies to the long term and the short term. In the short term, unity prevents the group from becoming ineffective under pressure and being prone to exploitation from other groups. In the long term, unity is needed simply to ensure that the group continues to be committed to paths of action that are consistent with the group's objectives.

10.3.2 Meeting these needs – group dynamics

So far in this chapter we have seen that a group can greatly extend the abilities of its members and put them at a considerable advantage. We have also seen that in order to provide these benefits there are various needs of the group that must be met.

Below, we will introduce some of the basic dimensions of group dynamics and explain a little of how these provide for the needs of the group. Each of the dimensions is an attribute that may be observed in a group. Group dynamics is an all-pervading subject and applies to ancient tribe and modern office alike. After reading this section you should be able to identify the group behaviours in your own social groups and observe them within your family, sports teams, friends, societies or associates.

Norms

Norms are standards of behaviour. They describe the 'normal' way in which the group behaves. It is possible to imagine all sorts of different ways in which these norms might come about.

Some norms stem from sheer practicalities. We all drive on the left in the UK, not because the left is inherently better than the right but because you have to choose one side and we have chosen the left. Other norms may be quite covert and unspoken. Examples of this might include a meeting that routinely considers topics in a certain order, or a group of employees who use their own made-up terms or expressions. The reluctance to explain such a system to newcomers enforces their oddity and in this way norms can be used to draw lines between those who are 'insiders' and those who are 'outsiders'.

These various types of norm support the likelihood of group success in many ways. Norms bring familiarity which protect individuals from feelings of fear or threat that might inhibit their contributions. Our natural desire to avoid insecurity therefore makes us take quickly to group norms and accept them in accordance with our level of personal security.

Some norms extend to preserving the social hierarchy and so illustrate to all the organization of the group. It is common, for instance, for the order in which a formal dinner is served to reflect status, though in fact any order would be just as practical. Where norms engender sound practice they may serve to make the group more long-lived. Certain cultural food restrictions are good examples of this. They may well have formed sensible hygiene practices when introduced, but such is the power of norms that they have survived long after the invention of the refrigerator and are now unnecessary.

Within a group of engineers it may be an office norm that the engineer who performed the majority of the initial project work gives a presentation to the Board and takes on the project. Such a norm might support optimal resource deployment since the jobs are probably best done by the same person. However, if an engineer was not asked to follow the route this would almost certainly lead to anxiety and the thought that the initial work performed was not acceptable. In this way, norms may be used as a vehicle of judgement.

Norms simplify things, they serve the belonging need we all have and they can be used as a ready reckoner against which to judge behaviour.

Group culture

All groups have their own culture. In group dynamics the term is used to describe the atmosphere or ambience of the group. Group culture has a major role to play in producing a climate conducive to the production of ideas. In Section 10.3.1 we saw that there was a need for creativity and that this in turn required a climate conducive to idea generation. Different groups might find different climates successful but all groups will want an appropriate one. Group culture provides this creative climate.

It is easy to see that, at least in some way, the culture of the group must be the product of the individuals that compose it. If the group in question is a design team, it may be that one senior member of the organization has a potent character trait which, before long, permeates the group. For example, if one person in a team is really enthusiastic about the work in hand, it can raise the spirits of the whole group. The whole team becomes bright and optimism prevails. Such outlooks can become entrenched as part of the culture. If we compared this group with a rival group in a different organization we may find the cultures to be completely different. The culture of a group is dynamic and will change with time. One cannot, of course, decide to engender a particular type of culture and simply create it. The culture of a group is a result of the group, not the other way round. Group culture can be thought of as an extension of the factual group norms into the more abstract. Issues such as beliefs, values and opinions shape group culture and they are not so easily defined.

Leadership

Leadership is a vast issue. Its many facets assist the group in every one of its needs. It is the most potent role of all within a group and it is the kingpin upon which the rest of the group depend. If you ask someone what their job is like, they will usually include a description of their boss. Difficult questions get referred to the boss. The cliché, 'take me to your leader', alludes to the overwhelming power of the leader's position.

Leadership is often misunderstood and seen as anything from the romantic swashbuckler leading a merry crew to fortune and adventure through to the lonely isolation of a compassionless military hero. Such fictional distortions have no place in engineering management. Leadership is an essential ingredient of group success provided by ordinary people. When did you last hear of a successful human endeavour that had no leader?

An important distinction must be made between leadership and a leader. Leadership is defined as those tasks which need to be done in order for the group to be led. A leader, on the other hand, is someone who is actually doing the leading at that time. Leadership tasks can be performed by anyone in the group, even by more than one person at once. Usually, one person is the leader for the majority of the time and is the person named when people are asked who is the boss. Much of the time, other group members take the lead with the full approval of the 'main' leader.

The style of leadership has a great influence on the rest of the group. In the long term the leader is judged by the other members against the group success in achieving the task. Unsatisfactory leaders are then impeached or abandoned. Many people regard leadership as the only group position worth holding because they have been conditioned by their culture to aim for the top. This does nothing for the individual's personal happiness and everyone has their own ideal group position, with very few actually preferring leadership. One cannot change one's basic preferences much, and it is important to remember that while skills can be improved through good training, one can never acquire abilities that were previously completely lacking. To some extent, the saying that good leaders are born and not made is true.

The leader of a group can affect the task. The leader is the ultimate authority and it is only the leader that can ultimately take the group away from the task. The leader's good judgement on group issues is vital for the continued success of the group. Group dynamics provides a safety valve for the authority of the leader and it is only while good performance endures that the group permit the leader to stay. It is the group's respect that regulates the leader's authority. The respect of the group gives the leader a mandate to lead and it can be withdrawn at any time. There may be no ballot boxes but a leader is democratically empowered. It is this need for a leader to be in position at the command of the group that undermines the authority of a leader empowered only by organizational decree or appointment.

The leader is the ultimate judicial authority in a group. In this role the leader settles disagreements and resolves disputes. A competent leader must already have shown fitness to lead and therefore commands the respect of the team. When disputes occur the team members have to be prepared to respect and trust the leader's decision. So powerful is the group need for a decision-making authority that when a new group is formed, choosing a leader for it is almost the first thing it does. Often this is done subconsciously. A group of three of four people asked to move a heavy object forms a good example. They normally sidle up to the object and start a reasonable but uncoordinated attack, then one of them says something organizing and constructive and is instantly made the leader. Usually it is this person who counts the group into the lift.

The leader is often the one who guides important discussions and makes the final decisions following group debates. In this way, the leader has more influence on the fertile generation of ideas than any other role. A group may have a very good ideas base but if the leader fails to seek the opinions of other group members, or is not convinced by a correct argument put forward by another, the ideas might as well not be there. Again there is a safety valve for the authority of the leader. If a leader subsequently finds a correct option put forward by a group member was discarded, the respect of the leader may suffer greatly. To command respect, the leader must constantly be making excellent choices. Because of this, it is harder to hold leadership than to challenge it and so potentially superior leaders are always eager to take over. Of course, when the authority of the leader comes by appointment, as is the case with modern employment, social norms prevent a rival leader from carrying out a challenge, but this does not mean that a loss of respect has not occurred. It is possible for a leader to lose command in all but name.

Finally, the leader is the taskmaster and figurehead of a group. The leader is ultimately responsible for who does what and is the person put forward to represent the group.

Conformance

People who have joined a group have already demonstrated some form of common aim by their very presence. Conformance in group dynamics refers to the group need for unity. The need is met directly by the individual's need to belong, or rather their need to conform. Of course, different individuals have different ideas about conformance and one group may be ideal for one person but appear hopelessly identity-crushing to another. From the group's point of view conformance is a force that must be appropriate to the needs of the particular group. If it is too weak in its binding effect the group runs the risk of falling apart. If it is too strong it may be inhibiting the creative abilities of the group members and so frustrating the very reason for the group's existence.

All individuals have conformance needs and all groups demand them in some form or other. Even the socially rebellious groups of teenagers, found in one form or another in all societies, cannot escape. Such groups claim to reject conformance and yet wear a uniform as recognizable as that of any army.

In general, the greater the commitment to the common aim, the more major personal differences can be suppressed and the more powerful the conforming force. In a group at work, the conformance of individuals to the group's task can become a contentious issue. The members of the group are bound together by the cohesive forces of the group and these must be maintained in balance in order for the group to stand the best chance of success in the task. Interpersonal compatibility, however, can interfere with this cohesive balance by becoming either too strongly attractive between two individuals or negative and repulsive. In both cases the mutual feelings may distort rational judgement and lead to a de-optimized decision-making process. This will be greatly exaggerated if the leader is affected. To prevent these problems from reducing group effectiveness, people tend to limit their social interactions within a group at work. Business is not usually mixed with pleasure and infighting is not normally tolerated.

Conformance has beneficial effects on the generation of ideas, two compatible minds can work together to form a formidable intellectual entity. Having like-minded members brings solutions to many group needs. The acceptance of norms, similar judgement of leadership, common purpose all make the group more likely to succeed in its task. Conformance should not be confused with similarity or lack of independence. A group of very similar people is a weak group, a group of very different and yet conforming people is very strong. Such a group will enjoy a good match of differences and is well equipped for all circumstances.

Conformance is the group behaviour that directly answers the need for unity. Conformance also provides benefits to the group by making the individuals relaxed within it and confirms that they are accepted. In this way, conformance can answer the deepest of all psychological needs, the need for identity. By wearing the uniform of your group, you state your identity as one of them. By using their expressions you pass the test of conformance. By resisting those who are different, you separate your identity from theirs.

10.4 Managing the creative process

Contemporary creativity comes from group work. The days of individuals providing the majority of breakthroughs in science and engineering are past. Modern problem solving is an organized, rational approach aimed at producing the required breakthroughs at the times they are needed. The techniques described in this section are all aimed at managing the creative process of engineering by assisting groups to plan their efforts.

In Section 10.4.1 we will meet some methods that can be used in an overall way to plan large blocks of creative work. In Section 10.4.2 we follow with the techniques of lateral thinking and brainstorming. Both are methods for assisting the production of new ideas. Finally, in Section 10.4.3, we investigate a particular skill in the creative process, that of good decision making.

10.4.1 Planning innovation

There seems to be an inherent paradox in planning innovation. Innovation is associated with inspiration, imagination and originality. Yet none of these things seem amenable to planning.

We have a very stereotyped view of a creative person at work. We imagine a great engineer thinking things over until inspiration strikes and a brilliant invention is born. In reality, even when individuals do make breakthroughs, they usually come after much hard work. People create the environment for innovation and breakthrough by constantly working at the problem. Of course, no-one would suggest that people can plan which ideas will be masterpieces and which won't, but this does not preclude the use of planning to assist creative results. By planning their work, so that they are constantly in fertile areas, people affect their 'luck'. There is, in fact, no paradox involved in planning innovation.

Chapter 8 described planning techniques for determinate problems. The 'special' techniques in this chapter are used when coping with indeterminate issues and should be used in conjunction with the methods discussed in Chapter 8.

In general, the techniques that we have at our disposal to assist in planning innovation may be divided into two categories. In the first are the techniques of planning itself. These are ways in which good plans may be prepared. In the second category are the different approaches to the work that may be taken. These are the underlying principles upon which the plan depends. We will start by looking at the different approaches and follow with the good planning techniques.

Approaches to planning innovation
The following techniques may be applied when planning the innovation process. They may be used singly or in conjunction with each other. Once a principle upon which the work is to be based has been selected, the planning techniques of the

next sub-section together with the planning techniques of Chapter 8 may be used to provide a complete plan.

● *Multi-directional attacks*: In certain circumstances a solution to the critical problem might come from more than one direction. In such cases it may be sensible to embark on developing parallel solutions based on different principles even though only one will ultimately be required. This 'insurance policy' approach can greatly increase the confidence of a successful outcome as well as doing much to ensure that the solution chosen is indeed the most appropriate. The amount by which these two benefits offset the cost of the additional work must be estimated and a decision made on the best information. This approach is particularly prevalent in the placement of military contracts where companies may compete for a tender long before the state finally chooses the winner of the contract. Meanwhile, each of the companies will have invested a great deal of time and money in demonstrating their ability to meet the needs of the contract in their chosen way.

● *Fixed-resource research*: In this approach a fixed amount of resource is directed at a problem with the intention of identifying the best solution available within a budget. This is particularly appropriate when the definition of performance is as yet unknown, perhaps when research is undertaken as part of a large programme or if the project is relatively small and is of an investigative nature, rather than a manufacturing one. A negative result is also useful in that the organization may then be sure that no competitor can solve the problem without spending more than the organization's own research investment. Some organizations attempt to pre-empt the creative work of others by employing levels of expenditure that are known to exceed those of their rivals. In this way, such organizations hope to be first to achieve the next hurdle, whatever it may be.

● *Rolling plan*: This technique can be applied where success is more important than timing. As the project progresses and the results of key decisions are known, new objectives for the next phase of the work are defined. The managers employing such a principle must exercise control in order for it not to degenerate into an unguided block of work in which the management has become mere reporting. The end goal of the whole programme must not be forgotten and timescales must always be imposed for each section. The technique is particularly appropriate for projects in which the future path is completely dependent on results that are as yet unknown, or to situations where a relatively large number of alternatives must be tried in turn until an acceptable result is found.

● *Undirected research*: In this approach the researchers are given a problem but are left to see what ideas and results they can generate. The free environment fosters their creative output and their natural enthusiasm leads them towards useful work. The technique can be very effective in providing those with creative skills an opportunity to make the most of their skills in a way that benefits the organization. However, the technique can degenerate into an unproductive ramble through the interests of the researchers and it should be managed in terms of desired goals and resources to ensure direction but otherwise left alone. Such work should be scheduled into existing commitments.

● *Key result research*: Clearly there is no point in conducting research into all areas of a particular problem if there is still a chance that the project will fail on some fundamental principle that has not yet been proven. For this reason, it makes sense to start with 'proof of principle' research. Care must be taken with such research projects, as the temptation here is to perform very cursory investigations whose results look positive. The research is then taken further and it is discovered that the principle does not really work in the situation in which it is required because of other fundamental design needs. This problem may be overcome by having further grades of research programme. After the initial surface investigation the work is taken a little further and may go through several stages before becoming a product. One should avoid the trap of thinking that simply because the work has gone a certain distance it must continue, and many successful companies claim a project drop-out rate of nine out of ten. Such organizations are still comparing the objectives of the projects with their strategic aims even in the last stages of a project's life. If a mismatch is found there is no point in continuing. Individuals do not pour good money after bad, nor should organizations.

Planning techniques for the innovative process

The following techniques may be used in conjunction with the planning techniques of Chapter 8 to plan the innovation process:

● *Upper and lower estimates*: No matter how complex or dependent upon future outcomes, it is always possible to put estimates on the upper and lower limits of cost and time. Such estimations can initially be used in a crude way to identify whether further planning is sensible. The first phase of examination can be done almost on the back of an envelope, giving order of magnitude estimates of the resources and principles involved. This will suggest whether it is sensible to consider the option further and will also make clear the critical milestones in the project. Once the first estimates have been prepared, further refinements may be made to increase their accuracy.

● *Milestone reviews*: Even though a new idea or device may depend on certain critical problems being solved, they rarely depend on a great number of critical issues. It is therefore possible to select a few milestones in the life of the project. These are stages at which, if the project has not achieved certain targets, completion within the original budgets is no longer possible. Milestones break unmanageably large blocks of work into manageable packages. Progress through the milestones is a good indicator of overall success.

● *Vested disinterest*: Individuals who are champions of their own projects or ideas are not necessarily good planners of the work. Their input may be essential to the process since they are the ones with the detailed understanding, but they are also ambitious for the work and want it to succeed. Such people tend to minimize the difficulties and overstate the benefits. When planning is undertaken people with opposite points of view should be invited to examine the proposals.

● *Strategic comparison*: Great benefit can be derived from comparing how the organization performed in similar previous ventures. Clearly, if a task is to be completed in an uncharacteristically short period of time an explanation of how this can be achieved is required. There is, however, a danger that people will use this 'experienced viewpoint' to decry perfectly viable options. We are all familiar with the phrase 'we've tried that before, it's no good'. When memories of previous tasks are recounted questions need to be asked to establish rationally whether the new idea should be tried. 'What is different now?' 'Why did the previous project fail?' Such questions deserve answers. This makes abundantly clear the complete waste of time and money that can result from an undocumented research project. If no write-up of a previous project exists no benefit from the experience can be fed into future plans. Regardless of success or failure, valuable planning experience can be gained from every single project an organization undertakes if it is documented.

In this section we have initially seen that there seems to be a paradox in planning innovation. How can one plan to have an inspirational thought or choose which piece of work will be ordinary and which brilliant? However, we soon realized that there is no paradox in planning innovation, and the most important point is to keep attacking a problem so that eventually we achieve the break we are looking for. People make their own breaks by working steadily at problems. The difference between planning innovative processes and planning other processes is that innovative processes involve indeterminate outcomes. In some way, the path that the plan must take cannot be known until the result of some future activity is known. This indeterminacy makes planning the innovative process difficult; difficult but not impossible. In this section we have also seen various underlying planning approaches that may be used when designing a plan to deal with an innovative process and some good planning techniques that may be used to ensure that such plans are effective.

10.4.2 Problem solving

Engineers solve problems. The organized production of food was an early achievement of the first agricultural engineers. As societies advanced, new problems were encountered. Engineers solved them by providing, among other things, buildings, sanitation, energy, lighting, armaments, transport, flight. The list grew as engineers strove to meet our incorrigible appetite.

Today engineers continue to propose solutions to solve these problems as well as many new ones. Cars straight out of yesterday's science fiction are in every showroom. For less than the average weekly wage machines can now be bought that effortlessly wash the clothes of a large family; such a wash used to take two whole days of hard work; today people complain about having to hang it out!

In all areas of human endeavour, problem solving is the gateway to progress. It is not surprising, then, that the very process of solving problems has been studied in a rigorous manner. Researchers attempt to reveal dependable strategies for success.

There have been many attempts to produce reliable problem-solving techniques. Often these take the form of a shortlist of tasks that must be performed before the problem can be solved. Any good technique will include the following steps in some form or other:

(1) *Examination*: One cannot start to solve a problem without being absolutely clear what the problem is. For instance, an engineer working on the design of a component in an engine may be trying to improve the life of the component. At first, the problem seems well defined, but in reality the problem might actually be that the oil pressure needs increasing or the material needs altering, or the speed of the component should be reduced. It is possible that what is required is another method to be used for the same function. It may be that all these things are important but which must be tackled? What actually is the problem? In some cases the problem may not be correctly identified at all. A company wishing to have more graduate employees may implement a recruitment policy lasting several years to correct the perceived imbalance. However, the shortage may be caused equally by low levels of recruitment as high levels of resignation. The real problem may be how to stop the graduates leaving rather than how to attract them more effectively. Clarity of thought and the application of rationality are the key to being accurate in identifying the real problem.

(2) *Proposal*: Good problem solving depends on being able to identify the best solution to the real problem. This might be achieved by having the ability always to think immediately of the best solution. Unfortunately, this approach is practically impossible for all but the simplest of problems. The ideal method to produce lots of different ideas and then weigh one against another to find the best one. Often, thinking of new ideas makes one rethink and improve old ideas, and so the process leads to an improvement in quality. In any case, one cannot know that the best idea has been selected unless all other possibilities have been presented and evaluated.

(3) *Implications*: A proposal designed to solve a problem of any complexity will have implications. These implications may even be significant enough to render a proposal unworkable. For example, if an organization finds that a valuable employee is resigning there is a temptation to offer more money as an incentive to stay. However, the implication of doing so is that the organization was not paying the individual as much as it felt the employee was worth in the first place. Clearly, this means that the organization was guilty of underpayment. In addition, how will all the other employees react? They cannot be blamed for thinking that the organization is underpaying them too. In such a case, the implications of the proposed course of action mean that it should not be taken up.

(4) *Implementation*: No solution to a problem is complete without due consideration being given to the implementation of the solution. A workable plan is required to demonstrate that the solution is indeed feasible. Once a problem has been defined, the best solution identified and its implications thought through, it is always possible to produce a plan for implementing the solution. Such a plan can then be used to manage the work. Implementing a solution to a problem is, by definition, a management issue and, depending on the solution in hand, any of the topics covered in this book may be appropriate.

Below we consider two methods for improving our abilities to propose solutions to problems. The techniques are excellent at unjamming thought processes that have become hindered by deductive reasoning and have reached a situation from which there seems to be no way forward. They are used in conjunction with deductive and rational reasoning to ensure that the best solution is found. These processes work in all problem-solving situations but are especially appropriate to a process as creative as engineering. They are called brainstorming and lateral thinking and are familiar, in name at least, to many people.

Brainstorming

Brainstorming is a method by which a flood of new approaches to a problem can be generated. In most cases the majority of ideas are discarded but from the generous flow only one idea may be needed to solve the problem and so the wastage rate does not really matter.

The technique of brainstorming relies on the assembled group seeing the problem with fresh eyes and using their imagination to produce new approaches to the problem. A purely rational approach is used after the brainstorming session to assess the ideas, but during the brainstorming phase the most productive individuals are those who let go of rational constraints and use their imagination to bring entirely new views to bear on the problem.

Successfully managing the brainstorming session means creating the environment in which this torrent of ideas flourishes. Various things can be done to foster good brainstorming conditions:

- Have a leader experienced in brainstorming.
- Have a group of less than about ten and more than three.
- Include some people who have little or no technical knowledge of the problem.

- Don't permit negative comments.
- Don't allow suggestions of solutions or evaluation of ideas while the ideas are being generated.
- Produce a group atmosphere that puts everyone on the same level. People tend to be inhibited by having more senior people present.
- Encourage positive statements. A negative statement like 'the problem is that we can't...' is a lot less stimulating for participants than saying 'What if...?' 'How may we...?' 'Hey, how about...?'
- Keep the session short – about half an hour.
- Record the ideas in sufficient detail to allow examination later.

After the ideas-generation phase is over the more technically qualified may consider the practical aspects of each suggestion and the implications of them. The reviewers should be aware that it is in the nature of brainstorming to produce some quite large and general ideas that normally require a considerable amount of work before they can be implemented. There may even be new ideas emerging from quite different directions.

The brainstorming process is particularly useful when an impasse has been reached or when the physical principle used in an attempt to solve a problem has been found to be unsatisfactory. The process tends to throw up rather wild ideas, many of which are interesting or even amusing but of little practical value. Consequently the use of the process is sometimes limited. In general, the technique is only applicable on an infrequent basis, but when it is used the results can be spectacular.

This is really an extension of a thought process we all go through when we are problem solving. While the problem is comprehensible and understood we apply rational thought to produce the optimal next step. However, we can often reach an impasse or get so close to the problem that we cannot take an overview and jump to fruitful alternatives. At this time we may have the sudden, inspirational thoughts that unblock the logjam. These thoughts often come at unexpected times and are often jogged along by association with some unrelated event or coincidental encounter. Brainstorming tries to manage and optimize this process.

Lateral thinking

We are all educated in classical problem solving. Our schooling makes deductive, logical thought habitual. It is certainly true to say that its application has brought impressive advances in our knowledge. In 'vertical' thinking one starts with some known condition and then applies steps of reasoning aimed at reaching a goal. The only rule is that the steps must each be true and self-consistent with the system of thought being employed. After many steps have been employed one is in a new place that might not have been foreseen without the logical process. As long as each step is valid, we may go as far as we like. Engineers make great use of this type of thinking and think nothing of starting with two or three physical laws, applying them to a physical system and producing, through vertical reasoning alone, an experimentally verifiable model of the system. It is little wonder that the process has earned respect.

Lateral thinking is the opposite to vertical thinking. In lateral thinking, one is not constrained to follow completely the formal deduction process. One has an inspiration or flash and sees a new angle on the problem which, if locally self-consistent, is a new model in its own right. The human mind is very good at seeing patterns, maintaining conformity of ideas and reinforcing concepts. Lateral thinking is used to free us from the limitations that these cause. It is not a replacement for our existing system of thought, it is a powerful tool for use at certain times. In many ways, lateral thinking is brainstorming on your own. A great deal has been written about lateral thinking, much of it by one of its most famous exponents, Edward de Bono. The reader will find a book of his, *Lateral Thinking for Management*, listed in the Further Reading. Lateral thinkers make jumps to new ideas and views, they are creative and imaginative, often supplying ideas that vertical thinking alone could not provide.

10.4.3 Decision making

Being consistently better at making decisions will place anyone at an advantage over their colleagues. Good decision making is common to all professions, and engineering is no exception. Good engineering managers are good decision makers.

In this section we will look at the problem of decision making with the aim of understanding the process and how to optimize it. Decisions vary greatly in their magnitude. Some involve trivial issues such as choosing an item from a restaurant menu, others are vast. When President Kennedy announced that the United States undertake the Apollo moon shots he is thought to have given the most expensive commitment ever made in a simple statement. At first sight these two decisions have little in common. However, they both involve choice and they both have consequences. In some ways at least, all decision making is the same.

We shall now investigate how to improve our decision making by dividing the process up into independent, elemental sections and optimizing each one.

Start with objectives
A prerequisite of making a choice is that one must have some objective against which the choice is to be judged. Objectives facilitate clear and corporately beneficial decision making at all levels. If the objectives are not clear, the decisions that result cannot be guaranteed to support the original intention of the objective. Worse still, if an individual is supplied with conflicting objectives, rational decision taking becomes impossible when a decision that cannot serve both objectives is faced. Section 11.3 explains how good objective setting always avoids such conflict and produces clear, unambiguous objectives which assist the decision-making process.

Collecting the right data

Clearly, a rational decision made in the knowledge of all the relevant facts will be a good decision. Sadly, we are rarely in possession of all the facts we need. In life, all but the most ordered and the smallest of decisions are made in a state of partial ignorance. The aim of the data-collection phase is to reduce the level of ignorance. Whenever a decision is to be made there are always options. The two important goals of this data-collection phase are therefore first, the collection of a sufficient amount of facts about each option and second, to ensure that all options are known. During the first phase the data is assessed to estimate the error in decision that may result from the error in the data. If this error is unacceptable, more data are collected until a more reliable decision can be made. There is no point in gathering unnecessarily accurate information to permit a decision. The important issue is that the accuracy of the information is consistent with the measurable effects of the decision.

Brainstorming sessions or more formal strategy meetings are often held to be certain of covering all options.

Decision-making techniques

When the decision maker is in possession of an acceptable quantity and quality of information, and has an objective to achieve, the process of selecting an option can begin. We will now look at some of the evaluative techniques that be used to assist the process.

It should be remembered that making occasional wrong decisions is an inevitable consequence of making decisions at all. No-one can expect to get it right all the time. The well-adjusted person will put down the mistake to experience and learn from it. No-one can do no more than make their best effort. One must accept one's own fallibility and not bear grudges.

- *Search for extrema*: In some cases the objective may require the decision taker to head for an extreme (maximum or minimum) of some variable. If a project is required to be completed in the minimum possible time, for instance, the decision maker need consider only critical path plans which allow this rather than fixed-capacity or fixed-resource plans which may be more appropriate but take longer. Alternatively, a camera designer may have the objective of minimum possible weight and so will have to select plastic for the lenses rather than the optically superior glass available. In general, extrema are very poor ways to specify objectives and almost always lead to unclear choices. The questions of 'How light must the camera be?' and 'What project duration is acceptable?' are unanswered in the above examples. The use of extrema has been included here for two reasons, first, because very occasionally it is the appropriate way to specify an objective and second, while it is bad practice, it is also common.

- *Penalty costs*: All decisions come not only with potential benefits but also with costs associated with the decision being wrong. These are called penalty costs.

Occasionally, even though there is almost no chance of the decision going wrong the consequences are so dire that an almost certain gain is left untouched. No-one in their right mind would gamble with their lives for modest financial gain, no matter how favourable the odds.

● *Matrix assessment*: This is a very simple and effective way of choosing from among alternatives when many different attributes have to be weighed against each other. The attributes are first listed and assigned a maximum score to reflect their relative importance. The items that are most important are assigned the largest maximum scores. Each alternative is then marked out of the maximum score for each attribute. A high score indicates that an option can successfully provide the desired attribute. When this is complete the columns for each option are added up to determine the best choice.

Table 10.1   Matrix assessment for selection of lens material

Attribute	Max. score	Glass	Plastic-N3	Silica	Plastic-N2
Clarity	20	19	17	15	17
Low cost	20	5	15	10	18
Weight	20	2	19	13	16
Refractive index	15	14	8	9	12
Uniformity	10	10	6	8	6
Stability	15	15	11	13	11
Total		65	76	68	80

In Table 10.1 a designer is selecting a material to use as a camera lens. The different options open are listed across the top. In the column down the side are the various attributes considered important by the designer. The next column shows the maximum score available for each attribute. The designer has allocated these scores and has, for example, decided that clarity is twice as important as uniformity and so given it twice the maximum score. The body of the table shows the score for each option. The bottom line clearly shows that plastic-N2 is the best choice.

In this example a decision has been made on technical grounds but the method is equally appropriate to any type of decision in which options have to be selected against several criteria.

● *Overriding constraints*: In some cases overriding constraints significantly reduce the number of options available. This serves to make the decision-making process easier and so it is helpful to look for such constraints at the start. Sometimes the constraints may be the quest for extrema as described previously, but there are other constraints too. For instance, a manufacturer may decide to upgrade an existing camera model and include various new features without changing the original body shell. The design engineers will therefore have overriding space constraints imposed by this decision. In reviewing the design options, the engineers can reduce the number of alternatives significantly by applying this overall constraint.

● *Use of maths*: Nearly all decisions involve an evaluation of probabilities. Wherever it is possible to select an option mathematically it is best to do so. Maths is an unbiased and dependable assistant. Even when the uncertainties are significant error bars may be allotted and maths still applied. No issues are able to defy the laws of probability, and understanding them gives one a significant advantage in decision making. For example, when asked how many people must be gathered in a room before two of them will have the same birthday, most people will give an answer somewhere in the hundreds. Only statistics can easily predict the answer, that you stand a 50–50 chance with just 17 people.

● *Consequence analysis*: In some cases the consequence of a decision interacts with the very issue upon which the decision was made. For example, a company which reduces its prices in an attempt to undersell its competitors cannot be surprised if they react by doing the same and thus restoring the status quo. The decision the company faces in the first place is not whether to reduce prices but rather, who can ultimately reduce their prices the most and how these reductions can be sustained.

Anticipating all the consequences of a given action is not usually possible and so efforts must be directed in the most important directions. One of the best-known and most comprehensive uses of this technique is the war game. Certain scenarios are fought out by staff taking the roles of all sides and the consequences of certain actions on the whole course of a war may be examined. Conducting such analysis becomes more difficult and less accurate, the further into the future the scenario is taken.

In Section 10.4.3 we initially looked at what a decision is and saw that while the substance of decisions may be very different, the processes involved are common. All decisions involve choice and, if implemented, they all have consequences. There are clear advantages to being a consistently good decision maker and the section progressed with a look at the three key areas of decision making: 'Start with objectives', 'Collecting the right data' and 'Decision-making techniques'. In the last section we introduced several very useful techniques for making evaluations that can be applied to all kinds of decisions.

10.5 Summary

In this chapter we have studied three major areas. These were 'Team Working', 'Group Dynamics' and 'Managing the Creative Process'.

In Section 10.2 we examined the potential benefits of team working and met holistic teams; a team whose whole is greater than the sum of its parts. This interesting and useful phenomenon was then studied and an experimentally verified model developed by Belbin and his colleagues was presented. The application of the theory in practice was then considered and it was clear that even when the opportunity for team balancing are limited, the model has useful management contributions to make. In Section 10.3 we discussed various aspects of group dynamics. The section started with an introduction and we then considered group dynamics by first examining the needs of a group and then seeing how the behaviours studied in group dynamics are able to meet these needs. We discovered that a great many behaviours are designed to answer the needs of a group. We do not just prefer to operate in groups, to some extent we are actually designed to do so. We are social animals and this result is in good agreement with experimental observations.

In Section 10.4 we examined the apparent paradox of planning a creative process and found that, in reality, there is no paradox. Planning creativity is possible. We described the problem-solving process and found that creative output can be increased by using the techniques of brainstorming and lateral thinking in conjunction with rational thought and deductive reasoning. After this we investigated the decision-making process and explained various techniques for ensuring that good decisions are made.

Taken as a whole, the material in this chapter has the potential to make a massive impact on our effectiveness as engineers. By using the team theories we have introduced, holistic teams can be composed whose productive output will far exceed that of unbalanced teams. We have introduced group dynamics and, as engineers, can improve our team-working success through understanding more of our team roles, why we have them and how they work. Finally, an ability to manage the creative process is clearly a benefit to the individual and so Section 10.4 is of use to the individual. However, when these skills are used in conjunction with good team-working skills a staggering level of effectiveness can be achieved. A team of engineers who competently apply the concepts of this chapter will never be limited by their own abilities.

10.6 Revision questions

1. Describe in your own words the following terms from this chapter.

   Apollo syndrome
   Brainstorming
   Group dynamics
   Holistic
   Norms
   Team balancing
   Matrix assessment

2. How did the team roles developed by Belbin come about?

3. What governs whether a team is more successful together than the members are singly?

4. Under what circumstances is team balancing particularly appropriate?

5. List five different 'teams' to which you belong.

6. Why might team balancing be of interest to engineers?

7. What features of Belbin's team model distinguish it from other such theories?

8. For one of the 'teams' to which you belong listed in question 5, get each member to answer Belbin's self-perception inventory.

   How well does your team fit together?
   Do you each agree that the team roles the questionnaire has assigned to you fit yourself?
   Do the others agree?
   What strengths does the questionnaire have?
   What weaknesses does it have?

9. What factors make improving the balance of your team difficult?

10. How does a group's task affect the group?

11. Why must successful teams have creative skills and the ability to resolve disputes?

12. How are these needs met?

13. For one of your teams listed in question 5, describe behaviour or incidents that are examples of the following:

    Norms
    Group culture
    Leadership
    Conformance

14. Why is there no paradox in planning to be innovative?

15. Think of an example of a creative process in which you have been involved and describe a process that either was used to plan it or that would have helped it to go better.

16. Get together in a group of five or so and brainstorm the following problems. Use the techniques described in this chapter and ensure that a timekeeper is appointed and used.

    (i) What sort of businesses could your team start?
    (ii) Choose a local traffic problem. How may it best be solved?

(iii) How many different purposes can your team think of for an electric kettle?

In your team think of another similar problem to brainstorm that is of benefit to the team or a team member.

17. What three steps are essential in decision taking?

18. Think of a decision you have recently made. What techniques did you use to arrive at it? Would using any of the techniques described in the chapter have helped or changed the decision you made?

19. In a group of three or four, choose a technical engineering problem with which you are familiar, perhaps the bending of beams or the design of a circuit. Select a goal to be achieved and then brainstorm four or five different approaches to the problem. Use the 'matrix assessment' technique to determine the best solution.

References and further reading

Belbin, R. Meredith (1981), *Management Teams: Why they succeed or fail*, William Heinemann Ltd.

de Bono, E. (1971), *Lateral Thinking for Management*, McGraw Hill.

Gawlinski, George and Graessle, Lois (1988), *Planning Together*, Bedford Square Press.

Maddux, Robert B. (1988), *Team Building*, Crisp Publications Inc.

Margerison, Charles and McCann, Dik (1990), *Team Management*, Mercury Books.

Stewart, Dorothy M. (1987), *Handbook of Management Skills*, Gower. A book in three parts. Part I, Managing yourself, Part II, Managing other people and Part III, Managing the business. Chapter 17 is a good general review of team building.

Torrington, Derek (1985), *The Business of Management*, Prentice Hall International. Chapter 14 deals with team building. The rest of the book forms a comprehensive investigation of the essential management skills required for work in business.

11 Personal management

Overview

Charity begins at home and so does good management. One cannot expect to be good at managing other people while unable to manage oneself. Personal management is more than simply knowing where one's things are, or having a tidy desk. It extends to planning one's career, setting objectives and being able to assess oneself objectively. In Chapter 1 we saw some examples of advertisements for engineering jobs. Personal management is particularly easy to scrutinize at interview time. Individuals who are weak at it are characterized by being poor timekeepers. They do not appear clear in their goals and are uneasy with questions that probe their reasons for wanting the job on offer because they do not truly know for themselves. The opposite characteristics apply to those with good personal management skills. Which do you want to be?

11.1 Introduction

Some people always seem to be organized and have their arrangements under control while others are muddled, often break engagements and are late for things. However, our organizational skills control more than just our arrangements, we use them to achieve our goals in life. If our personal organization is good we are more likely to be successful in achieving our life goals. The poorly organized drift, hoping success will come to them, the well organized name their destination and head straight for it.

This chapter is about improving our personal management and ensuring that we achieve the great things of which we are capable. It is of relevance to everyone since, whatever your level of personal organization, an improvement in it will always bring results.

The chapter is divided into three sections. In Section 11.2 we will introduce time management, good desk-keeping and boss management. In Section 11.3 the need for objectives and the process of writing objectives are considered. Objectives are the most basic building block of management and their use from a personal point of view is covered in this section. In Section 11.4 we shall look at appraising oneself. This section will look at career planning, the preparation and use of curriculum vitae, and finally, one's general wellbeing.

The learning objectives of the chapter are:

1. To understand how to be personally organized.
2. To understand time management and good desk-keeping principles
3. To realize the importance of managing one's boss
4. To develop the ability to set objectives
5. To understand the process of effective self-appraisals.

11.2 Personal organization

11.2.1 Time management

Time is a physical dimension that has puzzled scientists since humans first asked the question 'What is time?' Even today the question still cannot easily be answered by physicists. The average person is left with a concept of time as some medium through which things pass, changing as they go. One cannot affect time's passage, you cannot know the future and the only certainty is that eventually one's allotted span runs out.

People are very different. Some people achieve a great deal while others appear to achieve almost nothing. Some manage to find success and fulfilment, others do just the opposite. Making oneself happy and fulfilled in life is surely everyone's aim and yet some achieve it easily while others find it elusive. Although there are many differences between these types of people, one difference stands out above the rest. The way they use their time is different. Time is unlike any other resource. It cannot be stopped, it cannot be saved up in periods of plenty for use in shortage and absolutely everybody gets the same amount, just 24 hours a day. The difference is that the achievers use all their time effectively. People who do this are characterized by knowing what they are trying to achieve and managing their time to ensure that they get it done. They never seem to be involved in things that do not build towards their goals, and they frequently review their situations, making new plans and using opportunities to achieve their aims. People who are unable to achieve happiness have the opposite characteristics.

Time management is the rational way to ensure that our limited time is always used effectively. We shall now look at the mechanics of it.

Time management principles
As a first step to effective time management we have to identify those things that make us happy, the life goals we are currently seeking. These goals should include

all aspects of life – private, social, sporting, work – and finances should certainly be included. It is important that the statement of these goals is very clear and Section 11.3 will help greatly with the preparation of these. For example, a person moving to a new town to start a new job may want to do well in the new post but also meet new people and make friends, perhaps buy a house and run a car. These things potentially conflict, money must be saved for a house but spent on the car. Time must be spent on the job yet it is also needed for socializing. Such a person will have their own ideal balance between these needs and it will not just happen alone, they must be managed, time included.

The following paragraphs describe the essential steps required to manage your time. These are having clear objectives, prioritizing tasks, sticking to plans, making time to manage time, dealing with the unexpected, and controlling time wasters.

- *Clear objectives*: Time management starts with these overall goals which are then broken down into more manageable tasks. There may be smaller, individual objectives formed to deal with specific issues or on-going tasks that require regular blocks of time. Tasks that are sequential are divided into manageable chunks, planned and deadlines set. These are needed before time management can begin. You cannot organize yourself to achieve something without actually knowing what you are trying to achieve. To use a navigating metaphor, you cannot plot a course until you know your destination. If you do not have your goals clear, sort them out first.

- *Prioritize tasks*: First, break down the overall objectives into manageable chunks of work and collect all the tasks that are in progress. Second, assign priority to each task. Everybody has some tasks that simply have to be done and others that are less vital. The tasks on a 'to do list' should be prioritized and generous allocations of one's most productive times made to the most important tasks. At the other end of the scale, desirable tasks that are not vital can be scheduled into the background. Then, if other jobs permit, they will be tackled, otherwise they are left. It is inevitable that some low-priority jobs will not be tackled. This does not mean that time management is not working, rather that it is and you are achieving important tasks at the expense of unimportant ones. Good time management means that only jobs of a low priority do not get done. People who are bad at managing their time often complain that they do not have enough time to complete the important tasks. This cannot be: after all, they have just as much time as everybody else. What is usually happening is that some low-priority tasks are completed instead of the important ones.

- *Stick to scheduled tasks*: No system of organizing your time will actually do things for you! You have to follow your own plans, stick to them and get things done within your own deadlines if your system is to work.

- *Allow time to manage your time*: Setting out your time in an ordered way takes time. One must allocate a slot of time for time management. Even for very busy managers five or ten minutes a day is enough for the task. You spend money managing your money; you should spend time managing your time.

● *The unexpected*: Just because you cannot know what will happen does not mean you cannot plan for it. Individual jobs vary in size and complexity of course, but nothing we do is so unlike anything else that we cannot estimate time by some form of comparison. An organized manager will allocate periods of the day to background tasks which, though beneficial, can be delayed if the need arises to make way for today's crisis, and the undone work made up when the unexpected interferes less. If no such periods occur then the day is filled with too many tasks and it is time for a serious look at whether all the tasks should be underway. Perhaps too much is being attempted, or tasks that do not support the objectives have crept in.

● *Managing time-wasters*: All sorts of normal daily activities are time-wasters. It would be madness to think that they can be eliminated, indeed it would not even be good time management for two reasons. First, everyone needs to vary their mental activity rate and some periods of apparently wasted time. For example, a conversation in the corridor or the time spent photocopying can be beneficial due to the very diversion that they bring. The second reason is that sometimes these apparent time-wasters turn out to be goldmines of information. How many times do you pick up something new in passing or as an idle comment thrown in while talking about something else? For these reasons, it is not appropriate to remove all the traditional time-wasters. What is sensible, however, is to try to manage them so that the beneficial aspects are retained but the negative ones lost. We will now look at five of the worst time-wasters: procrastination, the telephone, meetings, the drop-in visitor and poor delegation.

Procrastination is, by definition, a waste of time, yet who is responsible for it occurring? One of the hardest lessons of time management is that it is not anyone else who wastes most of our time, it is ourselves. We choose what we do and so the responsibility is ours. Other people do impinge on our time but only as much as we let them. Procrastination is easy to do and hard to combat. Often people procrastinate because they do not relish the task in hand, they find it difficult, or it will bring them into contact with people they do not like. The best way to combat procrastination is first, to recognize that you are doing it and second, to be strict with yourself about the time allowed and the results required for the awkward task. Procrastination can extend to long timescales as well as short. Often people have 'pet tasks' that have been hanging around for months or even years and never get done. Some people procrastinate over major issues and the indecision can have a large effect on their lives.

The *telephone* is often cited as a time-waster which is surprising really, since, when used well, it is a very great time-saver. The important issue is to control the use of the telephone so that the communications performed with it are all contributing to the achievement of one's goals. A telephone call that brings understanding of how a goal may be achieved is clearly a useful call and may have brought the information more efficiently than by another method of communication. However, a half-hour call to explain something that takes seconds to understand in diagrammatic form or in which you achieve someone else's goals is clearly a waste of your time. The use of the telephone is a skill in its own right and is often taught on skills training courses to new recruits. The use of the telephone is discussed further in Section 12.5.

Meetings are another perennial offender for wasting time. As with the telephone, an effective meeting can be an excellent time management aid or a formidable waste of time. Meetings in which decisions are reached rapidly and actions are agreed by attendees who have the authority to do so can be as effective as they are brief. Conversely, meetings where the required authority for a decision is not present, where the purpose of the meeting is not clear, or where the attendees are unprepared can be as ineffective as they are long. The individual can ensure that time is not wasted in meetings by insisting on the good meeting practice explained in Section 12.6.

The *drop-in visitor* can be controlled at the times when they are unwanted by simply and courteously turning them away. Drop-in visitors generally waste time either by consuming time or by the very disturbance that turning them away causes. Strangers should have to make an appointment and a reason for the call requested. There is no need to be curt in such situations. The ubiquitous 'Hello, what can I do for you?' is both polite and to the point.

Poor delegation sounds more like an issue for personnel management than time management but in fact poor delegation is one of the greatest enemies of good time management. It means not being clear what outcome is required from a particular task. It brings three ways in which time is wasted. First, if a task has been poorly delegated its definition may not be clear, resulting in effort wasted in directions that do not relate to the task at all. Second, it may be that the task was not well chosen and does not contribute to the overall aims at all. In this case, even if the task is achieved, the time has been wasted since the task itself is of no use. Third, if a person does not delegate when really they should, they waste their own time since they end up doing the task themselves. An organization can lose substantially here, since if a long chain of delegation is affected every person involved is engaged in a task that truly belongs to their subordinate.

Time management systems

Many commercial systems are available for time management. It is important to realize, however, that you do not actually need specialized stationery – a loose-leaf diary is enough. However, the commercial systems do offer very practical solutions and many people who manage their time use them. The different systems all have their own particular format but typically they contain the items listed below. The cost varies considerably and they start at very reasonable prices but specialized, time management companies tend to offer rather more expensive solutions. Choosing a system is a matter of taste.

- *Binder*: It is essential that the system has a mechanism to allow the removal and insertion of pages. This is because as time progresses, past pages of the diary need removing and new ones inserting, and also since all the other items within the system are constantly being updated, the old versions must eventually be discarded.
- *Diary*: The largest section in any time management system is usually the diary. Diaries can be bought in almost any configuration from showing a year to a page through to one day to an A4-sized page. Most people prefer to be able to see a week at once but the important thing is that the one you have suits you by being large enough to contain the typical number of entries you wish to record per day. For people to whom this means a day to a page it is advisable to have another, synoptic, diary which shows how things will go over the next few weeks to keep an overview.
- *Address section*: A section for keeping the names and addresses of contacts is essential.
- *Blank paper*: Blank paper is often included for use in making up lists, notes, sketches, 'to do lists', writing notes for others, and any other activity best served by blank, rather than pre-printed, paper.
- *Report sheets*: Many manufacturers have available all sorts of report formats, including, for instance, monthly budgets or expense account forms. Some people find these useful especially since the binder keeps them all together. Often the use of the forms is restricted by the need to use the pre-printed stationary of your own organization.
- *General*: Many commercial systems include all sorts of other useful information such as maps, telephone area codes, common words in foreign languages, and even puzzles and diversions for idle moments.
- *Objective sheets*: These are inserts that are used for recording the objectives and overall aims. Separate sheets should be used for 'to do lists' or 'this week's tasks', since these will change while the objectives are relatively constant. These pages should be referred to from time to time to be sure of overall aims and, in particular, every time a new task has to be done the objectives should be used to decide whether to accept the task or not. Saying no to new tasks that do not assist the overall aim is the golden rule of time management.

Case study – Time management

A service manager for the subsidiary of a large manufacturing company selling washing machines has four service engineers working for him. The normal daily pattern is for the service engineers to perform the day's work and at the end of each day the team gather and arrange tomorrow's work. The meeting takes over an hour.

After his annual review the manager becomes aware that similar service departments in other companies service more machines per head. He is also alerted to the fact that his engineers feel overstressed, certainly their long hours of work must be tiring. Some of them complain that they are untrained for the jobs they meet. The manager's policy is to make the customer happy, but recently this seems to involve more and more areas of work. He discusses the situation with the service department personnel and the daily meeting comes under fire. After a long day on the road no-one wants to end the day thinking about tomorrow's headaches. The boss has a very long day because of taking his children to school first thing and is always run down by the day's end. The service engineers also feel that the meeting is in some sense a waste of time since it is impossible to plan a visit. 'Service isn't like that, you just have to get there and see what's involved,' said one of them.

After talking it through with the group and his own boss the manager introduces a new daily order. The department have the objective of solving all problems within two days. Overtime is to be used to provide more effort in time of need. Now each engineer phones into the manager first thing in the morning to collect the first job of the day. The manager gets in early, which suits him, and collects the messages on the answerphone and plans as much of the morning as he can. The engineers phone in between jobs and by lunchtime he usually has the remainder of the day planned for each of them. He lets the answering machine take over after about 3 pm with a message that the call has been logged and a return call to arrange a visit can be expected the next morning. The service engineers continue until the job is done and then get the customer to sign a form which includes the time at which work stopped. The boss uses the end of the day to do his office administration, catch up on the reading that goes with his job and plan possible future developments of the department. These less stressful tasks are done at the end of the day when he is least effective.

The manager now actually has time to keep up with the organization's developments and to install measures of effectiveness. Among other things, he realizes that the department are solving many more problems than rivals normally do and charging the same. He discusses these and other strategic issues with his boss and they agree a strategic business plan for the department to make service support become a marketing tool for the company. At the next annual review, the quantified success of the department is clear, the service engineers are happy and promotion looks imminent.

Case study observations

In this case study various important management issues can be examined. We shall only consider the time management issues.

The case study illustrates a common thread in management problems. Initially there seems to be no way forward. The manager knows that other departments in other organizations are more efficient yet everyone in the department seems to be overworked, including the manager. As a group they correctly perceive the daily meeting to be ineffective. However, the fact that they have no overall objectives or measuring systems, that they seem unable to control the flow of work to their own satisfaction, and that they do not seem to know what is required in order to become efficient are much deeper criticisms. However, even if these big issues could not be solved (in fact they did not even discuss them at the meeting), at least they can now manage their time better. The time-wasting meeting is replaced with a more appropriate communication system. Each engineer hears only what they personally need, not what everybody else needs to know too. The objectives for each service engineer are improved since they receive them during the day, when they already know how the work is going rather than the day before, based on pure guesswork. The change in communication method here is crucial and most service departments of this nature issue their staff with portable telephones.

Time has also been set aside for different tasks by the manager. Early in the day the engineers' workload is being managed, later the answerphone takes over and the manager is free to take on other tasks. It is not long before the manager is able to address the underlying issues of overall objective and measurement systems and is soon implementing them successfully. The new regime of 'management from above' has replaced the old one of 'management from below'. The manager now has departmental objectives with which to control the direction of the department. The needs that the organization has of the service department can be converted into objectives for it to achieve. The manager used to spend all his time serving the needs of his subordinates. They used to manage his time but he now does so for himself. A clear picture of what must be achieved and how it may be achieved has replaced the old order in which the group only responded to problems and were constantly battling against a tide of work without ever getting control of it.

The difference in management between the 'before' and 'after' case is very great and yet the way in which it was brought about was not radical or heavy-handed. Good management practice was introduced and it made way for more improvements. It is certainly true that scrapping the meeting saved time, but, as these observations have shown, that was a small change that eventually led to a completely new departmental time management system coming into operation.

11.2.2 Good desk-keeping

We all know messy colleagues. Some people run a desk so untidily that the desk surface cannot be seen. Such people would be outraged if they visited a supplier to find an order of theirs lying in a muddle on a messy desk, yet this is exactly how they deal with other people's jobs on their own desk.

A desk is the workshop of a craftsperson. Like all good craftspeople, the good desk user should only get out the tools necessary for the job in hand. The job will be completed and then tidied away before the next is begun. Tools and equipment are not left lying around to become damaged or muddled, and at the end of every day the workshop is cleared away completely. Most people judge a workshop from the way it is kept, and you can do the same with a desk and its owner. After all, if the owner finds working easy with so many things muddled up perhaps they have a muddled mind, perhaps they do not think the things are important, or perhaps the organization overloads the person to the point of their not being able to cope. In any case, the muddled desk indicates muddled work going across it.

Fortunately, there is a simple system to follow which will clear the mess. Once these rules become habit the desk is always used effectively and the craftsperson has more time to spend on creative work. We shall now look at the *clean desk rule*.

This states that the desk is to be clear of all things at the end of the working period. The only permissible exceptions are the telephone and the empty in-tray. One must be ruthless in despatching the clutter of a messy desk and even hardened managers find it almost impossible to accept that there are only four options for every item:

(1) *Bin*: Just because you have received something it does not mean it has value. The bin is the most underestimated aid to good desk-keeping we have. If an item does not fit one of the categories below then the bin is the only place for it. Periodically, scour the filing cabinet for binworthy material. Judge everything against your objectives and anything that does not fit in must go in the bin. Look at material not from the point of view of how useful it is but how it may be justifiably binned. Try to imagine the worst thing that would happen if you binned the item. Only reluctantly refrain from using the bin, try to wear it out with over-use!

(2) *Delegate*: If you cannot put something in the bin, perhaps it belongs to someone else. Ask yourself whether you are the correct person for the job. Delegation can be upwards, downwards or sideways. Just because someone else gave you the task it does not mean that you should do it. Only start to do something if you are sure that it is within your objectives and you are the best one for the job.

(3) *Act immediately*: Some items can be acted upon instantly and should be. There is no point in putting off a small action that might as well be done now or filing a trivial task, noting it on the 'to do list' and doing it later. A surprising number of things can be dispatched in almost no time at all, for example a single phone call, one letter, a payment, or an instruction.

(4) *File it*: Many things will need filing. Some will be for information only, to be used later when action is required, others will require action too complex to do immediately. These tasks are defined on the 'to do list' and then filed ready to be worked on at the allotted time. Some items may be of interest only, perhaps reading about your subject which should be done but will probably not produce particular actions. Whatever the situation, the first question is where to put it. People who cannot file things away to be completed later usually run an ineffective 'to do list'. They cannot trust the list and so dare not put something away for fear of forgetting it. Such people have the number of tasks they can simultaneously administer limited by the size of their desk. A successful manager will need to avoid such a limitation and must employ good filing.

There are many ways of filing – alphabetic, chronological, by size, for example – but for your own desk you need a retrieval system appropriate to your personal needs. The most effective way to accomplish this is to have an objective-based filing system. Start with the list of tasks that you are working on and have a file, or group of files, for each. Various benefits result from this system. First, all the information for a particular task is in a particular file and so the desk is not covered with many files for just one job. Second, when an objective is completed it is easy to archive, keeping the number of active files to a minimum. Third, since the files are laid out in accordance with the objectives it is easy to find the things you are looking for. Some items for filing may belong to a library of data relevant to the job, some may be reading material. To cater for these items two additional files, one for each, should be kept.

Finally, having set up a system, do not be tempted to let it lapse. Set aside a regular slot once a day, to maintain the workshop, and do the time management. In life there are few things more satisfying than a tidy desk after a productive day.

11.2.3 The boss–subordinate relationship

No-one shapes our view of work more than our boss. Bosses allocate work, they make decisions about us and our work, they have authority. All the things that motivate us in our work come via the boss. Is it any wonder that for most people, the image they have of their employer is really the image they have of their boss?

Bosses comes in many different shapes and sizes. Some people are natural managers with much skill while others may have accepted managerial responsibilities reluctantly in order to further their career. Some bosses need support from above, others are fiercely independent. Normally, people do not choose their boss and bosses rarely choose all their staff. Consequently, neither side is usually able to choose someone best for them.

A boss–subordinate relationship is a two-way thing. Bosses clearly want the relationship to be a productive one and so reflect favourably upon their own management. Similarly, the subordinate wants the relationship to be successful and to be fulfilled in the work done for the boss. Both parties are responsible since both contribute to its productivity and both can wreck it.

The following three areas of the boss–subordinate relationship are, sometimes wrongly, thought of as the responsibility of the boss. As can be seen, they are influenced by both parties.

(1) *Control of objective setting*: It is important that the objectives are clearly understood and Section 11.3 describes the mechanics of producing good objectives. Such techniques help enormously in ensuring clarity and are even more effective when done as a collaboration between boss and subordinate. The first step in managing your boss is to ensure that only tasks which have a minimum acceptable standard of objective clarity are tackled. After all, 'What, exactly, do you want me to do?' must always be a perfectly reasonable question to ask your boss.

If objectives are poor, the reasons should be examined. If they are simply poorly expressed, then the situation can easily be corrected. If, however, the weakness comes from a lack of clarity in the overall aims, or fundamentally weak premises, then there will be rather more work in correcting them. It is better to face the problems at an early stage and decide exactly what must be done rather than leave it until later and realize that unusable work has been in progress and so much valuable time has been wasted.

(2) *Manage expectations*: Much job dissatisfaction comes from a mismatch of expectations. The mismatch may occur either if the boss has an expectation from the subordinate who does not meet it or vice versa. Such mismatches are often associated with future career prospects or job content.

An employee who is ambitious and who performs well may feel ready to move upwards. It is vital that the boss in such a situation does not create false expectations of promotion. If a person is given to understand that success today will bring promotion tomorrow, the organization must honour its offer. You cannot blame someone for being disappointed if they feel they have earned a promotion but it does not come.

There are two issues here. First, the expectation, not just the promise, should only be given where the organization has the ability to meet it. Second, a clear definition of success must be understood. If objectives are used in the career management of employees there will be no disagreements over whether a given standard has been achieved.

Bosses and subordinates alike have ambitions and hopes, only communications between them and with the rest of the organization will ensure that a realistic view endures. Neither side is telepathic and it is madness to assume that the other party can know something if you do not tell them.

If a mismatch occurs it can be resolved only when it is confronted. The longer it takes to face an issue, the harder it is to resolve. For this reason, the management of expectations must be tackled whenever a divergence of opinion is suspected by either the boss or the subordinate.

(3) *Maintain the relationship*: There is no evidence to prove that getting on well with your boss as a person is central to a good working relationship. The converse is, however, not true, it definitely has a demotivating effect if you get along badly. From this we can conclude that it is necessary to avoid quarrelling but not to go further. In general, attempts to generate artificial friendships with bosses go down very badly. Of course, there are many friends who work for each other but there are clearly potential dangers and such people must constantly guard against the obvious conflicts of interest that might result.

It is not being likeable that makes a good boss, it is the quality of the boss's personnel management that is important. You can be happy working for a boss who makes absolutely clear what is required but whom you do not particularly like, whereas it is not easy working for someone likeable but who never actually tells you what is required and cannot guide you.

11.3 Objective setting

Objectives feature in all branches of management. They are one of the most basic and fundamentally important tools of management and they are constantly in use within organizations. Since objectives are so central, a good question to ask first is why we need objectives at all.

11.3.1 The need for objectives

There are two powerful influences that cause the need for objectives in engineering management. The first is the practical limits of communication. When a task is delegated one individual hands a task to another. The two must communicate the task and this allows room for misunderstanding. Objectives exist to eliminate misunderstanding from the delegation process. During the management of organizations complex pieces of delegation become necessary, and if during the process of delegation we are to avoid explaining every job within the task it must be explained unambiguously and succinctly. There must be no hidden assumptions. For example, if a manager says to a subordinate 'run the overseas section of the sales office for a while please', there is no way for the subordinate to know what is actually required. They may have a good idea from experience but that is not enough. If the person delegating had said 'over the next three months log 10 million yen worth of Japanese orders using existing resources', the picture would have been a little clearer. Objectives are also a desirable way to be managed from a personal point of view since they allow room to do things the way you want. You are not forced into methods that you do not necessarily approve of and you have the opportunity to bring your own ideas to bear on the problem.

In summary, the first need for an objective is to be an unambiguous answer to the question 'What, exactly, is it you want me to do?' This is a perfectly reasonable question to ask and a boss who cannot answer it should not anticipate admiration.

The second reason for objectives is to fulfil a deeply rooted part of our motivational psychology. Humans are 'goal-seeking mechanisms'. In Section 9.4.1 we reviewed some theories of motivation, and found that having something to aim towards is a very deeply rooted need. It seems we cannot be satisfied without pursuing a goal. This is often illustrated by the experience of people who strive for some particular goal for many years. The objective becomes all-consuming, and when in the end such people either achieve the goal or have to give up they often experience depression or even emotional breakdowns. Some may wonder how such people, who often have impressive abilities, can possibly fall apart so dramatically. The answer is, of course, that their objective system has been destroyed. For years every minute has been filled with work aimed at one goal, and then there is nothing. Although such extremes do not probably apply to you, most people are familiar with the lack of direction and anti-climax that follows a great achievement. This is the same thing but on a smaller scale.

People who are good at personal management will have many balanced goals. Some will be short term and others long term, they will have plans for today, next month and often for years ahead. In general, individuals who have such objectives and actively pursue them control their destiny, steer their lives, achieve much and are happy; those who don't, aren't.

We have now seen that there are two reasons for objectives: first, to provide unambiguous delegation and second, to serve a fundamental drive of humans, the need to achieve. By working towards an objective, these two things are brought together in one statement.

11.3.2 Writing objectives

An objective is an absolutely clear and unambiguous statement of a desired outcome. Objectives are used by people who delegate or by individuals who are managing their time. They may be corporate objectives applying to many people or they may be departmental objectives, applying to only a few. In all cases, there are various important characteristics of a well-specified objective that prevent ambiguity and foster accurate interpretation. There are many popular lists of attributes that well-written objectives should exhibit. In general, they include being quantified, time-based, measurable, feasible and compatible. We will now look at these properties in detail.

● *Quantified*: Objectives should be quantified, numerically if possible. If a given objective cannot be quantified then you can never know whether it has been achieved. Who would want to head for a goal such as 'reduce the failure rate' or 'make the price acceptable'? These goals do not explain what is required and consequently are demotivating for the individual and useless for the organization.

It is important that the qualification is appropriate. Words like 'all' or 'never' are special words, and should only be used to quantify objectives to which they genuinely apply. For example, the objective 'reduce the waste heat generated to zero' may be desirable and is certainly quantified but the use of zero is unlikely to be appropriate.

● *Achievable*: This component of an objective is aimed at two things. First, there is the motivation of the individual. It is clearly motivating to be given an objective that is exciting and, although tough, still achievable. The important factor is the individual's belief that the task is achievable. Motivation is a personal thing and it is therefore personal opinions that count. If an employee does not believe an objective is achievable, their boss will not sway them by simply stating that it is. The reasons for disbelief must be made clear and dealt with. The boss may be overlooking something that worries the individual – self-confidence or lack of experience perhaps. Second, achievable objectives make planning possible. Where complex, interrelated tasks are in progress planning is vital to optimize the use of resources and predict overall lead times. If the individual tasks that have been delegated prove unachievable, the overall plans will collapse as each failed task affects the progress of the next. If any of the benefits of planning are to be realized, there must be a known likelihood that each of the tasks will be achieved.

● *Compatible*: This attribute is related to feasibility but is distinct from it. Compatibility relates to a collection of objectives and describes the need for them to fit together as a whole. For instance, a design department manager might have the objective of 'answering 95% of telephone calls from customers with design queries within half an hour'. The same engineers in the department may also wish to improve their understanding of their customers' needs and have the objective of 'visiting 50% of homeland customers over the next year'. Both of these objectives may individually be feasible, but no one person can do both of them. One cannot always be in the office and always be out of it. Such objectives are not compatible. A set of objectives must be compatible no matter whether they are for an individual or a whole department. For this reason, objective setting always starts with one, central, goal that is supported by others that follow from it. Progressive subdivision ensures that conflicting objectives do not occur.

● *Time-based*: An objective which is not limited in time has no use whatsoever. Every result that is desired has some appropriate deadline. Open-ended objectives waste resources in rambling activities unrelated to central goals. If there is no particular time by which you need something, you don't actually need it at all.

● *Measurable*: This is related to being quantified. However, for an objective to be measurable you have to have access to an appropriate measuring system. For instance, it is quite possible for a computer software manufacturer to quantify the objective of halving the number of operating problems experienced by customers during the first year of operation of a particular piece of software. It may even be a very sensible objective to have chosen. But if the company does not acquire information from the customer about the operating problems the objective cannot be measured. Waiting for complaints or estimating the errors in tests are clearly not useful substitutes for the real data. Consequently, the engineers will not know if their efforts have been successful. If you cannot measure the changes you are making, you cannot know if they are having the desired effect.

11.3.3 Maintaining progress

When pursuing an important objective it is not surprising to find considerable interest in the progress of the work. The athlete aiming to win a race monitors progress in detail. Daily changes are used as a short-term pointer to the long-term success. This interest is much more than simple curiosity, it is used as a control mechanism to make success even more likely. If the athlete's progress is not fast enough, for instance, the training is stepped up, different training is used or advice sought from specialists.

Maintaining progress is a closed-loop control activity which starts with measuring progress, proceeds to comparing it with the desired amount and ends with action to correct any deviation. The loop is constantly in operation and has the effect of perpetually steering the work towards its objectives. The process has the same principle as control theory.

● *Measurement*: Various techniques are used to measure progress. Reviews are often held where the achievements of the last period are discussed and data from the measurement systems are presented. Progress meetings may be organized for suppliers at which demonstrations of progress are required. Whenever the individual is acquiring data that indicates in some quantified way how the task is proceeding, progress is being measured. It is important to bring the data together from time to time and ensure that the whole picture is satisfactory.

The measuring of progress is not over until it has provided unbiased observations upon actual progress whose accuracy is quantified. Only when you are in possession of sound data can you make a reasoned analysis.

● *Comparison*: This stage of ensuring progress involves the comparison of the data against the plans made to achieve the objective. In Chapter 8 and Section 10.4.1 we discussed different planning techniques. The comparison stage of maintaining progress is when the data of the measurement stage is compared against the plan to quantify both the progress made and the work remaining.

● *Corrective action*: Using the information gained so far it is possible to formulate rational paths of actions that offer the best route towards the objective from the current position. Once a certain position has come to pass the management question is not 'how did the situation come about?' but 'how best to proceed from here?'

These three steps of the closed-loop control process should be operated at appropriate intervals for the goal in question. It is through operating control systems that the decision takers can be alerted to danger or opportunity in advance of its arrival. The benefits of getting such information at the earliest possible opportunity are self-evident.

11.4 Self-appraisal

People vary greatly in their ability to manage themselves. Those that are good at it and are clear about their objectives manage their time to ensure that they achieve their objectives and they monitor their progress towards them. One of the key processes of good personal management is the ability to review yourself and see impartially how you are progressing. Such a process is called self-appraisal.

The main difference between appraising yourself and appraising someone else is that it is even more difficult to be objective about your own achievements and failures than about those of another person. In Section 9.5 we looked at the appraisal process as it is used between a boss and subordinate. Self-appraisal is very similar and still involves the review of the previous period, comparison against objectives and new targets for the next period. It is very easy to fall into the trap of selective memory and post-rationalization where our own performance is at stake. We do not find our own failure palatable and so we tend to remember only the good things and to think up rational reasons for our behaviour after the event to cover up any irrationality at the time. Consequently, serious self-appraisers will record it for future reference. Particularly where objectives for the next period are concerned, it is important to be clear and have a written record.

Of course, a self-appraisal may consider any item the individual feels appropriate but there are three topics that are of perennial interest. These are career planning, the curriculum vitae and general wellbeing.

11.4.1 Career planning

Good career plans have two aspects to them, the long term and the short term. The *long-term* planning should start with what sort of person you are and what you want to achieve. Section 9.4.1 examined the things that motivate people and they should be considered from your own point of view. Your long-term plans should provide you with a life that is fulfilling and that involves you in work you find deeply motivating. As with many aspects of personal management, doing this successfully depends greatly on your ability to be objective in assessing yourself and being able to separate out into different areas the reasons for having certain views. Those who can be objective about their feelings are much more likely to be able to make clear and rational decisions about themselves and their future than those who cannot. The *short-term* planning process involves tactical moves towards the long-term goals and selecting the best next step towards them. A short-term plan answers the question 'what am I going to do when the present career phase is over?'

Between these two plans we can be sure of constantly steering ourselves towards a fulfilling life. Clearly, taking the view that we can bring such achievements about and actually doing so is more likely to lead to fulfilment than simply hoping that it will happen. One must recognize that as life progresses, views may change and so the plan should too. If your outlook has changed, you must look for the basic change in motivation that caused it. Perhaps new long-term aims are called for. Many people do have changes of career, not because they choose wrongly in the first place but because they themselves have changed. It would certainly be a shame to miss out on such opportunities by simply not bothering to think about it. The important thing is that we always know where we are heading. For instance, an engineer may place importance on becoming a member of a professional body and perhaps a chartered engineer. If so, then planning will be required to ensure that the necessary post-graduate training is available from prospective employers, and choices of employer may be affected by this ambition.

11.4.2 Curriculum vitae

The curriculum vitae (CV) is a description of yourself, prepared for the benefit of another who wishes to view a synoptic account of you and your skills. Your CV is your envoy, it goes ahead of you and it represents you. It is particularly useful and should find frequent use when applying for jobs. Whenever a self-appraisal is conducted, it should include a look at the CV. Ask the question 'will my CV impress the next person who uses it?' If not, then plans can be made to gain the appropriate skills and experience to improve it. One's CV should always be balanced, meaning that it has a good cross-section of the skills needed in your field of work. Breadth of knowledge and experience are just as important as depth. From these points we can draw several conclusions about the way in which a CV should be prepared:

1. It should be appropriate for the organization to which it is sent. CVs should be tailored for each job application.
2. It should be synoptic, there should be no unnecessary verbosity, but it must provide the reader with the information required. People receiving CVs have many to read. The long and boring ones get skipped over, just as you would if you received them. If your CV is not a clear and complete summary of yourself the reader will infer that you are incapable of preparing such a piece of writing or will assume that there is something to hide.
3. It should be comprehensive and up to date. If you received a CV from an applicant that was ill prepared or out of date you would be disposed to reject

CURRICULUM VITAE

Name: Anne Example BSc (Hons)
Date of birth: DD.MM.YY
Address: XX your road, your address, your postcode.
Status: Married with one child aged eight.

EDUCATION

Date	Place	Qualifications
XX-XX	Crompton Park School	Eight O-Levels, three A-levels (Maths, Physics, Chemistry)
XX-XX	University of Lorrowbridge	BSc Hons in Electronic Engineering Class II (ii)

CURRENT POSITION
YY-Present-day
Senior Designer in High Current Division at Mainwaring Peylon plc.

I joined Mainwaring Peylon as the project engineer for the high current thermo-intelligent switch programme. I am responsible for all aspects of the group's management. I have seven people reporting directly to me including three graduates.

The remit of the group is the complete design and manufacture of a new product range. This includes the design of the switch amplification cavity and the three-dimensional multi-parameter finite element analysis. I personally designed the resonator control system. I have negotiated all aspects of the contract with the sales and marketing division which involved several trips abroad to provide technical support for the division and to visit customers. I am also responsible for all quality assurance aspects of the programme. The programme is nearing completion and I am engaged in identifying new work for the group at present.

PREVIOUS WORK EXPERIENCE
Last job YY-YY: Project Engineer at Halton Freionics Ltd: employed as a Project Engineer reporting to the group leader. I was responsible for all phases of product development from initial customer visits to commissioning. Projects undertaken included a Muon-catalysized fusion power cell for specialist hi-fi systems and a portable 27 TeV, 10A electron accelerator for pest control.

I also handled the commercial account with the company's second largest German customer. This included price setting, hosting visits, product support and the handling of all communications.
Job before that YY-YY: Dagaba Systems Inc. (UK): employed initially as a graduate trainee and ultimately as a salesperson. In my training period I was introduced to various functions of the business and learnt how the organization functioned as a whole. I liked travelling and was successful in my application for promotion to sales. In my four years in sales I travelled extensively in the UK and made three trips to the United States and the Far East. I became interested in the process of product development and spent more and more time on new product generation. Eventually I left to expand my opportunities and joined Halton Freionics Ltd.
Job before that YY-YY: Immediately after graduating I spent a year abroad travelling. I worked in Australia for nine months while travelling. The time included two months as a hospital assistant.

Figure 11.1   Example of a curriculum vitae

PROFESSIONAL TRAINING
Since graduating I have attended various courses in professional skills:
Time management (5 days)
Communication skills (7 days)
Management organization and teams (10 days)
Appraising employees (1 day)
Leadership skills (5 days)

SKILLS AND EXPERIENCE
I have a broad range of personnel skills including dealing with people at all levels within organizations. I have briefed a board of directors on many occasions but have also trained apprentices. I have interviewed candidates for selection and have conducted performance-based appraisals.

Upon delivering equipment I have presented training in its use and written instruction and service manuals for most of the equipment I have designed.

I have experience of various computer packages and have supervised Computer Aided Design draughtspeople on both electrical and mechanical systems.

INTERESTS
I am interested in the philosophy of design and have given various guest lectures at my old university college. I enjoy sailing and in 19YY I was entered in a national ocean-going yacht race. I was second rigger on an all-woman's crew. I attend night school in art at a local college where I study photography. I enjoy both listening to and performing music, I play the piano and also sing in a choir. I enjoy cooking and entertaining for friends.

PUBLICATIONS
In 19YY I co-authored a paper on the magnetic confinement problems of muon-catalysized fuel cells and presented it at a conference in New Jersey.

REFERENCES
The following people may be contacted in connection with my application and will be happy to answer any questions.

Name: Address: Relationship

Name: Address: Relationship

Figure 11.1 (continued)

it. If the applicant takes little care over something so important, how much less care will be taken with tasks the organization might set?

4.   Until you meet, the organization knows nothing of you except your CV. It goes ahead of you and represents you. It should therefore be neat, tidy, short enough to read fairly quickly, written well enough to make the reader to want to know more. The spelling, grammar and punctuation should be beyond reproach.

Although a CV should be prepared for each application, all CVs have points in common and so it is possible to use a format that will be generally applicable and then update the content as required.

All people who receive a CV will want to know some of your personal details to form a picture of you. It is therefore normal to include a section of your personal details and explain at least something of your education. In the case of a CV prepared for employment the receiver will want to know whether you have the ability to do the job and whether you are likely to be a personable employee who will fit well into the organization. To provide for these needs, your CV should include a description of what you are doing at the moment and what you have done in the past. Early in you life you will have far fewer things to talk about than later on, and so that the CV stays at a manageable size it is normal to say less about your early achievements as time goes by. For instance, in the CV shown in Figure 11.1 there is very little said about the time immediately after graduating since much has been achieved since. In earlier days this section would have been much larger and might have included things that the person learnt then or more details of the work undertaken. Personal skills and experience are of interest to someone receiving a CV since they say a lot about how someone views themselves. After all, why should you be considered for a post if you do not even have something to say about why you might be particularly suited? Finally, to complete the picture of what you are like, the reader will want to know something of your interests or pastimes. If you read some CVs it is more likely that you would remember those people that had something interesting and human about them. Not everybody has dangerous or exceptionally exciting hobbies to put down, but everybody does have something that interests them, or that they enjoy doing. Put it on the CV and the reader will have a better chance of knowing what you are like.

The example CV in Figure 11.1 is of an engineer with several years' experience applying for a job as a chief design engineer in the electronics industry.

11.4.3 General wellbeing

Every self-appraisal should include a review of general wellbeing – health, enjoyment of life, whether you are getting enough exercise, sleeping well and all the other things that compose a wholesome life. A healthy body and healthy mind will always go together.

People whose general wellbeing is good are characterized by having a happy home life, a group of friends they regard as permanent, good health, enough sleep, leisure pastimes and a satisfying vocation to fill the weekdays. All these things are equally important and they should all be reviewed, not just the last one. Any development that deprives the individual of their required balance causes stress. The more the imbalance, the greater the stress. Good personal managers are clinically accurate in identifying the source of their stresses. Their good self-appraisal skills give them the earliest warnings possible and they rapidly execute strategies either to defeat or to avoid the stress. They undertake actions that build up their general wellbeing.

Interestingly, people's ability to maintain their general wellbeing is often affected by it and the situation is unstable. Those who are doing well find it easier than those who are not. Stress causes a lack of wellbeing and the most common causes of stress are not having enough time, not having clear objectives, and not being able to take a clear, objective look at your own progress. In this chapter we have discussed ways of managing these three things and therefore have provided material of use to people trying to reduce their stress levels.

11.5 Summary

In this chapter we have introduced personal management. You can use personal management to help achieve life goals and be fulfilled.

In Section 11.2 we introduced time management, effective desk-keeping and boss management. Through these activities you can be sure that everything you are doing is assisting you to achieve your life goals; that you use your desk to its best advantage; and that you have the most productive relationship possible with your boss. In Section 11.3 we introduced objectives. We examined the need for them and found that there are five attributes of well-written objectives. The section ended with a look at how progress may be assured and the overall operation of objectives was seen to be similar to the operation of a closed-loop control system. In Section 11.4 we discussed three important aspects of self-appraisal. These were career planning, curriculum vitae and general wellbeing. In the section we saw how these may be reviewed. It is clearly a matter for you to decide that you will use these tools to ensure that you achieve your goals.

To find deep satisfaction in life, you have a choice. You can either just hope that some day you chance to come across it; or you can use self-management to head straight for it. The choice is yours.

11.6 Revision questions

1. Why is personal management important? What advantages does it bring? What characterizes people who are good at it? How may it be achieved?

2. What is 'time management'? How does it apply to the following groups of people?

   Students
   Qualified engineers
   Salespeople
   Shopfloor apprentices
   Executive Managers

3. Keep a log of your activities over the next week, accounting for periods of time no smaller than one hour. At the end of the week, total the amount of time you have spent on each category of activity. Does the way in which you allocate your time look sensible when compared against your objectives? How much time has been wasted? Could you manage your time better? How?

4. Could your filing system be improved? Can you always find the material you are looking for? Explain to someone else what problems you have with your filing system and how you will change it to overcome them.

5. What can you do to control the effectiveness of the working relationship with your boss?

6. What attributes do well-written objectives have? Find an objective of your organization and compare it against these attributes. Does it have them? Rewrite the objective and improve the quality of writing, taking the required attributes into account.

7. What are your objectives for the next year and for the next five years? Write down five things you want to achieve over the long and the short term. Work on the statements you have prepared until they have all the important attributes of a well-written objective. Show them to a friend and ask for comments.

8. How is progress towards objectives maintained? How might this occur in the workplace of engineers?

9. Prepare a CV for yourself. Ask a friend to go through it with you and compare it with the example in the chapter. Will it impress someone to whom you might send it? What things do you think you should do in order to improve its balance?

10. Explain the following terms:

   Clean-desk rule
   Compatibility
   Time-wasters
   Objective.

Further reading

Adair, John (1988), *Effective Time Management: How to save time and spend it wisely*, Pan Books. A well-written and informative account of time management principles. The book is easy to read and includes sensible guidance to improve your time management abilities.

12

Communication skills

Overview

In the previous chapters we have seen the various tools of management described as individual subjects. In the first two we explained the environment in which engineering businesses operate. In the subsequent chapters we saw how the needs that result from this operating environment may be met. We have seen how the personnel needs of an organization may by served, how teams operate, how objectives are set, how money and information is controlled, and how projects are managed. However, there is one skill that underpins all of these, a skill without which none of these things would be possible. This is the skill of communication.

The case study in Chapter 13 illustrates all the managerial skills we have described in the previous chapters. Team-working issues arise, financial issues are involved. Sometimes information control is foremost, at other times it is appraisal. Communication is there all the time. So all-pervading is communication that every single paragraph of the case study involves a communication of some sort.

12.1 Introduction

Communication is a vital skill for any professional. No-one can be effective in an important job without being a good communicator. It has even been said that there are no management problems in the world, only communication problems. In this chapter we will see the importance of information and consider examples showing how crucial particular pieces of information can be. Communication is the process by which we acquire all information and it is therefore worth studying in order that we may become good at it. Communication is important, it is fundamental to engineering, as it is to so many other things in life. Unless an engineer can effectively communicate designs, ideas, and instructions, the requirements of the job will not be met. Engineers who are ineffective communicators may also endanger life by not adequately defining critical processes or giving warnings. They are unlikely to be able to achieve satisfactory positions in their companies because they not only fail to 'sell' themselves but they also give a bad impression to customers, and they may cost the company money by not correctly determining the customer's needs.

All management tasks involve communication and it is not possible to be good at managing without being able to communicate well. In this chapter we will first examine the role of communications in the workplace. In Section 12.3 we will look at wider communications and at some of the techniques that can be used to gather the information which will allow us to do our jobs effectively. In Section 12.4 we will present techniques for ensuring effective communication in the common forms of writing. Section 12.5 deals with oral communications in a similar way. Finally, in Section 12.6 we will examine some of the ways in which meetings may be managed to ensure that they are effective.

The learning objectives of this chapter are:

1. To understand the need for effective communication in the workplace
2. To describe some of the methods that are commonly used to aid communication in the workplace
3. To provide guidance for achieving good oral communications
4. To show the accepted forms of written communications that will be used by engineers
5. To explain the role of meetings and to consider how they can be run efficiently.

12.2 Communications in the workplace

The workplace of even a modestly sized organization is a microcosm of the outside world. The complexities that exist in the world are usually in some form within the organization. Where communication is concerned this is also true. To understand how communications operate in an organization we need to do two things. The first is to understand what the organization needs from its communication system and the second is to look at the various methods that can be used to aid the communication process. If we understand communications properly we should be able to answer questions like 'how can an organization tell if the communication system in use is satisfactory?' 'What factors affect its efficiency?' 'How may the existing methods be altered to improve communications?' In Section 12.2.1 we will look at the needs of an organization that result in a communication system. In Section 12.2.2 we will consider the methods that are used in the communication system and the aids that assist the process.

12.2.1 The purpose of a communication system

Only by understanding what the organization hopes to achieve from its communication system can we expect to improve it or to design it well.

Understanding a communication system means first, knowing what purpose the system serves. We can then see how it might achieve that purpose and understand the limitations that a system might bring. We therefore turn our attention to why the communication system exists.

To understand what purpose a communication system serves we must, as with so many other aspects of engineering management, look at the organization's corporate objectives. In Section 2.2 we examined the nature of corporate objectives. We saw that an organization of any size needs somehow to make sure that the efforts of the people that compose it are all directed in some beneficial direction. We saw that using corporate objectives is the way in which this is achieved. An example of a corporate objective is given below.

> To return a profit of £75 000 by supplying centrifugal water pumps to European customers over the next calendar year.

In order to achieve this, or any other, objective the company needs a communication system.

The ideal communication system for an organization is the most cost-effective one that makes it possible to achieve the corporate objectives. Let us consider the communication implications for the above objective.

The first and most important implied communication need is for the directors who set the objective to communicate it to the managers of the company. With any objective there will be wide-ranging consequences and it is imperative that the managers not only understand what they are required to do but that they also know how the rest of the company is going to act. So we immediately have a requirement for downward communication, the directors to the managers. We also have a requirement for horizontal communication: the managers must communicate with each other. At this level of operation the communication must be accurate and formal, there must be no room for misinterpretation or lack of detail. The consequences of inaccurate or complete information are too great.

If the objective has been set so that departmental managers can develop their own strategies for meeting it, this again implies interdepartmental communications. However, it also implies that there must be a route back up the organization to the directors, particularly if there are likely to be any problems with meeting the objective.

If we now look at the other communication requirements for meeting this objective we can see that there is a need for customer contact. To trade effectively one must understand and know the world of one's customers. One cannot depend on their complaints to guide strategic commercial decisions, one must actively gather information, and to do this one must communicate with them. In doing this, the organization, and the individuals who communicate on its behalf, may use many different techniques but they are all employed to find out what actions will lead to satisfied customers. In the case of this objective the organization must know what the customers expect from a technical, commercial and service point of view.

Different customers may have different expectations, and one type of pump and method of sale may suit one customer and not another. If the organization is to be successful in making both satisfied with a purchase, different technical and commercial methods may have to be used for each customer. The market for the pumps may change with time, developments in technology may alter customers' expectations or changes in exchange rates may affect the price of competitive products. To be effective in selling to its customers the organization must be knowledgeable of all such developments in the customers' situation, and this means communicating to find out.

The next communication need implied by the objective is for knowledge of the financial performance of the organization. The organization cannot achieve this objective without knowing such things as how much each product costs to manufacture, how much are the overheads or the profit margin on each sale. It is only by ensuring that this important information is communicated to the managers who need it for decision making that the information can be effectively used. Similarly, measurement of the achievement of this objective requires adequate financial data. Many communication techniques are used in this process and there will certainly be a requirement for communication in all directions, with cost data being communicated across departments but budgets being set at a higher level. The directors will also want feedback on the costs of implementing the strategies for meeting the objective. As the information is communicated upwards in the organization it becomes more synoptic and describes the performance of increasingly larger parts of the organization. Computers play a major role in producing financial data for decision making and these certainly make the data to be communicated more widely acceptable and understandable. Ultimately, the communication of this financial information to the board of directors is the responsibility of the financial director, and such a post can even be defined in this way.

A further communication implication of the objective is the communication between the individuals of the organization. If the objective is to lead to actions for every person in the organization they will somehow have to made aware of what actions are required of them. The whole process of implementing strategies to meet objectives is one of delegation, and the managers at every level have to be capable of taking responsibility for their own particular area and then sensibly deploying the resources under their control to achieve their own objectives. The communication skills involved in managing a group of professional engineers are considerable. Communicating with those for whom they are responsible can absorb a large proportion of a manager's time, and for some it is the only task they do. The need to be good at communicating with people at a personal level cannot be overlooked, and the need to have an organized approach to handling departmental communications is equally important. Nothing is so frustrating for the employees of a department as not knowing what is going on.

Finally, the last communication issue implied by the objective is with the outside world. There will clearly be a need to identify and trade successfully with the organization's suppliers. There will also be a need to keep up to date with any legal requirements of the organization's trading position. Most organizations have some sort of relationship with at least the local community and often with a much wider public. Large organizations may employ a public relations department whose sole purpose is to ensure that communications with the outside world are handled effectively and that the public image of the organization is managed.

In all these communications there will be protocols that define the way in which the communications take place, and which set limits on the methods of communication. For example, it may be agreed that all communication with customers is done via the sales department and that everything is confirmed in writing. One of the most commonly used methods of communication is the meeting, and most companies will have rules that define when meetings should be called and how they will operate. Meetings are so important in the workplace that we have included a separate section that explains how they work (Section 12.6).

Having seen the need for communication in the workplace we will now look at some of the methods available.

12.2.2 Communication methods and aids

In most organizations many different methods of communication are employed and we shall describe some of them below. We will also consider some of the aids to communication that exist. The order is alphabetical and it is not intended to attach importance to any particular method.

Appraisal interviews

These are used as a communication device between employee and manager. They concentrate on a review of the performance and possible development of the employee. They can also be used to discuss specific problems, but they are not very effective for this as they usually take place at fixed times, annually or six-monthly. One of the main advantages of appraisal interviews are that they provide a forum for looking forward and agreeing personal objectives for the next period. There may be a two-tier appraisal system in which the appraisee also has an interview with the manager's superior. This interview serves as a check that things are well between manager and employee as well as providing the opportunity for the employee to discuss wider issues such as company development. Provided the manager is someone whose judgement is respected and is the sort of person that will discuss issues objectively, appraisal interviews can be very useful. Appraisals are a very important part of personnel management and are dealt with in detail in Section 9.4.

Conversation

The majority of workplace communications result from ordinary conversation. This is one of the most useful forms of communication because it is so flexible. In a world of memos, meetings, reports and procedures the skill of holding a successful conversation is often neglected. There is more to having a conversation than simply walking up to a person and starting to talk. The way a conversation proceeds depends very much on the sort of conversation it is and the expectations of the parties concerned. A chat with a friend about the state of the country is likely to proceed along different lines from a more formal discussion with a superior about some aspect of work. In all cases a good conversationalist will be able to measure how well the information being sent is received, they use their interpersonal sensitivity to see the conversation from the other person's point of view and adjust it to produce the most effective result. They control the direction of the conversation so that the main point is not lost. Normally they know what they want from a conversation and during its course they examine the possibility of achieving their goals and constantly aim to achieve the best they can from the situation. They can readily turn their speech to convince, persuade, alienate, sympathize, praise, chastise, caution, commend, or to any other message they want to impart. Above all, they always leave the conversation with everyone involved clear about what has been discussed and who is responsible for any action.

Drawings and signs

Most would immediately consider technical drawings as being the main use of these in the workplace. There are well-established standards and codes of practice for the representation of physical objects as drawings (for example, the British Standard BS 308). A good example of the use of signs is the standard safety logos for various hazards: hard hats, safety glasses, and corrosives. Their main advantages are that they are very descriptive, overcome many language barriers, can be designed to attract attention and are interpreted very quickly.

Facsimile (fax)

Faxing involves generating a copy of an original document at another location by sending the information via the standard telephone network. Faxes have all the advantages of the telephone for speed and distance without its limitation of being unable to see a picture or have a written record. The person you are faxing does not have to be in attendance in order to receive the communication. It is transmitted and is then ready for collection at the receiver's convenience.

Grapevine

This is the unofficial communication system of any workplace. It involves the sending of information from person to person in an on-going chain. Most of the information conveyed by the grapevine concerns individuals in the workplace. It may be slanderous, malicious, untrue, or any combination of these. Provided the information is from a reliable source and has not been through too many 'hands' it can give an invaluable insight into what is going on, especially in judging difficult-to-quantify matters like the mood of another department. Clearly, the grapevine is an intrinsically unreliable communication system, depending as it does upon the vagaries of human conversation. For this reason, it can never be considered as a formal communication technique.

Information technology

Information technology (IT) is the generic term used to describe methods of manipulating data or allowing communication by computer. There is wide use of computers in the workplace for applications from word-processing to mathematical modelling. The communication advantages they bring are speed, transferability without misinterpretation, and the ability to store a vast amount of information and retrieve it quickly. Information is often transferred directly from computer to computer, as, for example, when a buyer places an order directly onto a supplier's system, or when a flight reservation made by the travel agent is simultaneously logged with the airline. This offers advantages of speed and cost as there are no pieces of paper involved or laborious manual processes. The process is not limited to technical information as electronic mail systems allow messages to be written directly to a user's terminal or file anywhere in the world. There is practically no communication task for which software has not been written and communication has replaced computation as the primary role of a computer. Being competent in the use of a personal computer is now an expected skill for professional people. For more information see Section 5.4, and for an example of how IT can be used read the Stephen Lever case study in Section 12.7.

Meetings

These are convened to discuss all sorts of subjects. A potential problem is that as more people attend it becomes more difficult to achieve anything. Meetings are very useful for getting people to air their views and criticisms before a relevant audience or to review progress on a project, and they provide an opportunity for different skills and experience to be pooled. It is important not to consider them as a luxury for higher management as they give an opportunity for individuals to talk at length on a specific subject allowing a rapport or sense of cooperation to be achieved.

Memoranda

These are either communications with people outside the organization that are less formal than a business letter or they are used as a method of communicating within the organization which can be as formal or informal as required. They are characterized by being short communications that contain a few important pieces of information. Sometimes memoranda (memos) are copied to more than one person to indicate to the others on the circulation list that the information has been passed on. For this reason, it is important that memos are dated. There is no defined format for a memorandum but many establishments have pre-printed memo pads or sheets which are convenient to use and which can be folded to save having to use envelopes.

Notice boards

Notice boards have a variety of uses. They can be used to display general information of interest or to inform personnel of specific rules relating to work. It is important when using a notice board to put any information on display well in advance as very few people read a notice board more than once a week. Their main advantage is that they are cheap because only one copy is needed. However, unless they are properly managed they can become cluttered and information will not be noticed.

Pre-printed forms

These are used when something is required repeatedly: sickness, leave, expenses, stores issue. They give a standard format for anyone to use and it is easy to check that all the required information and authorizations have been included. Their great advantage is that they fix a protocol for communication in a specific case and provide a record for reference should any problems arise. Multi-part forms mean that a single written copy at once produces several copies so that all persons involved in its implementation can have their own copy.

Suggestions box

Often workers on the shopfloor have very good ideas for improving the way their tasks are done. As this is not usually considered part of their job there has to be a method of getting these ideas to the people who can change the system. Rather than appeal to individuals' better nature sometimes a financial incentive is used. Suggestions may be reviewed by a manager or a committee, and a decision whether or not to implement will be made. Although relatively informal, suggestions boxes can be quite motivational and shopfloor workers may see the box as their only means of communicating with 'management'.

Telephone

The telephone is the most important aide to communication that has been invented. It has revolutionized communication over the entire planet. The use of the telephone is a special skill and is so important to engineers and other professionals that we shall take a more detailed look at its use in Section 12.5.2 as part of our examination of oral communications.

12.3 Information gathering

Information is often the most valuable asset we own. By knowing where the enemy will strike, a defending commander can win a battle before it has even begun. A company that accurately knows the internal affairs of a rival is at a great competitive advantage. Even something as trivial as a purchase is affected by information. One cannot know that a purchase is good value for money without first having information about the options available. It can therefore be of immense value, but not all information is valuable. Information has three main parameters that give it its value. The first is quality. The quality is clearly significant, and where something important is at stake the truth of the information must be questioned. Before undertaking a major commitment one must be sure that the information upon which it is based is sound. The relevance of a piece of information also affects its value. If two people are bargaining over a sale, knowing the upper price acceptable to the purchaser is of great value to the seller but of no interest to a third party who has nothing to sell. Lastly, the timing of the information is also important. No-one is interested in what their commercial rivals were doing this time last year, it is the next year that is of value. Likewise, knowing where the enemy will attack is priceless beyond measure. Knowing where the enemy actually did attack is history.

In our everyday lives we are constantly in contact with potential sources of information. Every day we are exposed to far more information than we could possibly use. The important thing to do is to use one's objectives to define the information you require and then efficiently obtain it. With the invisible commodity of information, it is quality not quantity that counts. Good engineers use information effectively. Like great statesmen or military heroes, they make well-informed decisions based on a rational treatment of the relevant facts. For engineers there are many types of information – technical, personnel or commercial, for example – and there are many places from which the information can be obtained.

In this section we will look at some of the many sources of information that engineers have at their disposal and some methods that help assimilation. We will consider the customers of the organizations, a special group of people with information of particular importance to engineers. We will also consider suppliers, the boss, the use of libraries, personal filing, reading skills and note-taking.

12.3.1 Sources of information

Customers

Customers can provide four types of information. They will have technical knowledge which can be of use to a supplier. For example, a number of companies now use partnership sourcing and as part of this a customer will help a supplier to implement new management technologies or systems. Customers can provide information about their needs. This is clearly important when a customer order has to be met, but in addition to their own specific needs customers can provide an indication of general needs of customers in their industrial sector. This information can be used by marketing and engineering functions to develop markets and products that will meet future specific requirements. Finally, customers can provide market intelligence. They can give information about what competing organizations are doing, new products and their performance, price and delivery information. All these are useful to an organization that has to compete.

A common technique used to determine customer need is to listen to complaints and react to minimize them. While this is better than nothing, it is not enough, and can never provide a complete understanding of the organization's customers. A fact that is often quoted is that while only one in twenty-five customers complain of their dissatisfaction, on average, they will each tell ten other people about a bad experience. The organization is not only the last to hear about its errors, its poor performance is widely advertised. Complaints are therefore a very poor indicator of customer satisfaction. If you want to know about your customers, you are going to have to find out by listening to them and, where possible, by visiting them.

Listening to customers means much more than sending out a few questionnaires, listing the causes of complaints or looking at service department records. The need is for active listening, actually putting effort into getting the required information. One cannot aim to meet the needs of one's customer without knowing what those needs are, and one cannot expect customers simply to come along and describe them to you. Customers are often unsure of their needs themselves. An organization that assists them in accurately identifying their real needs will stand a far greater chance of generating customer satisfaction than a less attentive rival.

The easiest way to get a clear understanding of your customers' situation is to go and spend time with them doing what they do and seeing things from their point of view. The stereotyped image of good product development coming only from the application of research and development in a well-equipped laboratory, peopled with excellent academic brains, could hardly be further from the truth. Success in developing new products means meeting customers' needs, and an organization cannot expect to know what these needs are without being in good contact with the world of their customers. Visit them often.

Suppliers

Just as customers are useful in the provision of information so are suppliers. Organizations that supply products or services put effort into selling them. Like any other organization, they attempt to produce happy customers. A great service that suppliers can offer is technical advice. There are so many processes that are used and so many factors that limit these processes that no engineer can be acquainted with all of them. However, if the supplier is offering a specific process then there will also be technical advice available on that process. When engineers face problems with which their organization is unfamiliar, it is far better to get information from a knowledgeable supplier and to collaborate with them than it is to do it all yourself.

Boss and colleagues

One's boss and colleagues should be an invaluable source of information. A competent boss will be able to direct subordinates to information and guide them around the organization, but will often only do so if asked. The information held by the boss is especially useful to new recruits. However, while it is certainly appropriate to use the boss to gain information about the upper levels of the organization which cannot easily be found out in other ways, it is certainly not appropriate to expect a boss to provide general information available from other sources. The most important information a boss owns is a knowledge of which objectives are most important and why. In addition, a competent boss will be able to help an employee who gets stuck, either by giving information or by directing to an appropriate source.

One's colleagues are also sources of information and can be used to assist in learning the procedures of the organization, the technical issues or the history of the work currently underway. This should be a two-way process with the individual providing something in return; colleagues soon tire of repeated requests for help without getting anything in return. Learning by helping is both informative and fair. Colleagues that need increasingly to exchange information soon develop into teams. Teams can be especially efficient at information handling and Chapter 10 described this subject in detail.

Library

A library is a vast information storage and retrieval system. Libraries vary considerably in size but even the smallest usually contain far too many books to allow users to find their way around unaided. Many systems of referencing have been developed to assist the users of libraries in their search for information. At most libraries it is possible to search the whole collection on a computer index by author, by subject, or by 'keyword'. Good information gatherers are expert users of their library systems and regard their library as an extension of their own filing system.

Within the library there will be several sources of information. Journals such as those produced by the engineering institutions will provide articles and reports on new products and technologies, as well as access to more specific trade literature from advertising. Standards, such as those produced by the International Standards Organisation and British Standards Institution provide agreed technical specifications for many products in addition to protocols for design and manufacture. Information technology has transformed the task of searching for data on specific topics with the CD-ROM. These are now standard in most libraries and provide access to a number of on-line databases which can be searched by keyword. The searches will produce abstracts of articles and books on the selected topics, with each abstract indicating the information that is to be found in the full document, thus allowing the reader to decide whether or not to obtain that document.

12.3.2 Assimilating and organizing information

Personal filing

We all have our own personal filing system. The function of the system is to permit rapid retrieval of previously gathered information. This extends from the way we categorize and store our own books, magazines and collections of other printed matter through to how we arrange to store information such as when the local swimming pool is open. Most people have reasonably effective storage systems for their books and notes, but few have a comprehensive system that deals with all the information they collect and might want. Personal filing is a part of personal management and a simple system is described in Section 11.2.2.

Reading

Practically everything that is worth knowing is written down in some way, somewhere. Reading is therefore a particularly important part of information gathering and it deserves special attention. Printed matter has many advantages over other forms of information. It is permanently recorded and may be left and returned to later, unlike a conversation. It does not fade or change as verbal accounts may. By using the conventions of good writing such as contents lists, structure, and cross-references the author makes it possible for the reader to quickly search entire volumes for specific pieces of information. No speech or meeting can do this. Printed matter, or rather the characters, syntax and grammatical rules that compose it, is designed to assist the information-gathering process. There is more to using it well than at first appears, and people like engineers, who need good information skills, need to be effective readers.

Reading is a skill. We normally learn to read very early in our lives and many people cannot even remember being unable to read. As we grow up we improve our reading and use it to serve many needs. In our adult life we read at many different levels. At one end of the spectrum is the acquisition of factual detailed, even mathematical, material which is very condensed and contains a great deal of meaning. At the other end might be a collection of articles passed to an executive for glancing at to see if anything of interest is contained within them. In this case the reading is characterized by great speed and even skipping over complete sections. The reader here uses the structure of the material to decide what level of attention is appropriate. No-one should assess the relevance of an article by reading the whole thing. On coming across a new scientific paper, for instance, the reader will probably read the title, rejecting the work if it is not of interest. If the title is not recognized or sounds as if it might be interesting then the paper might be opened and the abstract or introduction read. The decision to proceed is constantly reviewed. This process emphasizes how important it is to structure and prepare written material well. A reader will conclude that poorly structured material is a worthless product of a muddled mind. In a number of cases the assumption will be correct. After all, if someone cannot even marshal their thoughts into a sensible and logical structure what hope is there that the material itself will be of any use?

Even when the reader finds a section of particular interest, the material will not necessarily all be read in depth – far from it. The skilled reader will use a speed and depth of acquisition appropriate for the particular reading need at any given instant. The following techniques show the sort of range of reading techniques any competent reader can use. Such a person will often use all the techniques during the reading of just one page.

● *High resolution*: In this technique reading is very slow, often going back to earlier sections and rereading sections in detail. When the reading is finished the structure of the whole section can be remembered. The structure and the line of reasoning are memorized, allowing the reader to be able to reproduce the material and explain it to someone else in a different way. Perceiving the structure of a body of knowledge is the gateway to deep understanding. Such understanding, not surprisingly, takes time. The ability to re-explain material in a different way is characteristic of deep learning and a good test of it. In general, material, once deeply learnt, will be retained for a long time.

● *Normal speed*: This is the speed at which the reader naturally takes printed matter. It is the speed usually used for recreational reading. The normal reading speed varies greatly from person to person but it does change as the reading ability changes. It can be used to measure reading development in children and, in general, a slow natural reading speed goes with poor general reading skills and a lack of ability to vary the reading style. A high reading speed is associated with the opposite. This is due to the fact that good reading skills are a matter of practice.

● *Speed reading*: In this mode the reader does not acquire detail, and after reading the section may not remember large portions of the text. The purpose of the technique is to allow the reader to get quickly to areas of interest. Once interesting material is found, the speed goes down and the detail acquired goes up. In this mode the reader may not settle the eye on every word. Often only a few words per sentence are actively read. Sometimes groups of words or whole phrases are acquired at once. This technique takes advantage of the intrinsic structure of the language being used to allow the reader to gain understanding of the material in an overall way. Practice can greatly improve this technique and it is often the case that people under-use this mode.

● *Scanning*: This word 'scanning' means to examine in detail, but when used in the context of reading skills it means an extension of speed reading. More material is left out. Sometimes only the first and last sentence in each paragraph are read. Whatever is done, the important thing is that the reader gains an overview of the document, knows what sort of thing is in it. As with speed reading, the purpose is to allow the reader to make a decision about whether more or less detail is required.

Note-taking

Note-taking is a process of synoptically recording information whose delivery speed or situation does not permit the receiver to control the acquisition process. The notes may then be used for detailed learning at the individual's own pace.

Good notes reflect the structure, key elements and unique facts gained during the receipt of the communication. The best notes are made by collating rough notes, made at the time, into a complete and well-presented set. Usually this involves researching the information from other sources such as books to augment what was presented and make clear any areas of uncertainty.

Notes are used in many ways. They may be used to make a temporary record of directions and be thrown away afterwards. Alternatively, an engineer visiting a customer may take notes during the conversation and later they would be written up in detail, with care taken to record every item that might be of interest. Notes are most commonly used for recording the contents of a lecture or presentation.

12.4 Written communications

We all need to communicate, but some of us have a greater need than others. Engineers and technologists are among those whose requirement to communicate, as part of their job, is absolutely essential. There is little point being a creative designer if you can't sell your idea to someone, can't explain how to make it, or can't explain how it works or what it does. Other professionals will need to read your explanations, your manager may need to read your reports, the financial staff of your organization may need to have the financial implications of your work explained to them. How will you achieve these aims without being effective at written communication? Similarly, we need to be able to understand other people's communications. As engineers, it is often necessary to understand many different forms of communication such as design reports, technical specifications, test reports, and market reviews.

Outside the workplace we still need to communicate. For instance, applying for a job involves communicating ideas that suggest that you are better suited to a job than anyone else. More simply, you might just want to make a complaint about a product that you have purchased. In all these activities written communication is of the utmost importance.

12.4.1 Factors affecting written communication

Some of the information that humans communicate to each other is of a transitory value only. Knowing that the departure of a particular flight has been delayed by two hours is only of interest to passengers waiting to catch it. However, other blocks of information will always be useful. The work of Isaac Newton, for example, is constantly referred to and the texts that are published are simply contemporary accounts of the original principles. Written communications have evolved to serve the communication needs associated with information that has a significant lifetime. When speech is in use it is possible for feedback to occur. If the receiver does not understand the communication, it can be questioned and the meaning explained by the sender in a different way until understanding has been confirmed. Written matter, however, does not answer back. For this reason, any system of written communication must have firm rules of construction upon which it is based to enable understanding. It does not matter what these rules are, only that they are known to the communicators, and humans have developed many very different languages that can all cope effectively. Some languages have significant differences between written and spoken usage. In China, for example, some regional differences in dialect and interpretation of the language are so great that people from different regions can easily communicate in writing although they cannot understand a word of each other's speech.

The rules of language exist to make communication easy. The only things that set limits upon these rules are the limitations and needs of humans themselves. Our voice box and acoustic interpretation skills have a limited speed of operation, our physical size and hand–eye coordination dictate ideal dimensions for characters upon a page.

In essence, written communication is only a permanent record of spoken communication and so the rules of language apply equally to both. However, it is within written communication, where the very structure of the language is open and visible on a page, that the rules of language are most easily scrutinized.

Many things impinge upon the ease with which a section of writing can be understood. In the following six sections we shall explain the role of some of the most important.

The parts of speech

Language is a code. Each word has its own unique meaning but some groups of words have similar functions. The function that a group of words serve in language is called a 'part of speech'. The parts of speech exist to make the language clear and complete. Examples of commonly known parts of speech in English are the noun, the pronoun, the verb, the adjective, the adverb, the conjunction and the preposition. By studying the functions of these parts of speech we can see how the language operates. For instance, consider the verb 'to run'. Phrases such as *he ran*, or *they were running*, or *don't run* all have clear, unambiguous meanings because of our knowledge of how the verb 'to run' should be modified in each case. In other languages there are different rules. These rules are particularly important in written communications because we have no opportunity to say something again if we make a mistake. Written text must be correct. The purpose of dividing the language up into these parts of speech is that with the description of the language that they provide, we can use them to define the absolutely correct usage of our language.

Punctuation

Blocks of words that do not have punctuation are impossible to understand easily. Punctuation is a written substitute for the pauses and non-verbal communication that accompany spoken language. However, punctuation can be more than just this, and if used properly it greatly assists the rapidity with which a piece of writing can be understood. Correct use of punctuation is not trivial, and people usually underestimate its importance. Even the humble comma can be used to alter the complete meaning of a sentence. For example, '*I pressed the button, not meaning to endanger life*' has a very different meaning from '*I pressed the button not, meaning to endanger life*'.

Spelling

To a good reader, poor spelling is an offence against the language. Good spelling exists not only to avoid ambiguity in meaning, as between '*too*' and '*two*', but also to allow speed in reading. Good readers go too fast to stop and construct each word from its letters; the whole word is often acquired at once, especially small ones. If a particular word is misspelt the experienced reader's eye stops on the word because it is noted not to be the word at first imagined. This interrupts the flow of reading and efficiency of communication and must therefore be avoided. Some people find spelling mistakes annoying simply because they indicate a lack of care and attention, others feel that the language is being weakened by colloquial acceptance of alternative spellings and jump on the errors as proof. Whatever the reasons, you are unlikely to impress your reader by poor spelling. The implication will always be that someone who cannot get such simple things right cannot have spent much time on it, thus the content is also suspect.

Sentence construction

The sentence is the basic unit of the language. It is a collection of words that has a complete meaning in its own right. Any sentence abstracted from any piece of good writing will have meaning. This does not mean that one can always know what the author meant by the sentence. For example, '*They are always doing that*' is a sentence but we have no idea what '*they*', actually are, nor what they are '*doing*'. Nevertheless, the sentence is complete on its own. Many issues impinge upon correct sentence construction, 'numerical agreement' between subject and verb, for example – '*two bolts are steel*', not '*two bolts is steel*'. The 'tense' of the verbs within the sentence must agree. The 'case' of a sentence must be correct. '*They were the ones*' is correct, not '*them were the ones*'. Most rules of sentence construction come naturally through using the language. We instinctively know which version is correct and many people don't even know the rule (in this case 'that the pronoun should be in the nominative case if it is the subject of a verb') that they are applying. At its most sophisticated the science of sentence construction is extremely complicated. Once again, however, this complex system of rules, conventions and exceptions exists to make the language easy to follow, and being skilful in its use is likely to aid your ability to communicate effectively through the written word.

Style

Everyone has their own style of writing. Language is such a versatile thing and people so different that it is possible to identify authors simply by the style of their work. The most important issue about style is that it, like the factors above, exists to improve the reader's ability to understand the material. Even the distinctive style of a great novelist or playwright exists for this purpose and the enjoyment that well-written text brings is an important ingredient of style. Good advice about style stems from the factors that improve the effectiveness of written communication. The following are some points for consideration.

● *Avoid using many words when few will do*: For example, '*the audibility of oral delivery should be received by the listener with sufficient acoustic energy for reception*' is an unnecessarily complex way of saying 'speak clearly'. No-one is fool enough to think that overcomplication is a sign of intellect.

● *Avoid ambiguity*: Many simple ways of expressing things allow for misinterpretation. Ambiguity can creep into text in the most unexpected ways. '*Nothing is more effective in relieving stress concentration than increasing the radius of the concentrating feature.*' Does this mean that increasing the radius *is* the most effective thing, or is '*nothing*' the most effective thing, meaning doing nothing is better than changing the radius.

● *Avoid saying nothing*: '*An appropriate value of ballasting resistor is selected by using a technique which permits the determination of an ideal resistance.*' The sentence says only that '*to choose a resistor, you use a resistor choosing technique*'. No informed reader will be impressed with such an insult.

● *Avoid accidental untruths*: It is quite easy to make untrue statements in writing while trying to illustrate a point. For example, '*No-one who has simply written a business letter and posted it immediately can have prepared it properly*'. This may be true for an individual case but the statement says that it has been true throughout all time, and this is clearly not something the author could know even in the unlikely event of it being true.

● *Avoid jargon*: Science and commerce are pervaded by jargon words that restrict the audience of the text. It is very annoying for the reader to find a TLA slipped into the text without explaining what it is! The correct use of jargon is an efficient way to avoid unnecessarily repeating a long word many times over. Deoxyribonucleic acid is commonly replaced by DNA, and this one simple abbreviation alone has probably saved acres of paper. Everyone knows what it means (unlike TLA above, which stands for three-letter acronym) and the effectiveness of a written communication is not compromised by its use. If a particular text involves repetition of a particular group of words for which there is an abbreviation then use it, but define it the first time it is used, and either include it in a glossary or index the first entry where it is explained.

Handwriting

Most written communications are not handwritten but even the wide availability of word-processors has not meant that handwriting is dying out, far from it. Handwriting is a very versatile technique for making marks on a page, and those who can do it quickly and legibly will always be at an advantage. The way the characters that compose the English language have evolved is by no means arbitrary. The characters have conflicting design constraints impinging upon them. On the one hand, they need to be very different in shape from each other so that they are distinctive on the page and the eye can easily distinguish one from another. On the other, they need to be very similar to one another so that the hand can easily produce them in a flowing, easy-to-control way that permits speed. The arrangement we have today represents the culmination of much evolution. The character shapes are all based around the same simple shapes and the curves used to join one character to the next are all of the same form, making them easy to reproduce rapidly. The differences between words come from surprisingly small differences in the characters, and some of these differences are added after the word has been written. In writing the word '*section*' most people write out the whole word and then come back to cross the '*t*' and dot the '*i*', allowing them to produce the word at greater speed. It is also true that the current design of the characters allows for a great variation between individuals, which is a disadvantage, since it makes the character set open, and the more an individual departs from the basic font, the harder it becomes for the reader to understand the text. Just as with the other important factors in written communication, there are books describing handwriting in detail and, contrary to popular belief, it is quite possible for adults to make big improvements in their handwriting abilities if they so desire. For those with poor handwriting this is certainly advisable. Being unable to write something that another person can read easily is a great life disadvantage.

Copyright

Copyright comes into existence as soon as a work is published. It applies to written material, to music and theatrical works. The purpose is to ensure that the originator of the work is credited with it and this can be very important for a person's reputation. For example, a comedian who finds another person using his or her material has a case for copyright infringement. Clearly, the first person has been materially damaged by the acts of the second. Similarly, the text of a book is covered by copyright. Another person cannot legally reproduce the text without the author's permission. It is illegal (and immoral) to suggest that someone else's work is your own. The law of copyright is the law that governs this offence. It is, of course, perfectly acceptable to quote the work of another, indeed it is a necessary part of most scholarly activity. The only condition is that the quoted work is clearly identified and that the copyright holder is acknowledged.

In this section we have considered various factors that affect written communication. Some of them, grammar and parts of speech for example, often seem to be tiresome rules that serve to make written communication difficult. In fact these rules and protocols exist for the opposite reason and are very effective in their task. As professional engineers we must stop thinking of them as a source of loathing for generations of schoolchildren, and start to consider them as the science of communications, a science at which we can excel.

12.4.2 Preparation for creative writing

A written communication must be effective. You must be able to state what you want in a way that your reader will be able to understand. With oral communication, testing this understanding is easy – you can see by the body language of your listener, you can ask questions. Because you are dealing directly with the receiver you can react to feedback and change the delivery to make it appropriate. You cannot do these things with written communication and therefore it is particularly important that you prepare well beforehand. When a reader uses your material you will not be on hand to explain it, and the text must be self-explanatory. Something written by you is your envoy, it goes before you, people form opinions about you based upon it. Because of these issues, most people want their writing to be looked upon favourably and so they employ some effort in making sure that the communication is effective. In this section we shall look at the seven most important issues that should be addressed before pen is put to paper or finger to keyboard. A careful consideration of these issues should be made before embarking on any type of written communication. After we have introduced these important general issues of writing, Section 12.4.3 will look at some specific types on communication that are important to engineers.

Your reader
It is important to think about who your intended reader is. You need to have a picture of what the reader will want, or need, from your communication. You must also consider the knowledge level of your reader, particularly when using technical terms and jargon. You do not want to alienate your reader by detailing every basic point, but, at the same time, you want to ensure that the terms you use and the things you explain are understood. Your reader is your customer and, like any business, your success depends on happy customers.

Your objectives
You need to be absolutely clear about what you are trying to convey. What are the limits of your subject matter? What topics should be included? Which should be left out? Are you trying to inform your reader or explain something? Do you wish to persuade or influence your reader? You have to decide what your motives are so that you can choose the most appropriate style and language. The reader will look at the text and formulate an opinion on what the work is about. If your writing is unstructured and often leaves the central aim, drifting into things without

explaining the need for them, the reader will soon tire of the text and consider it too poorly written to deserve attention. You would not read something so poorly written, why should your reader?

Authority

The authority behind the communication should be made clear. A discerning reader will want to know why the communication should be believed. It may take the form of a reference list showing the sources of information or it may be a statement of the author's position. It may even involve a description of the author's life, previous publications and contemporary reports. Not all communications, however, have such a prestigious validation. A personal letter from a friend may contain a recommendation we choose to follow. In this case our knowledge of the individual gives the letter its authority. The authority or power behind a communication is so important that almost whenever we receive an important text we ask '*Who is it from?*'

Readability

The language you use must be appropriate to your intended audience. You should not write a report for your boss in the same way that you write a letter to a friend, and no-one would write to their bank manager in the same way that they would write instructions for a child. The general issues covered in Section 12.4.1 should be considered from the specific point of view of each communication you make. The issues of spelling and punctuation need always to be beyond reproach, but the complexity of the grammar, the vocabulary, the style, and the structure you use should be varied to suit the reader.

Media

You need to decide what medium is most appropriate for your communication. You need to be sure that writing is the most effective way of conveying the message, but then you need to decide whether it should be handwritten, word-processed, or produced on a desktop system. How large should the communication be? An advertising leaflet and a specification leaflet may be describing the same thing but they usually have a very different size, number of words and approach to the communication. Will a photocopied sheet of text be acceptable to your reader or is a glossy pamphlet expected.

Pictures, diagrams, and tables

You need to decide whether these will enhance or detract from your message. 'A picture paints a thousand words' but if you use a poorly produced diagram or inappropriate picture you will confuse the reader and make understanding difficult. As with language, your use of pictures and tables must be thought through to ensure that they are appropriate to your reader and that they convey your message effectively.

You must always ensure that when you are using these to show points or describe something that you make reference to them in the text and clarify what you want your reader to gain from them. Diagrams can take a great deal of effort to get right and, as with text, the reader will conclude that if the diagrams have not received attention it is likely that the content did not either, and the quality of the whole communication may be questioned.

Checking understanding

It is difficult to check understanding in a written communication since there is usually no immediate feedback between author and reader, but it is not impossible. Redundancy is the term used when points made previously are re-explained using different language. The intention is that if the reader did not understand the first explanation they will understand one of the others. The danger with redundancy is that, in excess, it will bore the reader, but if too little is used, understanding may be compromised.

Other techniques for checking understanding include making statements that explain what the reader should have understood by particular points in the text. If you come across '*We have therefore demonstrated that only the principal stresses play a role in failure prediction*' it is clear what should have been gained from the text. If the reader does not understand then a return to earlier sections is needed.

Questions are an excellent way to test understanding and it is common for textbooks to contain progressive questions that test successively deeper sections of the material. A good test of deep understanding is whether the reader can express the information in a different way. Of course, the author is not there to check the answers and so either the answers must be provided or available within the body of the text. In this way the readers can monitor progress for themselves.

It is common for text to be written for a range of abilities, and this makes it difficult to ensure that all the material is appropriate to everybody. If it is difficult to do this it is better not to try. Rather, include sections for people with little understanding or knowledge and other sections for those with previous experience. If the sections are clearly identified readers will easily be able to use the contents, cross-referencing an index to find the way to the sections they need. If, however, these issues are poorly treated it will be impossible for the readers to find their way around and the text will be practically useless.

If there are large amounts of prerequisite understanding, too much to reproduce in the text of the communication, it may be appropriate to define the knowledge required. This may be done either as a start to the text or within the body of the text by guiding the reader to other helpful texts at the point where they are likely to run into difficulty.

12.4.3 Specific writing techniques

We will now consider some specific forms of writing that are widely employed in engineering. These are reports, specifications, instructions, and business letters.

Report writing

In the world of engineering you may be required to produce many types of reports – technical specifications, progress reports, feasibility studies, laboratory reports, project management reports, personnel reports, or customer visit reports. A report is usually a fairly formal document used for consideration by others in the formulation of a decision. Reports may have information to present, they may need to highlight issues raised by the information and they may be required to draw conclusions. All reports are self-consistent rational bodies of text that permit the reader to understand the issues at stake. Reports are characterized by introducing information, discussing its meaning and limitations, and then drawing logical and relevant conclusions.

The following are some rules for preparing management and project reports. We will consider two issues; first, the format of the report and second, its contents.

- *Format*: The format that you use for your report defines how it will look. It is important to present a consistently formatted document since this assists the reader in following the structure of the report. The format of this book, for example, has been very carefully maintained throughout and is the same from cover to cover. References are always cited in the same way and the chapters all follow a similar structure. These things make the text easy for the reader to follow. The issues that need careful standardization are chapter and section headings, how many lines of space are left between each section, whether there will be more lines of space to distinguish major sections from minor ones, how lists are presented and how references are cited.

It is particularly important to think about these and decide what you want to do before you start writing, since you can then write in the correct format and reduce the time that would have been given to 'tidying up'. If you are writing on a word-processor the format is just as important, as you can often set up many of the format's parameters within the software and so save time. It is useful to prepare a format document which acts as an *aide-mémoire*. This will ensure that time is not wasted reformatting and allows the establishment of a 'house style' which will help readers to identify your work. An example of a format document is given in Figure 12.1.

Tables and figures also need careful format attention. Clearly, you will have to decide what they are going to look like and what typeface you will use for labelling. You will also have to give them numbers, or references, and titles. You will have to decide where they are going to go and how the text will reference them. Diagrams are usually included in the body of the report as the text references them but sometimes they are all left to the end. If you want to put them in the text you have to decide how much space to leave above and below, i.e. how you break the text. For figures and tables less than a half a page long it is possible to include them within a page of text as near as possible to the reference. If they are much longer it is usually best to put them on their own page as soon as possible after the reference. If you place all the tables and figures at the end of the report then they are normally located after the references and before the appendices. Often all the tables are put together first and then all the diagrams.

FORMAT DOCUMENT

1.   FONT, LINE SPACING AND PAGE LAYOUT
All report type to be in Times Roman 12 point with single-line spacing. Figures to be labelled using Geneva 10 point.

All text to be justified with no indents. Paragraphs to be shown by a line break. Right and left margins 2.5 cm, top 3 cm and bottom 2 cm.

All pages to be numbered in centre of bottom margin using Times Roman 12 point.

2.   SECTION HEADINGS
All headings to be numbered followed by a point and a five-character indent. Two blank lines to be above the heading and one below. The title should be in capitals and bold.

2.1 Subsection Headings
All subsection headings to be in leading capitals and bold. Subsection numbering to follow after the section number with no final point. Title to start at five-character indent. Leave one blank line above the title, no blank line below.

Figure 12.1   A format document

Finally, always proofread your work to ensure that you have eliminated spelling and grammatical errors. These will only detract from your work.

●   *Contents*: In its longest format a report may contain all the sections listed below:

> Title page
> Abstract
> Contents
> Nomenclature
> Introduction
> Text
> Discussion
> Conclusion
> Recommendations
> Acknowledgements
> References or bibliography
> Appendices

Although it is not normal for a report to include all these sections, most contain many of them. We will now examine the typical contents of each.

●   *Title page*: The title page should be the first page that you see when you open the cover of the report. It should clearly give the title of the report, the name of the author, the organization for which it was commissioned, the date, and other information appropriate to the situation. The title page may also include the abstract, depending upon the size of the report and of the abstract.

● *Abstract*: The abstract is a summary of what the report is about. It should be short but must provide enough information for readers to decide whether or not there is anything of interest to them in the report. The abstract of a management report is called an 'Executive Summary'.

Abstracts of journal articles and published reports are maintained on library databases. A prospective reader has only the abstract in order to make a decision on whether to obtain the full article. Thus, the abstract must contain sufficient information to allow that decision to be made.

● *Contents*: The contents list should give chapter numbers and titles with the number of the page on which the chapter starts. In a long report, or even a short but complex one, it may also be appropriate to list sections and subsections. A contents list for figures and tables should be given, after that for the text. The way in which titles are written in the contents list should be the same as they appear in the text.

● *Nomenclature*: The nomenclature section defines terms, symbols and abbreviations that are used within the text and which you want to be sure that the reader understands. It saves you having to repeat definitions throughout the text and prevents misunderstanding, particularly where common terms or units are used.

● *Introduction*: The introduction is the background to the report. It may be an introduction to a specific problem that your report attempts to solve or a more general introduction giving some of the background explaining why the report was written.

● *Text*: The text is what your report is all about. It should be divided into logical chapters that follow through the subject matter. If appropriate, chapters should be subdivided into smaller sections and subsections. The titles that you use for each chapter and section should give an indication of their content. It is normal to include data in the text, although if there is a large amount these may be summarized and then included in full in an appendix. If the data is too large even for an appendix it should be referenced as a separate entity.

● *Discussion*: If your report gives results or work that should be compared with other similar work then you will need a discussion section. You may give it a different title but within this chapter you will be discussing the relevant and pertinent points of your data and arguments. Discussions can be very lengthy and, if so, they should be divided with their own structure to make reading them easy.

- *Conclusion*: The conclusion to any report should not be very long. It should provide an end to your report summarizing your findings and your arguments. The analysis and background will have been provided elsewhere in the report and the conclusion should not duplicate them.

- *Recommendations*: If you have done work that requires further research which you believe could be logically carried forward, or if you expect some action following the submission of your report, this is where you make it clear. The recommendations of very large reports are often very short. For example, at the end of a 200-page marketing report you may find '*In order to achieve its strategic goals Haldone Freionics should open a US subsidiary company, introduce a new model of medium-power fridge for the home market, discontinue its off-site manufacturing and secure a loan to finance the redevelopment of the main factory*'. As with the conclusion, all the argument for this should be contained within the body of the text.

- *Acknowledgements*: You do not have to include acknowledgements although it is usual if anyone has provided you with specific help that has allowed you to do the work upon which your report was based.

- *References*: Throughout any project report it is likely that you will have drawn upon data, ideas, and material from other people. Your report should clearly indicate when this happens by making reference to the source material. The reference section of the report should provide a list of all the material used and it should be possible from the way in which the text is referenced to find to which particular document is being referred. If the number of references is fairly low you may choose to distribute them throughout the text rather than collect them into one section. A typical method of referencing is to insert a bracketed number which then points the reader to the reference that has been used: '*The necessary understanding of hydraulic pumps may be gained from Sayers (14), and his explanation of pump options makes clear the need for…*' Later in the report, possibly at the bottom of the page or in a separate reference section at the end of the chapter or report, details of the reference will be given. There are many standard formats in use and a common one is shown below. The important point is that the reader has enough information to obtain a copy of the referenced text.

(14) Sayers, A. T., *Hydraulic and Compressible Flow Turbomachines*, McGraw-Hill Book Company (UK), Maidenhead, 1990. ISBN 0–07–707219–7

When material has been used to provide general information, allowing the report to be written, it is not usually referenced but is listed in a bibliography. This list must also contain sufficient information for a potential reader to locate the book or article that has been used.

- *Appendices*: Appendices should be listed at the end of the contents section and included at the back of the report, in the order in which they have been referred to in the text. If you have more than one appendix it is usual to label each appendix as Appendix A, Appendix B, or I, II, etc. It is not usual to number the pages in the appendices, although if you do this each appendix is numbered as a document in its own right; the report page numbers do not continue. An appendix should be used whenever you have some background material that some readers may wish to look at but which might be known, or not required, by other readers. For example, if you were reporting on a project that had required a lot of statistical analysis you might include statistical tables, or relevant theory, in an appendix. This would mean that your text contained only your results and the results of the analysis.

Specifications

A specification is a detailed description of all the important aspects of the subject it describes. Sometimes the specification will cover technical issues and will be a complete technical expression of a product's performance. Other specifications include marketing specifications, in which the commercial profile of a product, often imaginary, is defined. Specifications are prepared by engineers to describe designs, commercial performance, purchases, or assembly procedures. At any time when important parameters need definition specifications may be prepared.

Engineers are perhaps most frequently involved with specifications for design work. This is a two-stage process which includes, in the first instance, preparation of a design specification. In this case the specification will define what is actually required of the design but will not suggest how the criteria are to be met. This allows other people to make an input to the specification but ensures that the designer can apply maximum creativity in meeting the requirements. After the design work has been finalized the designer will then produce either an equipment or process specification which defines the criteria that the equipment or process actually meets. It will usually be much more detailed than the original design specification. For example, the design specification might give a maximum size and weight for the equipment whereas the equipment specification will say what the actual size and weight are. An example showing the difference between a design specification and a product specification is shown in Figure 12.2.

Design specification:
1. height must not exceed 1.2 m
2. manufacturing cost to be less than £5.00 per unit
3. weight, fully loaded, not to exceed 50 kg

Product specification:
1. the height is 1.15 m
2. manufacturing cost is £4.00 per unit
3. weight, fully loaded, is 43 kg

Figure 12.2   Specification

Chapter 5 contains a description of how a technical specification may be used in the product development process and should now be read since it is only by understanding how such a document is to be used that it is possible to write one effectively.

Some general guidance for the preparation of specifications is as follows:

1. Specifications must not be ambiguous; it may be difficult to achieve this if you do not use technical language and terms.
2. Aim to use the minimum number of words and clauses. For example, some requirements may be presented in tabular form and contain many numerical descriptions.
3. Language should be simplified and care taken to avoid subjective statements such as 'of adequate strength'.
4. When preparing the specification, aim to define optimum quality for the job rather than highest quality. Overspecifying can increase the design time and cost to a point at which the equipment ceases to be economically viable.
5. The specification should define essential features and therefore should not contain phrases such as 'if possible' or 'preferred'. The obvious question is 'at what cost' are such things wanted? It must be possible to decide whether these things are needed and the specification should define the answer, not pose the question.

Instructions

Instructions in an engineering environment can be given in a written or pictorial format, or a mixture of both. They are the means by which you get someone to produce exactly the result you want by following a defined procedure. This does not mean that the instruction has to define every last elemental step the operator must follow. This would make the instructions far too lengthy to be practical and would limit their use to a very specific task. Good instructions make good use of the skill level of the person who will use them. An engine assembly manual might contain the instruction '*tighten the head bolts to ...*'. This instruction is not enough to describe the whole process to the layperson but a skilled mechanic will know that implied in the instruction is the tightening of each nut in turn, the use of a torque wrench, the correct 'pinching up' of the gasket and checking of both faces for burrs.

Some companies will have standard forms for various types of instructions (for example, test, manufacture, inspection). Although the forms may be different the guidelines for their production are the same:

1. Break the task down into a series of small, logical steps.
2. Describe each step in understandable language.
3. Refer to the equipment, or draw suitable analogies, so that there is no misunderstanding.

4. Use diagrams and models to illustrate what is required.
5. Ensure that the reader knows what the aim of the task is, when the task is complete, and what is supposed to happen next, if anything.
6. Include tests for correctness at important stages. Statements like '*You should now be able to see the flange inner-rim bearing on the swash-plate*' help the reader to confirm that things are going as planned.

It is particularly important to be clear for whom you are writing instructions so that you use appropriate vocabulary and appropriately detailed steps. You should also check the instructions carefully to make sure that you have not omitted anything important and that you have given any necessary warnings.

Figure 12.3 shows an example of an instruction. Notice that it makes clear reference to more detailed documentation.

INSTRUCTIONS FOR ASSEMBLY OF CIRCUIT BOARD W23

CARE: You must use a static protected workstation for this assembly

1. Locate components as shown in drawing W231
2. Solder all joints in accordance with soldering standards
3. Clean all joints and carry out continuity checks
4. Attach connector strip to board as shown in W231
5. Wrap board in bubble-wrap and place in Stores

Figure 12.3 Example of an instruction

Business letters

Business letters are constantly in use by professional people. You may wish to apply for a job, contact a supplier, complain to a customer for non-payment of an invoice or comment on the service received from a supplier. All these situations need well-written letters. It is a mistake to think that composing a good business letter is quick or easy; some of the most important ones can take hours to deliberate over and get exactly right. Often a well-written letter will produce a response when a poorly written one will not. If you were on the receiving end of a poorly written complaints letter in which all the problems were your fault and there was no acceptance of responsibility on the part of the letter writer you would, like most people, probably feel less inclined to act in favour of the complainant.

A business letter should be concise, giving as much information as required and no more. It is usual to open with a statement summarizing the objective of the letter and to finish with an explicit statement of what you require to be done. People often mistakenly think that because business letters are about business they should be cold, formal and cannot offend. Businesses are operated by people and it is still a person who will read your letter. Therefore be courteous, fair, clear, straight to the point and give all the information required but don't be long-winded.

The conventions for business letter forms and the style of language that should be used vary from country to country. It is always polite to try to adopt the accepted form where possible and therefore if you are writing abroad you should consult a large dictionary (e.g. *The Collins Robert French Dictionary; English–French*) which will have sections on letter forms. Figure 12.4 shows a widely accepted format for a business letter.

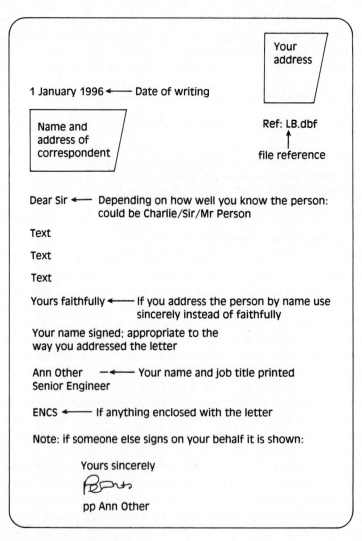

Figure 12.4 Business letter format

12.4.4 Using a computer to produce written communications

There are many packages available for making the production of written communications easier. The best known is the word-processing package, and a few notes on the use of this are provided below. However, data presentation usually requires more than a word-processor, and therefore spreadsheets, drawing packages and reference managers should also be used. All these can be merged into word-processed files to produce professional-looking documentation without the use of a desktop publishing system.

Word-processors are widely used, and it is becoming the norm for people to produce their own material directly via a keyboard. However, as a non-secretarial person it is easy to be scornful of the word-processor and to use it as a typewriter, thus losing the advantage. When approaching a package it is worth finding time to determine how it can be used to help with whatever you have to prepare. Most packages will store standard formats, will allow you to tabulate data, produce indexes and tables of contents and enable you to set headers and footers. In addition, they will check spelling and grammar. Spelling checkers are invaluable for the non-typist if the free flow of words is not to be interrupted by endless corrections. However, they will only check spelling and not usage, so unfortunately it is impossible to avoid proofreading. Spelling checkers and grammar checkers exist for most languages.

12.5 Oral communications

Oral communication allows an exchange of information between people who are able to see or hear each other. It probably developed in the social groups of our primaeval ancestors and was used initially to alert members of the group to the arrival of danger or the location of water and food, and to impose the hierarchy of the group. However, before long, a communication system had developed and the exchange of more general information became possible. The tool developed for this purpose was the question. Early questions might have been sounds to ask 'which direction is the danger?' or 'how far is the water?' The greater the ability to communicate such things, the better the groups were likely to fare. Consequently there is an evolutionary effect which favours good communicators. From these simple, early oral communications, and the evolutionary pressure that favours those who are good at them, language developed.

Oral communications have limitations, however, and other types of communication have developed to overcome them. The greatest weakness is that the whole communication has to be remembered, and this limits the complexity of the communication to the memory capacity of those involved. Thus to communicate lengthy messages or ideas that would be required in the future an alternative to the transitory memory was needed. The permanent marks of mud or charcoal on cave walls were probably the first use of a permanent record and their modern counterpart is the printed word. The pressure to succeed by communicating still exists today. Communications are still evolving and within organizations there is no doubt that success comes preferentially to those who communicate well.

Oral communications are used in many ways, from public speaking to private scheming, bringing a board of directors to a consensus, dealing with an argument or conducting a job interview. We all communicate orally, every day. Nevertheless, some people are much better at communicating than others.

Apart from the memory problem, another major difficulty with oral communication is that the quality of a communication relies on the communication skills of two people, a receiver and a sender. The receiver is the person receiving the communication, listening, and the sender is the one talking. As a sender, one needs to exert some influence over the receiver to ensure that the message is being both understood and given the weight that it deserves. This requires skill, but there are some ground rules which will help, such as structuring the delivery, checking understanding, summarizing periodically and delivering in a way that is likely to interest the receiver.

In the workplace we use many different methods of talking to each other. There is the primarily one-way communication, as exemplified by the oral presentation. There is the two-way information-gathering type of communication as exemplified by the job interview and the more discursive negotiation process. Finally, there is the group communication that takes place in meetings.

In this section we will start by looking at the factors that affect oral communication, the sending and the receiving. We will then describe specific techniques for achieving effective communication in presentations, interviews and negotiations. Finally, we will consider some ways that can ensure effective use of that much-abused instrument, the telephone.

12.5.1 Factors that affect oral communications

In any oral communication there will be at least two parties, the sender and the receiver. The success of the communication depends on the sender's ability to make the message clear and on the receiver's ability to listen to the message and understand it.

For the sender, there are a number of areas that will affect the clarity of the message. The first point to consider is that the message should be audible. Oral communication means using the mouth to communicate. Usually this means making a noise with it, although there are many non-audible communications such as mouthing, kissing, pulling faces, and sticking your tongue out. When the mouth is used to make a noise it is clearly important that it makes enough noise to be received and understood by the target of the communication. Words should not be mumbled, they should rise above the background noise. Adjusting the volume of your voice is difficult, particularly when dealing with unfamiliar surroundings and you do not know how well the sound will travel. It is almost as important not to shout at people. Generally this is perceived as aggressive and may deter people from listening.

Different audiences have different speeds at which communication is best. An engineering course which extends earlier knowledge might open with a review lecture, which would cover the material only once and at quite high speed. On the other hand, a message delivered using a public address system at a railway station announcing a change of stations for a connecting service would be given very slowly and repeated often. Determining the speed at which the message is conveyed can be done by knowing something of the audience. However, a sender should also be able to vary the speed according to audience reaction, either speeding up or slowing down.

Making the message audible and controlling the speed with which it is delivered obviously improves the ease with which the message can be understood. However, on their own they do not ensure that the message is likely to be received and understood. To meet these needs we must hold the attention of our receiver, and we need to make sure that the receiver understands what is being said.

To hold attention we need to make the communication interesting. Interest in a verbal delivery can be stimulated in a variety of ways; humour, anecdotes and enthusiasm are powerful techniques. In some cases the interest may come from circumstances, as in the case of the station announcement above, in others it is important to engender interest. For instance, a group of communication students will not respond well to a communications course unless it is delivered by staff who fully appreciate the importance of the subject and construct a course that includes practical seminar experience together with theory and relevant examples. The language used and the style of the delivery must be appropriate to the target audience if you want to maintain interest.

Making sure that the receiver understands what is being said is a continuous process. First, we need to use devices that will help understanding, such as using diagrams, repeating information, and making analogies. Second, we must check understanding. The outcomes of conversations are often very different from those imagined by the participants. The development of misunderstanding is a constant risk during conversation. For this reason, good oral communicators always check understanding at key points during the communication regardless of what variety of communication it is. This is most effectively accomplished by asking probing questions that require the receiver to process the information in some way in order to prepare the right answers. In ordinary conversation this may be done simply by asking if the other person understands or is happy with the conversation. However, the sender should also look for signs that indicate incomprehension – the glazed look, doodling, doing something other than listening attentively.

Summarizing so far: to send a message effectively we need to make sure that it is audible, that the speed and style of delivery is appropriate to the audience, that it is interesting, and that the receiver understands what is being said.

In addition to considering the content of the communication and general style of the delivery there will be other factors that affect the way in which the message is received. Consider the effect of your behaviour, dress, appearance, actions. If you are communicating with someone shy and introverted it is not appropriate to be forceful and dominating, as might be appropriate towards a loud extrovert. By being sensitive to the personality of others you can greatly increase the chances of effective communication.

Often it is necessary for communications to involve criticisms. The other people in the conversation are just as entitled to their points of view as you are. If you respond to their criticism with a retaliation, either open oral warfare or silence will result. Both these are barriers to effective communication and so must be avoided. As emotive subjects such as criticism enter the conversation the sender should respond by increasing objectivity and further reducing the margin for ambiguity. Depersonalizing issues is an excellent way to avoid confrontation. If someone feels they are under attack on a particular issue they will be very defensive. If, however, they feel they are being offered constructive ideas aimed at solving difficult issues they are more likely to respond positively.

Active listening

Active listening is an oral communication skill in its own right. In essence, it is a way of ensuring that errors of oral communication are eliminated, but it requires the active participation of the receiver, not just the sender.

When conversations are under way the participants often have a list of things that they 'want' to hear. Imagine an employee who really does not want to do a certain task and is in a conversation with the boss. It may be that a boss finds it hard to ask an employee to do this task, perhaps because even though the employee should be doing it, the boss didn't make this explicitly clear at the start and so it reflects badly. Subconsciously, neither party wants the discussion to turn to the sensitive area. More often than not, in such circumstances, the participants will avoid the subject and the two people will have different recollections of what was said because their hidden 'wants list' has coloured their objectivity.

When active listening is to be employed the following points should be considered. Remember that this is a technique used by the listener to ensure that the message is received accurately:

(1) *Use open questions*: Attentively acquire all the information on offer. Where there seems to be a shortfall, use open questions to fill in the gaps. For example, instead of saying 'Was the unit delivered on time?' use 'How do you feel now that you have the unit?' Open questions lead the sender to provide full information.

(2) *Ensure that you make all the points*: If there is a problem to be discussed or an issue debated then the active listener can take some control by ensuring that all the points relevant to the issues are agreed. If this is done prior to the wider debate then it will be clear to both parties what has to be addressed.

(3) *Handle criticisms effectively*: Often it is necessary to make comments that may be perceived as criticisms. All people are entitled to their points of view. If you respond to criticism with open oral warfare you are unlikely to be effective in the communication. Your arguments will not beat another person into submission, rather, your 'opponent' will take up the challenge and the discussion becomes a heated argument. The alternative to this may be total withdrawal, and the other person does not then divulge important information that might be of use to you. If you argue points with someone you may force them to give in on some factual matters but you will not alter their own opinions. Again, actively listening and watching for signs that show people's sensitivities can ensure that these situations do not arise.

(4) *The customer is not always right*: People in conversations are not always right about things, even when it is their specialist area. Handling this part of a conversation is especially important if you don't want the other person to feel threatened and so stop communicating. Don't confront the other person and so force him or her into a position from which they cannot win something. Use gentle questioning to examine ideas using his or her own knowledge to find the solution. Saying things like ' Yes, I see that approach, how would we then cope with ....?' will be helpful in this situation. Calmly make firm statements which do not attack the other person's conversational territory. In this way the conversation will carry along a lot further and more information will be exchanged as the other person learns to trust you. When a difficult part of a conversation is reached it often helps to depersonalize the issue. For instance, if you want a customer to explain the details of a rather complex design they are suggesting, don't say 'using your system is bound to introduce errors', say 'yes, that is always the most difficult area to get right, how important are the errors of that technique in your situation?'

All the points discussed so far are relevant to any oral communication. We will now look at how they apply in specific circumstances and consider some techniques that can be used to improve effectiveness for different types of communication. We will start by considering the oral presentation.

12.5.2 Oral presentations

Presentations are a particularly important form of oral communication which engineers, and other professionals, use to brief others on particular topics. Engineers may present their work to colleagues, personnel managers may make presentations about recruitment plans, sales engineers may present solutions to customers.

There are two aspects to making a good presentation: preparing it and giving it. We will look first at the preparation.

Preparation of the presentation
Preparation is essential to a good presentation, although if you are a member of the audience the preparation may not be obvious. What is obvious is the person

who has not prepared and overruns the time, finishes too early, cannot find the right slide, or starts to wonder what topic should come next. The time to prepare for even a short presentation can be significant and should always be allowed for in the planning process. There are five main areas to consider in the preparation of a presentation: timing, objectives of the presentation, the structure, visual aids, and questions.

● *Timing*: This causes more problems with presentations than anything else. If you overrun it gives the impression you have not prepared the material. If you are too quick then it appears either that you raced through the material or that you did not have enough. To get some idea about time, start by taking the time slot you have been allocated, then take away a time for questions at the end, a time for getting everyone settled at the start and a time for operating any visual aids or demonstrations that you may be using. You can then prepare the delivery to fit the remainder. Estimating how long it will take to deliver a given piece is not easy and even experienced presenters can get it wrong. The best thing to do is to give a dummy presentation to an empty room, speaking the words aloud, just as you would for the real thing, and time it. Talking aloud is important because you always read faster than you speak. You may feel silly giving the presentation to an empty room but you will feel far worse if the timing goes wrong during your presentation because you did not rehearse it well enough.

● *Objectives of the presentation*: When preparing your material consider what your objectives are for the presentation and then ensure that they can be met with the material. For example, if you are trying to sell a design to a board of directors, consider what aspects of the design they will want to know about and make sure that these are covered in your presentation. Also, consider your audience and their level of understanding so that you prepare material at an appropriate level, particularly in terms of using language that will be understood by the audience. Finally, you must ensure that your material has a coherent structure. Plan it as you would a report with an overview (abstract), an introduction, the text, and finally a conclusion.

● *Structure*: You should make a clear plan of the presentation, showing its structure with timing markers. The best thing to use is a single side of large text that is easy to read during the presentation and which contains the headings of each section and the points that are to be described within each. You can then easily pick up your place by glancing at the sheet. You will be sure not to miss anything and if you are not in the margin of time by which each section should be completed it is easy to correct as necessary. Your plan should clearly have on it the one or two critical messages you wish to convey in the presentation. You must be absolutely clear what the message of the presentation is and resist the temptation to include material that is interesting but not really relevant.

● *Visual aids*: These can be a very effective way of reinforcing points or demonstrating what is meant. They should help the audience to understand your message. Many forms of visual aid are available: models, slides, posters, demonstrations, video, overhead projection slides, and handouts.

Unfortunately, too few people give as much weight to the preparation of their visual aids as to the extent to which they are used. Visual aids can only be effective if they can be seen and understood by the audience, and if they form part of the material. They take just as much effort to perfect as the body of the presentation. Diagrams should be simple and easy to understand, yet they may convey a great deal of information. Clear diagrams and visual aids can make a presentation, they offer the audience new interest and stimulation and, if well chosen, can explain points with great clarity. On the other hand, they can be a real embarrassment and when poorly prepared can ruin an otherwise good presentation. For example, a final-year engineering student wanted to show that a device he had designed was more effective than a similar one that was available in the shops. Unfortunately for the student, he forgot that his device required being secured to a bench and so when he came to demonstrate it in his hands it didn't work. The golden rule is to try them out first.

The choice of the form of visual aids is usually limited by the place in which the situation is to take place and the equipment available. The nature of the presentation will also affect the choice. If a lot of information is to be transferred which may require note-taking then the presenter can ease this burden by providing handouts with summaries, or copies of complex slides.

Overhead or slide projectors are the most common form of visual aids. Certainly, overheads are the easiest to prepare and consequently the easiest to make a mess of. How many times have you looked at an overhead like the one in Figure 12.5?

CONTRACT REVIEW

The objective is to check that the requested technical specifications can be met prior to accepting an order. Two methods for achieving this are described; there are many others that could be used.

Method 1
All orders are reviewed at a weekly management meeting prior to their acceptance. If the managers do not have the appropriate technical expertise then the design manager is given the responsibility of reporting back at the following meeting, having consulted the design engineers in the intervening time.

Method 2
The design engineers have technical responsibility for a given range of products. When an order is received the Sales Department contacts the engineer directly to receive a decision on acceptance.

Figure 12.5 How not to prepare an overhead

To prepare clear and effective overheads a few rules are followed:

- Keep the number of words per slide low.
- Keep the number of ideas per slide low.
- Characters should be large enough to be seen clearly from the back of the room.
- Keep within the bounds of the screen.
- If you are handwriting a slide then print the letters.
- Label diagrams clearly and use the same drawing standard that you would if it were part of a report.
- Use colour and highlight key words and phrases, but avoid pale colours that may not show up. If the colour is significant then avoid red and green, as there may be someone in the audience with red/green colour blindness.
- If preparing with pens either use those with permanent inks or avoid touching the slide after using water-soluble ones, to avoid smudges.
- Keep slides in frames or clear wallets so that they can be picked up without sticking to each other or without the slide being damaged. Clear wallets have the advantage that they provide a ready surface for drawing/writing on when the slide is in use without marking the slide.
- Test slides in the situation in which you wish to use them first, if possible.

● *Questions*: These are usually taken at the end of a presentation, although this can vary depending upon the style of the presenter, the topic of the presentation, and its length. If the presentation is long and many ideas are being covered it is unreasonable to expect the audience to have to sit through it and remember their questions at the end. Wherever you decide to take questions it is always worth preparing for them so that you are not thrown 'off-guard'. Similarly, if no questions are forthcoming you may wish to proffer one yourself to prompt the discussion. When you have completed your material preparation consider a few questions that might be asked and prepare answers to them.

Making the presentation

When making a presentation there are five areas that will have to be considered: timing, presentation of material, using visual aids, reacting to the audience, and helping the audience.

(1) *Timing*: Throughout the presentation you should monitor progress and make sure that you can see a clock or watch without obviously looking at them. If questions are unexpectedly long, adjust your talk to ensure that the main points are still delivered within the time. The hallmark of a skilful presenter is the ability to be flexible and change the details of the presentation as it progresses without losing the flow or skipping material. This is difficult to accomplish but will come with practice and experience.

(2) *Presentation of material*: When presenting the material the aims should be to present it clearly and in a way that is interesting to the audience. First, begin with an 'icebreaker', rather than leaping straight into the body of the presentation. Introduce yourself and explain what you will be talking about; if appropriate, check attendance so that missing people can be accounted for. When you start talking you are in control and authority is with you, it is up to you to take control, shut doors, get things going, settle the audience and so on. The icebreaker is the time to do all this and it serves many functions at once. Primarily it establishes you as being in control, it settles the audience and raises the concentration level. It confirms what the audience are about to hear and it gets everyone focused before any of the important material is transmitted. A presenter should make a time allowance for the icebreaker, which increases with the size of the group and what is to be said. For a group of ten colleagues it may only last a minute, with a mixed group of a hundred or so in an unfamiliar place it may take several minutes.

When making the presentation, talk clearly and talk slowly. Some presenters talk to individuals in the audience, changing person from time to time. It is certainly good to make eye contact with your audience since this will help to keep you in contact with them. Making a presentation is not like talking to someone standing next to you. To be effective in a presentation you must be heard and understood by everyone you wish to address. This means that you must go at the pace of the slowest and be loud enough for the most distant. Your audience cannot control the speed of your delivery and so you must do it for them. It is good practice to ask your audience if they can hear you, although this must be done at the start during the icebreaker to avoid having to repeat material if some cannot hear.

Unless it is really necessary, avoid reading from a script. Nothing makes presentations more boring than having them read like this. The implication is that the presenter is so bored or unfamiliar with the content they are likely to forget what they have to say. The members of the audience are left wondering why they weren't just given a copy of the script and left to read it in their own time. Giving a presentation involves adjusting the speed of delivery to suit the audience, going over particular things as that audience needs to, taking feedback and making changes as you go along. Rigid scripts afford none of these and, indeed, if this is all a presentation becomes then it gives no more than the audience would receive by reading the script for themselves.

There are occasions in life when a script is called for, times when every last word of a delivery will be subsequently dissected and its meaning closely scrutinized, perhaps a presidential speech, a statement of the company's position at a shareholders' meeting, or a very technical lecture attended by world authorities where the ambient level of understanding is very high. These are suitable subjects for the prepared speech and it should be reserved for them.

(3) *Visual aids*: If using visual aids the first check must be to make sure that the room in which they are to be used is suitable. Matters to check would include whether the sun is reflecting on an OHP screen, whether the audience can see the video, if the lighting is suitable for slides.

The visual aids should complement the material and add to, not detract from, the overall presentation. In particular, they should be used when required and removed when they are finished with. If a slide is kept in view long after it has been discussed the audience will be distracted by it and won't give due weight to the rest of the presentation. Always refer to the aids, incorporating them into the presentation – the audience should not have to wonder what the point is. You should also make sure that you do not obscure any visual aid, and that it is clearly visible to the audience (for example, checking focus and that a slide is the right way up).

(4) *Reacting to the audience*: During the presentation a good presenter breaks down the barrier with the audience by reacting to them, rather than remaining aloof and untouched. The presenter should not stand still and deliver a monotonic slab of material but walk around, address individual people in the audience, establishing eye contact with them. In this way the audience will feel that they are being spoken to directly as individuals. This will hold attention and make the presentation more interesting for them. It is also important not to alienate your audience. Dress like them, speak like them, talk like them and be like them. People can be very prejudiced against differences and take the view that others unlike themselves cannot have anything useful to say. If such an atmosphere prevails during a presentation your audience will be less receptive to the message and more damning of the material. Therefore to minimize such effects and give your presentation the best chance your audience should be able to identify with you.

(5) *Helping the audience*: To help your audience you should summarize at the end of each section and, especially at the end, summarize what happened, what was done, what it shows and what was the point of it all. The summary is a very powerful teaching tool. People remember things in a structured way and reinforcing the structure by restating it synoptically at the end of a section is a most effective technique for assisting learning. Summarizing should not be confused with repetition. Repetition can be used to ensure that key concepts or buzzwords are remembered while a summary is employed to ensure that the structure of the knowledge is remembered. A common piece of good advice for making a presentation is '(1) tell them what you are going to tell them, (2) tell them, (3) tell them what you have told them'.

Finally, do enjoy it – enthusiasm is contagious. If you are interested and dynamic about your subject your audience will be too. If, on the other hand, you, as the champion of the subject, appear bored or indifferent your audience will conclude that the material itself must be boring and dull and they will not pay attention. Be confident.

12.5.3 Interviews

Interviews cause more apprehension and stress than practically any other type of oral communication, which is surprising really since they are simply a discussion between two parties about whether they can make a deal which both parties find fulfilling. Usually interviews are associated with employment, with an employer attempting to ascertain whether an applicant is acceptable to the organization and the applicant trying to assess what the job will entail, what the prospects are and what it would be like to work for such an organization. Interviews also take place in many other circumstances when one person, or group of people, need to obtain information from another person or group, such as trying to establish a customer's requirements.

The success of interviews depends mainly on those who run them. When they are run well they are an excellent vehicle for communicating the position of both sides and identifying critical information. Sadly, the converse applies to badly run interviews. Interviewing is a skill and there are many books and training courses that exist to enhance the effectiveness of both interviewer and interviewee. In this text we will review the most important aspects of the interview, looking at the interview process and the role of the interviewer.

The interview process
The process must begin with a definition of the aims of the interview and, where appropriate, these should be agreed by both parties to the interview. From the aims the interviewer can draw up an agenda to ensure that these are met. For example, if the aim is to choose a suitable person to join a team of designers then the agenda should cover areas such as previous experience, technical ability, what the person is like as a team member, and what he or she expects from the job. It might also include some informal meeting with the design team.

Establishing an agenda will also require determining who should attend. However, the more people attending, the greater the problems of coordination and communication and therefore the need to make sure that everyone is clear about the aims of the interview and the procedure that will be followed. The procedure will define the times of the interview, where it is to take place, if any demonstrations, presentations or tours are to form part of the process, and how decisions will be agreed and ratified.

The next stage requires that the two parties are brought together in a suitable environment so that the interview can take place.

Bringing the parties together means arranging facilities, agreeing dates and times, and ensuring that travel arrangements are practical. The greater the distance between the two parties and the more important the interview, the more difficult this arrangement becomes. Certainly, when trying to get many people together it is may be necessary to plan a long time in advance.

If a number of people are to be involved in carrying out the interview then they should prepare beforehand, ensuring that they agree and understand each other's role. In particular, agreement should be reached about who is to run the interview and who will cover which aspects of the interview. Most importantly, all should agree on the criteria that will be used to base any subsequent decisions. One thing that should never occur in this situation is that the interviewers start their own conversation, excluding the interviewee, or argue about issues in front of the interviewee.

A suitable environment is very important. It has to be able to cater for the parties and must be conducive to the free flow of information. In the recruitment situation this may mean having informal surroundings that are screened from external interruption. If dealing with a major customer this may mean having good communication links and catering facilities. These domestic arrangements are an important aspect to the interview. No-one is going to relax if they are in desperate need of a toilet, or perhaps if they haven't had a cup of coffee in the last four hours, or have just finished a long car journey. Certainly, when dealing with sensitive issues there should be no disturbance. The telephone should be switched off and people notified that the room is unavailable.

As the final part of the process, the interviewer and interviewee should agree on the follow-on procedure, how the interview will be recorded, or how the interviewee will be notified of actions and decisions.

The role of the interviewer

The interviewer will be responsible for ensuring that the process is defined and understood. However, in addition to this the interviewer has the responsibility for ensuring that the requisite information is extracted from the interviewee. There are a few simple techniques for ensuring that this happens. Initially, the interviewee should be welcomed and put at ease. If it is a job interview then the candidate could be engaged in conversation about some of the interests given on the application form, or there could be a more general conversation about the weather or the journey. Following the general introduction, the interviewer should 'set the scene' by discussing the aims of the interview, the procedure and either agreeing the agenda for the interview or indicating the structure that it will take.

During the interview the interviewer is trying to gain information upon which a decision can be based, therefore it is necessary to get the interviewee talking. In fact, in any interview the interviewee should do most of the talking. Obtaining information is done by asking questions and there are two rules that should be followed:

1. Ask open questions.
2. Ask single questions.

Open questions are ones that will allow the interviewee some opportunity to give detail – they cannot be answered in one word. For example, if a manager is trying to find out if a subordinate was settling into a new job the question could be asked 'Are you settling in?' If the answer is yes, then the manager may feel that his or her job has been done, even though they may have no evidence to substantiate this. If the answer is no, then more questions have to be asked and these must be developed from zero input. However, if the original question was 'How do you feel about the new job?' then a more detailed answer is likely to be forthcoming. More questions may still be required but at least the interviewer has a direction to follow. Open questions start with 'how', 'why', or 'what'.

Asking single questions is also important. Asking double, triple, or worse, questions can result in both interviewer and interviewee forgetting the original question, and it becomes very difficult to answer any question. An example of a triple question is 'What is your experience with that programming language and don't you think it's very hard to learn? I used it a long time ago but I think it's just been upgraded. Am I right?' Where would the interviewee start?

The interviewer can have a very hard job to do, particularly if the interviewee is unwilling to contribute. Unfortunately, as with so many management skills, the only way to increase effectiveness is to practise.

12.5.4 Negotiations

Negotiation is the process of reaching agreement between two parties that do not share the same view. Negotiations may be very large and protracted as is the case with nations negotiating defence or economic settlements or they may be small and rapid as with two people bargaining a price for a good. In either case the parties are negotiating and this involves various stages. The purpose of negotiation is to arrive at some agreement which is acceptable to both parties. Anyone can reach an agreement with someone else when a mutually beneficial possibility exists but only experienced negotiators can be sure that they have gained the most and given the least that the particular situation afforded.

There are four stages to effective negotiation: preparation, examination, confrontation, and resolution:

(1) *Preparation*: Experienced negotiators know that this is the most important stage. Without taking time to consider what is actually wanted from the negotiation one cannot be certain of heading towards it. On a simple level an engineer might prepare for a meeting listing the desired outcomes and actions together with responsibilities. Alternatively, a trades union negotiator may discuss with colleagues all positions to be taken, set upper and lower limits on issues, decide how the cases will be argued, even consider how the other side will respond and prepare countermeasures. If aggressive tactics such as sudden mood changes, unannounced adjournments or decoy concessions are to be used it is clear that good preparation will be required. It is also necessary for a negotiator to establish what is valuable to the other side. This is a continuous process and must be maintained throughout the negotiation as more information is unfolded, since the success of negotiating comes from trading in valuables.

(2) *Examination*: The early part of negotiation is concerned with an examination of the position taken by the other party. In particular, the negotiator is looking for indicators to the upper and lower acceptable limits on issues. At the same time, negotiators will try to keep their own position secret. A lot can be deduced in the examination phase about the ranges of acceptability and various tactics are available for examining them further. Suggesting values indicates an expectation and so tends not to be used. This phase also includes a 'sizing up' of the opposition which can be useful later, particularly where the choice of tactics can have a critical effect.

(3) *Confrontation*: Inevitably, during the negotiation confrontation will occur. Good negotiators never let it escalate into an open battle and control it long before such a point is reached, avoiding an impasse. An impasse means that dialogue has stopped and unless one side gives way negotiations cannot continue. This is a very unpredictable and potentially ruinous position and is therefore best avoided. Nevertheless, if two parties have incompatible views then they must be confronted and negotiated away. Of course, compromise is the way forward, and during the confrontation phase the disagreements have to be dismantled one by one. If a large impasse seems likely it may be best to tackle simpler, less contentious issues first and build good relations.

(4) *Resolution*: As the negotiations progress, compromise is established and principles upon which concessions can be made are discovered. The negotiation becomes the resolution of these differences once the confrontation has been passed. In the end all the contentious issues are either resolved and both sides have a way forward or they cannot be resolved and the impasse is insurmountable.

The success people have in these different stages has a lot to do with their experience and negotiation is a skilful process. Experienced negotiators nearly always outmanoeuvre their opposite numbers and where there is much negotiation to be done in a job special training is given. Sales personnel, for instance, are usually trained by their organizations to be effective in commercial negotiation. There are many tactics to use in the negotiating game – bluffs, making much of small concessions, negotiating with the most favourable person opposite, seeding dissension, undermining power, adjournments, tactical behaviour, stating one's position early, stating it late, opening much too high. The list is very long.

Negotiations are always about important things and so it is not surprising to find that they take some time. People new to negotiating often feel that the process is unnecessarily lengthy and if all sides simply stated their case a solution could be found much more simply. However, important issues need to be examined from all angles to make sure that nothing is missed and especially that an agreement will endure. Agreements will collapse if, subsequently, important issues prove to have been inadequately covered.

12.5.5 The telephone

The mass communication device invented by Alexander Graham Bell is now totally integrated into life on earth. One can now phone practically anywhere in the world; a vast network of satellites, cables and fiber optics make conversations with people on other parts of the planet a normal part of life at work. The power of the telephone is more than its long-distance capability. Communications over short distances, within single companies, even within single departments, are very different as a result of its existence.

Using the telephone is not as trivial as it is often considered. It is common to train people if their job requires them to use it a great deal. Because the telephone permits verbal but prohibits non-verbal communication, successful use of it cannot be achieved by having a conversation in the normal way. During telephone conversations it is necessary to provide verbally the information normally supplied non-verbally. Often this is achieved through an exaggerated range of voices and verbalizations that are not speech.

People often mistakenly think that because the telephone has been answered the person at the other end is ready for a conversation. They usually answer a ringing telephone even if they are quite busy, simply because the telephone call may be even more important than the task in hand. Therefore good telephone callers will always ensure that the following are included in their calls:

- A statement of who you are and where you are calling from
- A check of to whom you are speaking
- Finding out if it is a convenient time to call
- Stating clearly and precisely the nature of the call: social, business...
- Making sure that phone numbers are exchanged if necessary
- Being clear who must do what as a result of the call
- Summarizing at the end of the call
- Signing off courteously.

Finally, when transacting any business via the telephone it is good practice to keep a record of the call.

12.6 Managing meetings

Meetings are notorious for the extreme levels of communication effectiveness that can be experienced. Some meetings achieve vast amounts in short periods while others ramble on, decide no actions and might as well not have happened. In this section we will look at the meeting and at some simple steps which can be taken to avoid poor performance in meetings.

Meetings have a wide variety of functions. Some are progress meetings where engineers report on progress to their colleagues and superiors and receive information about the progress of others or news that affects them. Others are briefing meetings where many people attend to benefit from the questions asked by colleagues and by the interactive nature of the meeting. There are also meetings with collaborators or customers where difficult issues may need to be discussed and agreed. Such group conversations cannot easily be conducted by telephone or letter and the only way is to bring all concerned together. The following paragraphs indicate some of the most important things to remember when setting up effective meetings.

● *Is a meeting appropriate?* Meetings are called to bring several people together and make the best use of their collective talents. The more a proposed meeting makes use of the assembled talents, the more successful it will be. For instance, a project planning meeting conducted by a group of engineers who have a project to complete and the skills to do so between them is likely to be productive. However, a design meeting in which one or two delegates have their own opinion about methods to be employed while the rest are excluded or superfluous is likely to waste time. Meetings allow sharing of ideas, rapid communication, interactive reasoning and the ability to solve problems that span far more than the knowledge-base of one person. These are impressive attributes and they should be reserved for difficult problems.

● *The purpose of the meeting*: Every meeting should have a purpose, there must be some desired result or decision that needs to be reached and the meeting must keep it in sight. It is the responsibility of all those present to maintain a focused view and not sidetrack the meeting. However, the chairperson has particular responsibility for this task.

● *Circulate agenda first*: One cannot expect people to contribute their best if they are not forewarned of what is required. Apart from anything else, delegates may be busy and the meeting will then have to be rearranged. Finding a time when all the required parties can attend can be a problem in some organizations. The agenda is a list of the items that will be discussed at the meeting. It should also contain the start time, location, details of subject matter, the election of the chairperson and the duration of the meeting. Having the agenda will also allow people who are unable to attend an opportunity to make a written contribution.

It is usual to have three standing items on the agenda of regular meetings:

Apologies for absence
Minutes of the last meeting
Matters arising.

The second item is the ratification of the minutes where the group have the opportunity to point out any inaccuracies before the minutes are agreed as a true and accurate record of the previous meeting. The third item is where the actions that were agreed previously are followed up and reports are made to the meeting. Often there will be a final item of Any Other Business (AOB) which provides an opportunity for people to raise small issues that do not warrant a full discussion.

● *Have a chairperson*: During meetings people often put forward creative ideas and intense debate can develop when important issues are at stake. Quiet personalities with much to offer can be overpowered by loud ones. If the meeting is to provide the benefits of many peoples' views and produce constructive ideas these pitfalls must be avoided. This is the purpose of the chairperson.

The chairperson has a special role in the meeting and is responsible for the process of the meeting rather than its content. It is the duty of the chairperson to ensure that the meeting is kept to the subject and to alert the group when the discussion wanders. The chairperson must ensure that all voices are heard and that the loud do not dominate the quiet. He or she must steer the discussion towards making conclusions at the appropriate time and ensure that the record of the meeting contains the important points made. The chairperson must not volunteer opinions on the matters at stake or enter into the technicalities of the discussion. Such a combination of skills does not come easily and people who are good at this task are not common.

● *Take minutes*: Minutes record what was decided at the meeting. Some minutes are very brief such as those from a regular meeting and record no more than the actions that were agreed and who is responsible for them. Others, such as the minutes from court cases, are more comprehensive and record the arguments as they were put forward and how the discussion went. The style of minutes chosen must be appropriate for the meeting in question. A meeting to decide whether or not to take a particular action will not normally have the arguments recorded, only the result, whereas an appraisal meeting will normally record all the views and opinions of both the parties present.

It is always advisable to have someone taking minutes who is not involved in the discussion and decision making. This ensures that the minutes are objective and that someone who should be contributing is not distracted from doing so. After the meeting minutes should always be circulated promptly while the meeting is still fresh in the memory.

● *Ground rules*: In meetings where tensions are likely to be high, time is of the essence or where interpersonal conflict is likely to occur, ground rules are an excellent way of keeping things positive and constructive. Examples of ground rules that are normally taken for granted are (1) only one person can speak at once, (2) the chairperson's decision is final and (3) all members are free to participate. In particular types of meeting other ground rules may be appropriate. For instance, in a brainstorming meeting it may be sensible to make the ground rule that seniority is not important. The ground rules should be established and agreed before the meeting starts.

- *Good meeting behaviour*: Everyone at the meeting is responsible for their own contribution. It is a common mistake to think that the chairperson is responsible. People who come away from meetings criticizing the decisions taken but have not contributed significantly themselves are just as much at fault as those who force their own points of view through at the expense of others. Incumbent upon every member is the responsibility for giving their own point of view succinctly and unambiguously, and not wasting the time of the other delegates with ill-considered trivia. It is also no more than common courtesy to remember that everyone has their own point of view and is entitled to express it without censorship. There will always be people at different levels of understanding in a meeting and those who ask questions that appear basic and obvious to some team members are to be respected. Not only are they learning during the meeting but they often bring the more erudite members back to earth with basic questions. After all, if an idea is sound, it should be easy for the idea's author to answer the question. If it is not, something is clearly wrong.

12.7 Case study: Stephen Lever

Stephen Lever works for Richards Mouldings. The company was set up 50 years ago by old Mr Richards and it makes plastic components. In the early days the company made non-ferrous castings and then moved into plastics. Today it manufactures a range of goods related to plastic mouldings. It sells both moulding machines and moulded plastic components. Some of its customers own the dies and Richards uses them in its own machines to produce the products; other customers buy from a wide range of plastic components off the shelf. At the other extreme are customers who buy in very low quantities of ten or even single units for very specialist applications such as defence and research.

Stephen works in the sales office and is one of a sales team of fifteen people. Much of the work centres around order progressing. Customers ring up Richards wanting to know when their order would be ready. Traditionally, Richards has, like many other companies, had sales personnel go to the production department, find out how the order is progressing and then ring back with the details. This process continued to be a general function of the sales office until about three years ago, when a subsection was formed with Stephen and an assistant who handled all the progressing of orders. The aim was to provide consistent service of accurate information communication to all customers. Some of the customers have long-standing orders and deal with Richards in the very long term. They need to know accurately when deliveries will be made and sometimes expect very good treatment in exchange for this arrangement.

Stephen reviewed his work and that of his colleague. He reasoned that it must be possible to cut out the phone calls altogether. This was possible because the production control computer had all the information they required in any case. The MRP system that the company operated showed when each works order was due for completion. One could log-on to the stores system and check the availability of all the off-the-shelf items, and once you had learnt your way around the system there was really no need to ask anybody for the information.

Stephen investigated the area, attended an exhibition on information technology and by using journals, colleagues, computer promotional materials, and his colleagues, he became aware of EDI.

EDI stands for electronic data interchange and has been in existence for years in the computer world. Almost everyone who has used a computer knows about logging-on and connecting to distant machines and then using the data and programs that they contain. Stephen found that some organizations sold software designed to permit transfer of general production data across phone lines. The software would take an input file of virtually any specification, convert it into a particular format and then transmit it on request through a modem. The idea was that if anyone had a compatible system and wanted to exchange information they would simply ring up the source of the information, log-on, select the files for transfer and then instruct transfer to commence. At their end they could then use the software to convert the transmitted file into any desired format and so have compatibility with their own system. The only matter that really had to be agreed was the order in which the data would be stored on the file (for example, which piece of data would represent the order code, the delivery date, etc.).

Stephen had discussions with other company personnel, in particular the production and managing directors. They were concerned that customers would be able to see the internal workings of the company. For example, if a valued customer suddenly wanted a large order processed quickly or a delivery brought forward it was standard practice to let the orders of lesser customers slip a little in order to give the best service to the most valuable customers. If such a customer had effectively been given access to the shopfloor production schedule then they could deduce why their order was late. The problem didn't stop there, however. If a smaller customer observed the orders of another company being brought forward or receiving a good lead time it would be impossible to quote them a longer lead time, and soon every job would be on the minimum lead time and any amount of internal juggling with the schedule would become impossible. Lastly, it was pointed out that some customers would not wish the state of their orders to be made available to other customers of Richards since they may be in competition. Richards in fact supply several groups of two or three competing companies. Yet other customers, such as defence, might have confidentiality needs that would prevent them from being entered onto the system at all.

Armed with the views of other company personnel Stephen implemented a system that replaced telephone support for routine customer enquiry with an EDI-based system. Customers with compatible computer systems (most of them) could log-on to the system and see a limited picture of what was happening. This was discussed and agreed at a very senior level within the company. They had access to the expected delivery data, the order codes and tax point data. They could change the order within certain prescribed limits and see immediately what effect this would have on delivery, they could request changes and later be informed of the possibility and also log-on to a quality index that showed the average lead time, the number of late orders, the number or order change that had been made, etc.

Now Stephen is working on extending the system to order placement and interfacing to the accounts department so that the administration associated with invoicing can be reduced in volume but increased in accuracy. Two years ago no customers made use of EDI, now nearly 70% of the value of orders are sold to customers who use EDI to do all their routine order chasing.

12.8 Summary

In this chapter we have introduced the subject of communications and explained its importance. We have examined the need for communications and seen how it occurs in the workplace. Most of the chapter was concerned with an examination of written and oral communications. In these sections we described some of the most important communication methods used by engineers in their work. Finally, we looked at a special example of communication, the meeting.

It has been said that there are no problems in management, only in communication and it is certainly hard to think of a problem that has occurred that could not have been solved by a timely communication. The importance of being a good communicator is repeatedly emphasized by employers, by customers, by suppliers, and by our colleagues and friends. The individual who ignores this need does so at their own peril. It is possible to make great improvements in our communication abilities and there is no-one who does not stand to gain by doing so.

Communication is self-evidently the most important skill we possess, not only for management but also from our need to communicate with other members of our own species. It underpins our abilities to perform any other management or engineering function. Those who disagree with this view would be well advised to steer clear of these.

12.9 Revision questions

1. Make a list of the communication techniques to which you have been exposed over the last two weeks. For each, describe the advantages and disadvantages as experienced by you.

2. What communication issues are important in the process of appraisals? Start by considering what an organization needs from its appraisal process and then describe the communication issues associated with each area.

3. How do computers affect communications in your life? List all the communications that you have experienced that were computer assisted. What impact have computers had on communication techniques and communications within organizations? Use a library to do your research.

4. Why might memos be 'copied' to people other than the person to whom the communication is directed?

5. Obtain a copy of the 'mission statement' of your organization or of an organization you know well. (If no statement exists, compose one that might be appropriate.) How do communications in your organization exist to serve the statement? Are there communication needs that are not addressed? Is there more communication than is necessary? Can you explain the success of your organization is term of its communications?

6. Choose a set of instructions or directions that you have found to be poorly written in the past. Try to use examples of domestic products, like a power drill or coffee percolator, with which you are familiar. Criticize the text and rewrite all or a portion of it so that it is perfect. Discuss your text with a friend and get them to criticize your text. Then answer the following questions. Why is writing such text so difficult? What issues must be considered in its preparation?

7. Explain the following communication terms:

   Feedback
   Redundancy
   Jargon
   Abstract
   Reference

8. Describe the communication issues that might be important within an engineering organization between the manufacturing department and (i) the sales office, (ii) the personnel department.

9. Discuss the communication issues raised by the implementation of a quality system within an engineering company.

10. Discuss the communication issues raised by the implementation of a training programme within an engineering programme.

11. Discuss the communication issues that influence the effective operation of a team of engineers.

12. Prepare a letter of application to an engineering organization for a post taken from a real advertisement.

13. Describe the interview process from both points of view. What communication techniques can be used to ensure that the interview goes well?

14. What is active listening? How does it help a conversation?

15. You are to chair a meeting in which the engineering department of a company will meet an important customer for a discussion about progress on a contract. Progress has not been good, the customer blames the company and the company feel that the selling price of the contract was competitive enough to begin with, but since the customer has started to insist on changes and additions to the original contract it has become impossible to meet the contract. How would you prepare for such a meeting?

What communication issues would be addressed? How would you go about achieving a resolution between these parties?

16. Explain how information, which is intangible, can be of value to an engineering organization. What gives a piece of information its value?

17. Why do customer complaints make a poor method of gaining market information? What else can an organization do to gain market information from its customers?

18. What other organizations may be of use in supplying useful information to a company? Describe them and the sort of information they can provide.

19. Examine a colleague's reading skills in the following way. Ask the colleague to read a page you have chosen at random from this book, in their own time. Ask them about the contents of the page afterwards and record your comments about the level of detail they remembered afterwards together with the length of time they took to read it. Ask them to read another page in only three quarters of the time and record the same data. Finally, do the same but allow the colleague only a quarter of the original time. Repeat the test for several other colleagues. What conclusions or observations can you make about reading speeds and its effect on recall?

Further reading

Fowler, H. W. (1965), *A Dictionary of Modern English Usage*, 2nd edition, Oxford University Press.

Waldhorn, A. and Zeiger, A., advisory editors South, R. and South, J. (1986), *English Made Simple*, William Heinemann.

Part 4

ENGINEERING MANAGEMENT IN PRACTICE

13

The vocation of engineering management

In the introduction to this book we explained in general terms the way in which management topics affect the jobs of engineers. Throughout we have described the different areas of management and in fairly limited examples have shown their application in engineering situations. However, these examples cannot give an overall picture and they were specifically chosen to illustrate discrete areas. As engineers writing for engineers we will now redress the balance of the book.

In this chapter we have a case study. It is a complete engineering project, a job on which James Balkwill was employed for nearly two years. The type of situation described, that of a small project team, is quite typical. Many engineering graduates will work in such teams. The technology may be unfamiliar but engineering is so varied that it would be impossible to choose a technology that would be widely known. We have, however, included the technical detail so that you have some understanding of the technical issues that were faced by James and his team.

The project is the development and delivery of a new laser product. Naturally, a project of this size cannot be described quickly and so the case study is a lengthy one. To assist the reader in learning from the case study we have therefore divided it into eight sections.

The project is presented chronologically with the emphasis on the task. However, we can see the management issues and skills that are required to tackle an engineering job like this. After the case study it is debriefed. We have prepared a review of the case and the management issues raised. This is structured to follow the format of the book so that you can look back to the relevant section and reconsider the theory in practice. You will see that engineering management is not only important, it is an integral part of the job, and you will notice that none of the areas previously discussed works without affecting other areas.

13.1 The Cu100 project at Oxford Lasers Ltd

(This case study is reproduced by kind permission of Oxford Lasers Ltd, Abingdon Science Park, Barton Lane, Oxfordshire, OX14 3YR. England, and of the Cu100 team members themselves: John Boaler, Richard Benfield, Steff Inns, Rhys Lewis and James Balkwill. We are also indebted to Keith Errey, a director of Oxford Lasers for his assistance in the preparation of this case study.)

13.1.1 Background

Oxford Lasers is a very self-explanatory name for a company. It is, after all, in Oxford and it manufactures lasers. There are many types of laser, but Oxford Lasers concentrates on copper laser technology and its applications.

The laser action in copper was first demonstrated in 1967. The first lasers were based on ruby crystals and even these date back only to 1962. In England much research about copper lasers was carried out in Oxford at the Clarendon Laboratory which is part of the University of Oxford. Over a period of many years, research students carried out experiments that examined the operation and physics associated with copper lasers. It is an inevitable consequence of this type of work that those engaged in the research come to understand a great deal of the engineering associated with the object of their research. After all, copper lasers were not available commercially and therefore performing research upon them meant building them first. The research group continued to expand and the question of whether there was any commercial demand for the lasers beyond the confines of the university arose. Naturally, those involved were keen to exploit their work and the big question hanging over any commercial possibilities for the laser hinged on why people would use it.

There was one particular application that appeared to be of great commercial significance at this time. This was a process called AVLIS, which stands for Atomic Vapour Laser Isotope Separation, and it is used to separate one isotope of an element from another. The copper laser has a wavelength that makes it ideally suited to pumping a dye laser tuned to separate the isotopes of uranium. Uranium is used in nuclear power stations and is very expensive to separate. When uranium is mined the proportion of useful isotope is far too small to use it as fuel and so it must be 'separated' or enriched so that there is enough of the radioactive isotope to make a useful fuel. AVLIS appeared to offer an estimated tenfold reduction in separation costs.

In the United States, Japan and France there were already national efforts aimed at establishing a laser-based isotope separation facility. These were secret installations and there was no possibility of obtaining the technology directly from them. The UK had no such facility but clearly since the UK is a supplier of enriched nuclear fuel the investigation of AVLIS was almost inevitable. An opportunity therefore arose for the supply of lasers for initial investigations of the AVLIS process. As a result of the university research and its existing commercial status, Oxford Lasers was able to respond to this opportunity. The US and Japanese efforts used nationally funded laboratories to produce the lasers, but copper laser development in France was less successful, a fact which was readily exploited by Oxford Lasers.

Oxford Lasers had been formed in the late 1970s to commercialize the first excimer lasers which had been developed in Oxford. These are pulsed lasers giving high-energy pulses of ultra-violet light. However, it was the development and commercialization of the copper laser which really offered the chance for commercial success, and the first full-time employees were taken on in 1982 to build lasers in a former butcher's yard just north of the centre of Oxford.

While there was a clear focus on the development of copper lasers for AVLIS, there continued to be a need to sell lasers in other fields, and so the problem was how to sustain sales in these other areas and develop the company to be ideally placed for a major AVLIS contract when it arrived.

Copper lasers have, like all lasers, a particular wavelength at which they operate. In this case it is 511 nm, which is a deep green colour, and 578 nm, which is golden yellow. The laser is also pulsed, meaning that rather than emitting a constant stream of light, it is on very brightly for a very brief period of time (about 30 ns) and then off for a comparatively long time after that (about 0.1 ms). In other words, it is off for three thousand times longer than it is on. Despite this, the pulses are so bright and the whole process fast in comparison to human eye reaction times that the beam appears continuous.

The heart of a copper laser is a gas of copper through which a discharge current is passed. The electrons in the discharge then excite the copper atoms, by collisions, from their low-energy state, called the ground state, to a much higher energy level. The region in which all this occurs is called the gain chamber and this is contained within a tube of exotic material capable of withstanding the arduous conditions caused by the high temperature and corrosive action of the discharge. This is called the plasma tube.

Once excited, the atoms then start to shed energy in accordance with the probability distributions dictated by quantum theory. Before long many of them have fallen to a lower energy level and have emitted a photon in the process. The energy level to which the atoms fall is not the ground state but another energy level just above it. The central principle of the laser is that if a photon emitted from one such decaying atom happens to collide with another atom still at the high energy level it causes it to fall immediately to the next energy level down and therefore emit a photon as it does so. This is called stimulated emission – a photon 'stimulates' the excited atom to emit a photon. The initial photon is unchanged in this process and there are therefore now two photons travelling in the same direction, and, as it happens, the two photons are always in phase. This action continues until nearly all the atoms have fallen to the lower energy level. During this phase there is a massive conversion of energy into light, and mirrors at each end of the lasing chamber channel the light created back into the gas of copper atoms and so increase the chances of a photon colliding with an excited atom and stimulating even more light.

It is an acronym for this process that gives the laser its name, *Light Amplification by the Stimulated Emission of Radiation*. After all the atoms have emitted a photon and are in the lower energy level one has to wait for a comparatively long time for all the atoms to relax to the ground state, and only then can the process can be repeated. Because of this action, the copper laser is a pulsed laser, and although the output beam appears to be a constant and intense beam of green light it is in fact made up of a series of javelin-shaped arrows of light, each being only 10 or 20 feet long and each separated from the next by about 40 miles (milliseconds at light speed).

The physical characteristics promoted investigations of possible applications based on the unique properties of the copper laser. Applications therefore focus on the very high gain or amplifying capacity of a copper laser, its colour or wavelength, its high pulse rate and the fact that it is the brightest of the visible lasers. These characteristics led to applications in:

1. High-speed photography where the short pulse of light is used to illuminate and film events with a resolution of about a ten-millionth of a second
2. Dye laser pumping, in which the particular wavelength of the copper laser makes it especially suited for injection into another sort of laser called a dye laser that has the property of being tunable. This allows the user to obtain exactly the wavelength required rather than having to accept whatever the closest naturally occurring lasing material happens to be
3. Display applications, where the very bright visible light is used for entertainment at, for example, rock concerts or for projection television
4. Materials processing, where the beam's properties are used to modify or machine materials on a very small scale (for example, cutting a microscopic fillet in metal or drilling holes for fuel injectors or ink jet printers)
5. Medicine, where the light from a dye laser is used to modify a previously injected chemical on cancerous locations within the body. The light from the dye laser travels a good distance through human tissue and so permits the treatment to proceed with a relatively low amount of bodily invasion
6. Lastly, of course, AVLIS.

These very technical applications became the sales base for Oxford Lasers and the company continued to grow throughout the early 1980s. It is wrong to think that these markets were secure for the company since they were new markets being addressed by new products. Such a situation is inevitably difficult, and there is generally competition between the relatively small number of contending new products which leads to confusion in the marketplace.

The company had its own declared objective to grow to a certain size and turnover within five years and this goal was reviewed annually. The greatest obstacle to this growth was the very marketplace itself. Without massive marketing resources almost nothing could be done to stimulate the market size and the number of orders received depended on factors beyond the company's control such as the current direction of laser research, the trends in medical research and practice, and the uses of dye lasers.

However, the company was moving forward, growing rapidly and soon became 'functionalized' with two separate sections – manufacturing and R&D. Manufacturing was really assembly; there are thousands of components in a copper laser that range from proprietary electronic components to exotic materials and chemicals. Manufacturing was carried out according to a record of how previous lasers had been built, and consisted of sketches and drawings that had developed over the years. Technicians assembled the lasers and tested them. Production ran at about one laser a month. There was an open stores area, purchasing was done through the supervisors and product development was on-going to eliminate reliability problems and reduce costs. The responsibility for this whole area was that of the Engineering Director.

The other area, R&D, employed a roughly equal number of employees and was responsible for all kinds of new initiatives. These involved research into new designs for assemblies in the laser, improving the efficiency of the laser, keeping abreast of developments in the laser physics world and dealing with the technical problems of manufacturing and customers. One-off systems, particularly those involving customized optical accessories, would be tested and installed by R&D personnel. Since the power of the laser is something that sets it apart from its competitors, there was always an eye on applications that would involve more power. There was also the feeling that by announcing a continuing escalation in power on offer the rather feeble competition to the company would be left further behind since high power for a copper laser is technically more difficult to achieve. Also by this means the company sought to place itself in an advantageous position for supplying high-powered AVLIS lasers at some future point.

By the mid-1980s the company had a separate sales force and service department, there was the start of a computerized production control, the stores were controlled, and there were about 40 employees. Support continued from the Clarendon Laboratory through formal arrangements with the university and several researchers became employees of the company.

The company continued to grow throughout the 1980s by providing lasers for the applications described above. However, the background aim was to be successful in securing the supply of lasers to any European facilities for AVLIS. Company contacts and links in the scientific and commercial community identified the organization that was eventually to build such a full-scale pilot facility. Some time was spent considering the power requirements of such a system and a value of roughly 100 watts was determined to be optimal. Oxford Lasers ensured that its 60 watt laser was well publicized and that bench tests were conducted on a device large enough to produce 100 watts. These tests were expensive and without an order to pay for them they were simply used to demonstrate that the physical principles upon which the copper laser depends would continue to function in the way so far determined up to this power level.

13.1.2 Customer requirements

After many years of diligence and attention to the marketplace, negotiations with the customer finally arrived at a commercial proposition for lasers whose use was to be AVLIS. A European effort was to be developed in the UK and this customer would be building the facility in the UK. Ten years of waiting was over and the first step towards AVLIS supply was underway. This customer wished to separate uranium in a pilot industrial plant set up to evaluate the process. It was understood that if this facility went well there was the potential to supply a very large number of lasers to a much larger facility that would follow. With such a tantalizing commercial opportunity on offer, the senior management reacted enthusiastically but with some caution. There are, after all, dangers of becoming too closely allied to just one large customer and the other areas of sales should not be neglected.

Detailed negotiations opened and it became clear that a laser of about 100 watts would indeed be required and one which was capable of operating very reliably over long periods. This requirement was at odds with the current state of the product, which was not suitable for long-term, unattended use. This apparent problem was not a weakness in the existing product since it was sold primarily to scientists who had no problem with becoming expert laser operators and whose need for long periods of trouble-free operation was almost non-existent. Their primary requirements were low cost and ease of maintenance, something quite different from a very high mean time between failures (MTBF).

The need was for industrial use characterized by simple control, robust operation, easy servicing, long periods between routine services and large MTBFs. The existing products, in contrast, were complicated to control since most customers were well acquainted with the operation of a copper laser and wanted to perform and control all the different sequences and modes of operation themselves.

The customer wished to place the lasers in chains, the light from one laser entering the next and being amplified in the process. In essence, they wanted a chain of three or four lasers, and the first one in each chain was to be a special version of the laser, called an oscillator. The purpose of the oscillator was to produce the initial pulse of light. The pulse would be about a 100 feet long and

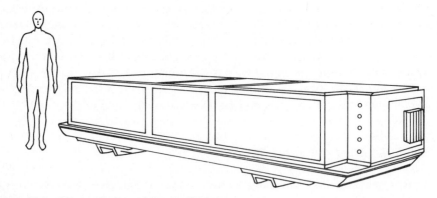

Figure 13.1   A general view of the Cu100 laser

composed of parallel rays so that they would not be too spread out by the time they reached the end of their flight path.

The other lasers in each chain existed to amplify this initial pulse. After a few nanoseconds the pulse would enter the first amplifying laser in the chain. This laser would be discharged and start to amplify just as the front of the javelin-shaped pulse of light arrived. The pulse would increase on its passage through this laser and would then enter the next laser in the chain. This, in turn, would be discharged at the right moment to reach peak amplification as the light pulse entered its gain chamber. In this way, the initial pulse would be boosted to several hundred watts of light.

The customer also intended to have many such chains operating simultaneously in a large rack containing many lasers. The rack would be built from 101 mm square section box girders of steel and sounded more like a shipyard project than an optical bench. It was imagined that each laser would weigh about a tonne and be placed in or out of the car-sized slots in the rack by forklift truck. The lasers were to be optically aligned with a dimensional error of less than a quarter of a millimetre over the 60 metre light path and each laser in each chain was to be synchronized to light speed and start to amplify the pulse of light only when it arrived. The amplification period of a copper laser is over so quickly that this additional complexity was necessary if the whole system was to have any chance of working.

The output of this warehouse-sized rack of lasers would be about 1500 watts of green copper laser light that would then be directed into an enormous dye laser to modify the wavelength. This light would then be directed into the separation chamber itself where it would interact with the atoms of gaseous uranium and be selectively absorbed by only one species of isotope. This selection occurred because of the precisely controlled wavelength of the light from the lasers. Once absorbed, the isotope had somehow to lose this surplus energy by shedding an electron and so becoming charged. The result of this is that one of the isotopes is electrically charged and the other is not. From here they can be separated easily by electrostatic means.

It was clear that a special project would be required to complete the work in accordance with the customer's specifications. Rhys Lewis was selected to run the team since his PhD and background meant that he knew more than most about copper laser gain chamber design and he had experience in R&D on other development projects. A contract was negotiated with the customer and a specification agreed upon. The specification covered the major technical aspects of the laser, particularly the power, the beam size and the length of time the laser should run without interference. The rationale behind the details of the agreement was based on experience with other lasers and a knowledge of the physics that goes on within them.

The contract was agreed and a 96-week project began. In just under two years the company was contracted to have installed two 100 watt copper lasers at the customer's lab and have completed a comprehensive set of acceptance trials.

13.1.3 Recruitment

I had been contemplating a job change for about six months when I read the advertisement but had not, until then, seen anything that had really suited me. There were other jobs that I had perused but I had decided against them for various reasons in each case. I replied to the advertisement with a CV and covering letter and was invited for an interview. I knew nothing about lasers or Oxford Lasers, all I knew was what was described in the advertisement, which was not much. I took a day's annual leave for the interview, had my suit cleaned and checked the address on a map. I spent the night before the interview going through the details of my reasons for leaving my previous employer, my ambitions, the sort of job I was hoping was on offer and trying to prepare for the questions I thought it was likely I would be asked.

When I arrived at Oxford Lasers in my interview suit I realized that the interview was to be conducted by a panel. This was a first for me and it did little to calm my nerves. Two of the panel were directors and so at least I knew something of their role within the company, but the third person was an unknown quantity. We walked round from the reception area to the side entrance of some smart new industrial units from which the company was operating. At the top of a staircase was an empty room which seemed to serve as a library, perhaps a meeting room and, at this time, an interview room. I knew there was a post available on a large project to build a new and powerful laser, something that would break new ground, but I was short of any real facts about it.

The interview began with a description of Oxford Lasers. The company had been trading for nearly ten years, manufactured copper vapour lasers at this one site in Oxford, and sold them worldwide. There were about 50 employees, the turnover was in the millions and Oxford Lasers was the market leader by a large margin. The lasers were sold for various applications – high-speed photography, medical practice including cancer therapy, research, television projection systems and isotope separation. The post for which I had applied was indeed on a large project, the aim of which was to produce a new a more powerful copper laser than any previously built by the company. There was a customer with a special need and a contract was already in place. The project was to last 96 weeks, and it was already week 4. Some specifications had already been completed and the diameter and power of the laser beam itself had now been agreed. The third person on the panel explained that he would have overall responsibility for the project. He then spoke of the importance placed on the project by the company, commenting that he even had his diary marked out in project week numbers. We joked, I was used to doing the same.

I explained my reasons for wanting the advertised project engineer's appointment. I was already employed in a similar role elsewhere but wanted to tackle one very large project rather than many smaller ones. I wanted my own team and to increase my role with customers and be asked to produce equipment whose specification I had assisted in negotiating. I felt confident on the technical side and explained the sort of projects I had undertaken previously, which included many specialist one-off scientific instruments. I thought I'd lost it then, everything went quiet and they started asking lots of other questions, including technical ones.

Much later I found out that the fit between my wants and their needs was so close that they were taken by surprise and just didn't want to seem too enthusiastic. About two weeks after the interview the papers came through and after a minor altercation over the salary I was hired. I started a month after the interview.

I didn't know what the traffic would be like on the first day so I left a lot of time and arrived really early, the place was locked and I spent the first hour of my new employment in the car park! Rhys arrived and introduced himself again. He looked at his watch, 'you're early' he said. Normally I'd been at work for over an hour by now. We walked upstairs into an office and I was shown my desk. There were five of us in the one office, the size of a large living room. They were clearly the desks of engineers. Papers were piled high, in-trays groaned under stacks of drawings and files, dotted about the room were several machined flanges, interesting shaped components and some other proprietary parts I did not recognize. There were books on laser physics and electronics catalogues everywhere together with two computers. Anything that provided a flat surface was piled high with papers, journals, and handwritten design ideas. There was a general atmosphere of people doing things.

Over the first few days I started to get some idea of what was going on. There was a company induction programme and I was even introduced to every other employee. Rhys started giving me lessons in laser physics.

The intensive induction programme and introduction to laser physics continued for the first month or so, punctuated only by other introductory talks and visits. All the physics described by Rhys clearly posed great engineering problems. 'How does one achieve a contained gas of copper?' 'How can one obtain the massive rate of current rise required to make the process work?' 'What sort of materials and components can be used for the arduous conditions inside the laser?' Rhys explained how these issues were solved in the present designs of copper lasers produced by the company. This knowledge formed the basis of Oxford Lasers' success and some of it is a jealously guarded secret. Once I had developed a working knowledge of copper lasers we moved onto the project itself.

A 100 watts of light does not sound much when compared to an ordinary light bulb of a 100 watts which may easily be bought at any hardware store. The reason a 100 watt copper laser is so different and so much more powerful is that the 100 watts used to describe a light bulb refers only to the power consumed by the light bulb. In fact, only a small fraction of this power is turned into light within the bulb and the vast majority is converted into heat. With the copper laser, however, the 100 watts refers to the power of the light and is vastly greater than the light output of a bulb. The second factor is that a light bulb emits its light radially outwards in all directions whereas the copper laser emits its light as a narrow beam, in our case 60 mm in diameter. This factor also greatly increases the light density in the laser beam compared to that from a conventional bulb.

The aim of the project was to produce the first two amplifying lasers each capable of producing a 100 watts of light when run alone. At present we had no order for an oscillator but we did expect that a similar contract for an oscillator would be negotiated.

The first amplifying laser was due for delivery in week 68 and the total project duration would be 96 weeks. Eleven weeks had already passed in the preparation of plans and contracts and the hiring of a project engineer. We also had a fixed budget. We were to use a team of five people and a new industrial unit was to be given over to the project. Rhys introduced me to some of the plans for the project that had already been drafted and it became clear that we already had things to do that were urgent even with more than a year and a half of the project to go.

Most of the first three-month period had been taken up with enclosure design. The laser would require a special enclosure to house and mount the internal components and the design of this assembly was an urgent issue since it would have to be competed before manufacture of the lasers themselves could begin. In addition, Rhys and I did a lot of work on the product development specification (PDS). I also spent some time with the technician who had been working with Rhys on research aimed at producing a 100 watt laser before I joined the team. Richard had a test bench in the same building which was furnished with the latest laser design with the goal of achieving reliable operation at 100 watts of light. Richard gave me many lessons on lasers and fast pulse circuit design, and the opportunity to spend some time working on a real laser gave me a much deeper understanding of the problems involved.

After I had been employed for about three months Rhys and I held interviews for the remaining two people we felt were required for the team. With my increased knowledge of the project we felt able to be more accurate about personnel requirements and the skills we required. It was clear that another technician was needed to look after the mechanical side of the laser and to do much of the laser assembly. It could be that the same person would also do some paperwork and ordering. There would be a lot of this later on and a technician who could grow into the job would be preferable. Certainly someone who could handle change was needed. A job description was prepared.

The second person was a draughtsperson. We had already had to use subcontract drafting services and the need for drafting was going to increase rapidly, especially to produce drawings for all the components that would be finally designed and detailed once the design report was out of the way. The new unit was available very soon and so with all these changes it seemed sensible to hire the two staff, set up in the new unit and officially change the reporting structure so that they reported to me, since I was to be their manager. I still reported to Rhys.

John joined almost immediately as junior technician. He was an internal applicant and certainly fitted the needs of the group very well. His old boss, the production manager, was somewhat vexed to lose a very good technician who had been trained by his department using his own budget. The view was that the production department spent all the time and money training technicians only to have the best ones appointed to other internal jobs. I could see this point of view, but then John wanted the job and Rhys and I wanted to give it to him.

Hiring a draughtsperson took longer, and in the first round of recruitment no-one who fitted the needs of the team and who would fit in could be found. We therefore continued using subcontractor services and carried out another round of recruitment. To meet our customer's request that certain drawings be prepared only by Oxford Lasers personnel I had to prepare some of them myself. Eventually a draughtswoman from a local electronics company responded to our advertisement and interviewed well. She was older than the rest of us, had all the skills we needed and struck both Rhys and I as someone who would help to make the team more balanced in outlook. We offered her the job and three weeks later Steff joined Oxford Lasers.

Steff visited Oxford Lasers during the week prior to starting work to attend the monthly meeting of the whole company. Often this informal meeting was held in the pub just 50 yards from the company and it always consisted of news from each of the departments followed by drinks. Rhys had made a short speech about our progress and other people gave reports about other departments. I introduced Steff and explained that she had come along to meet everyone and would be starting work the following week. Then we all had a drink. A team meal was felt appropriate and curry was hailed as the only food with which to welcome a new member of the team. Consequently Steff's introduction to Oxford Lasers was beer and a curry. My only comment was that it was better than an hour in the car park, as had been mine!

13.1.4 The design report

Like most companies, Oxford Lasers wanted to control its new product development costs. To do this a detailed specification of the product called the product development specification (PDS) was prepared. This was also a customer requirement. This document was more than the technical specification. It covered all relevant attributes of the laser. It did not specify the design, rather it said what the design would do. It included testing schedules, a quality assurance plan, and what inspection schedules would apply. Even measurement techniques, colours and test equipment were specified, and in one section the effect and operation of every single control and indicator were detailed. The report was a complete specification for the project and it ran to 120 pages. The successful completion of the PDS had an associated stage payment and the money received at that time would cover the cost of many of the expensive materials used in the laser construction.

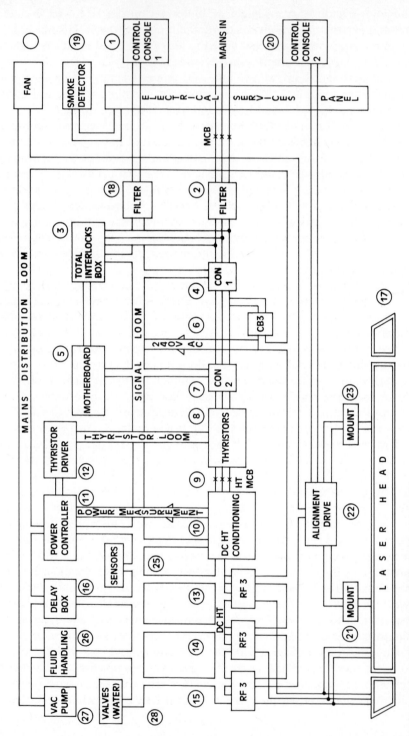

Figure 13.2   Block schematic of laser operation

Several visits to the customer were necessary to get the contents of this document agreed and 32 weeks of project time were dedicated to getting it right. We visited our customer again one week before its official signing. We still had a number of issues outstanding, some of them tricky, and had prepared extensively for the debate by considering our reaction to the alternatives that might come up. We also had a 'shopping list' of issues to be resolved and the way we wanted them to go. The customer must have had a similar shopping list and disagreements certainly occurred. The points were negotiated and agreement was reached with compromise on both sides. It was certainly true that even this far into the project both parties failed to appreciate exactly why some issues were important to the other side. The delegation from the customer's side was quite large and people kept coming in and out of the meeting, which was very disconcerting. Only Rhys and I represented Oxford Lasers. We stayed there all day. With mutual trust established it was ultimately possible to resolve our differences but it was very hard work. There were compromises made on both sides, admissions that certain issues had not been anticipated accurately, and some issues difficult for one side seemed straightforward to the other. By the end of the day we had a list of action points that reached practically all the way back to Oxford but we did have agreement that if these were included in the draft of the PDS it would be acceptable. We booked into a hotel overnight since it was too late to get home and flexed a little company plastic on sustenance.

In the course of our discussions it had become clear that the customer was already building the structure that would support another sixteen lasers. The present contract was only for two and we felt that this showed a considerable commitment on the part of our customer. In the hotel that night we made notes of the meeting and discussed what was said, making a report for Rhys to give back home.

One of my responsibilities as a manager was to conduct appraisals of my staff. The company held annual appraisals and the time for them came round during the preparation of the PDS. The appraisal scheme was two-tier, each person being appraised first by their immediate boss. This appraisal considered the facts of performance over the last year and existed to give objective feedback to the individual and offer ways in which the individual might develop. The second appraisal was with the boss's boss and covered longer-term issues such as career and company developments. I appraised the team individually. Each had their own particular points to make and some problems, although no-one had anything particularly difficult. The appraisals acted as a chance to re-examine my feelings about how the team was working and whether everybody within it was being fulfilled by their work. The second appraisals were with Rhys and he and I met to go through the appraisal forms for each team member before he held the second appraisals. Appraisals are time consuming and are pointless if the appraiser does not know the appraisee's work very well. At the same time, they can be extremely effective for the individual if done well and from my point of view that particular round worked very well in providing me with an objective look at each team member through a long conversation with just the individual and no interruptions. Naturally, in a situation like this, where one sees those working for you all day every day, there were no real surprises but one or two important issues came up that might not have otherwise surfaced.

The day for the signing of the PDS came and a delegation from our customer arrived. We felt well prepared but there were new faces round the table and nothing was certain. Discussions progressed all morning, there were corrections to minor points and these were made by a secretary who had moved into the next room to be ready for any such occurrences. After about three hours two new versions of the report were printed off with the minor corrections and these were duly signed by both parties. This released about one third of the total project payment to Oxford Lasers, much-needed cash that would fund the next stage.

The PDS became a useful document on both sides. There were many subsequent occasions when points of design were raised and the PDS showed unambiguously what was agreed. It also helped in planning the work since it detailed exactly what had to be done and could be read by all team members. No-one was in any doubt about what was required.

13.1.5 Detailed design and manufacture

The general pattern of the next few weeks was rapid construction according to the project plans defined in the PDS. When necessary, we would all gather round the laser to sort out particular issues, otherwise we just carried on with the tasks in hand. The signing of the PDS meant the project entered a new phase. From now on we would have to deliver what we had promised, and any deviation from the milestones would result in penalty clauses operating.

Richard had already been using some of the floor space in our new unit for a very experimental version of the power control system he wanted to use. A laser as large as the Cu100 uses a lot of power and the simple but bulky power control systems on the other smaller lasers would not be practical for our purposes. We were concerned that there might be stability problems with the proposed solution and, in any case, we had no experimental evidence for believing that it would work. We therefore, or rather Richard, built a three-phase version of a new, and as yet unproved, power control system intended for use on another new product being developed elsewhere in Oxford Lasers. Apart from one flashover, when a loose wire arced to the concrete floor inside the safety enclosure and took out three fuses and blew a hole that could accept a pencil in the concrete floor, the system worked fine and Richard was confident of proposed design. I got him to write up the experiment. He hated writing things up but it had to be done. We now had many technical issues underway and the simplest way of losing the information you need is not to keep accurate documentation. We cleared away all the equipment and made the unit ready for manufacture.

The laser enclosures were to arrive in about two weeks' time. The task in hand was the ordering and assembly of the first laser. All attention was directed towards the next milestone, to demonstrate the laser discharging in the presence of our customer. The discharge would not be at full power and the laser would not have copper in it, but it would be a continuous run of several hours. This milestone was due in week 52, only another twenty short weeks. We were also giving thought to the milestone after that, which was due only four weeks later. This was a full-power continuous demonstration of the laser at over 100 watts.

Project control was achieved with one progress meeting each Monday morning. We would all meet and I would take each person's comments on last week's achievements in turn. We would then discuss what each should achieve over the next week in order that we would meet the next milestone. The plans in the documentation agreed with the customer were freely available for the team to view. There were two major areas of work and these were divided among the team.

The first was the design for the final geometry of the water-cooling jacket and the flanges that mounted the laser tube itself. Rhys and Richard had responsibility for this. The second area was the alignment system for which there was no design at all. John and I had this responsibility. In addition, there was the ordering of components and the completion of the design of various relatively straightforward but nonetheless time-consuming assemblies.

A new filing system was set up to cope with our activities. We started to use some of the money we had generated from the design report. There was a set of files, one for every major subassembly of the laser. Each file contained several sections, and some covered the design of the subassembly others, the administration. One file was reserved for the bill of materials and another for the purchase orders. Running totals were kept and the cost profile of the laser was reviewed at the weekly meeting. The contract had particular quality control standards that had to be respected. For instance, we had to control the stock for the lasers separately from the rest of the company. However, we still wished to take advantage of the company's existing ordering and receipt of goods procedure. For this reason, orders were processed by the company operations department, but as soon as they had been received they were sent to our unit for inspection. The store shelves filled up and we started manufacturing various subassemblies, those that could be built in advance of the complete laser construction. The first enclosure appeared likely to be late and so a progress trip was made to try to prevent any further delay. Even so, the delivery date was delayed by ten days. We altered the manufacturing schedule to bring other jobs forward but ten days was about the limit. There was flexibility in the plans and although we had used up some float, progress was still on track. The enclosure finally came eight days late.

Rhys had to make monthly reports to the board about progress and the weekly meeting before a board meeting was very important to him. Questions would come back from the board but, in general, they were satisfied with progress

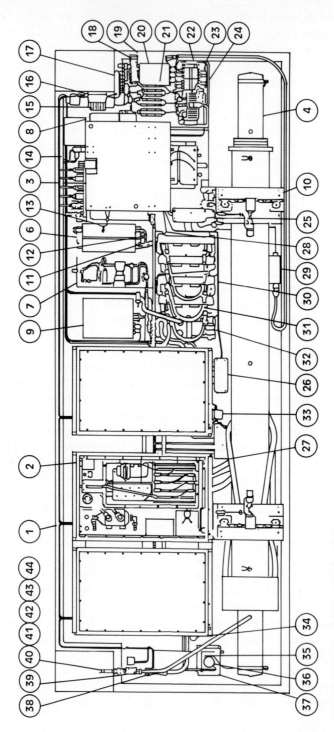

Figure 13.3   General assembly of the laser

and left us to work without burdening us with any other tasks. Some company-wide issues came to us from that route. For example, the way in which appraisals were to be held was regulated across the company, so were all general personnel matters such as holiday arrangements and internal promotions. As well as these, other important issues were communicated in this way. We learnt, for example, about the implementation of a companywide quality programme that would have an effect on business with our customer in the future and might mean not being able to have the separate stores. Policies on the type of computer equipment we were to buy came from the board and when we bought computer equipment for Steff, for example, it was the same as used elsewhere in the company and was negotiated centrally.

A major effort was underway in the preparation of the drawings. Steff had a difficult task in that much of the design was not finalized and yet drawings were required to enable John and Richard to build the laser in a controlled way. We agreed the objectives of the drawing preparation work. Primarily, the purpose of the drawing package was to provide a detailed description of the first laser so that the second one could be built identically. It was imagined that there would be design changes to be made between the first two lasers and any follow-up orders, and, of course, it was necessary to have a complete and accurate record of what was built. The second requirement was to permit manufacture of the first two lasers to progress without delay. This meant drawings for components that had to be ordered would have to take priority.

We established a hierarchy of drawings, starting with a general assembly of the whole laser and then divided it into progressively smaller subassemblies. The drawing structure copied the bill of material structure and over the weeks this emerged in progressively more detailed form. With a fixed amount of time available and so many drawings to prepare (there were to be about 400 in all) the possibility of not achieving the two drawings' objectives was a constant worry. In the end Steff had to take notes, sketches and photographs of some of the assemblies in the first laser since there simply was not enough time to complete all the drawings by then. This did not matter in fact since, between these notes, the design files and Richard and John's knowledge of what had been done it was possible to get the complete drawing set prepared before the second laser was shipped. I was especially glad that we were all together without interference from other projects at this time. The constant ability to ask questions of each other, to look at the progressing work, and meet together instantly if required made this phase possible. If we had other projects making calls on our time I don't believe we would have been successful.

13.1.6 Problems and delays

No project goes without a hitch of any sort and the Cu100 was no exception. Two particular ones are described here.

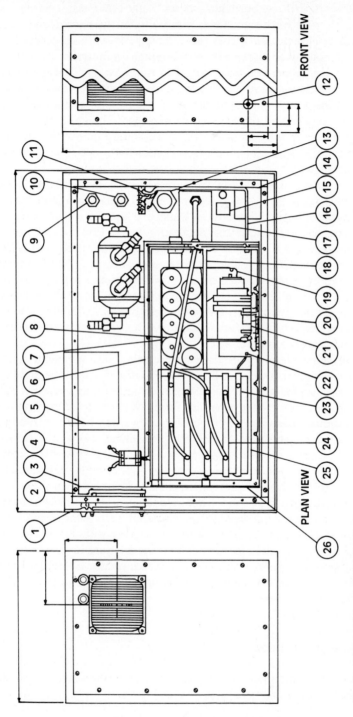

Figure 13.4   Subassembly of the laser

The enclosure

Shortly after starting I was concerned about the design of the laser enclosure, and more specifically, the amount of time left in which it had to be designed and ordered. This was the only piece of equipment that would have to be sorted out before the design report was finalized. This need arose from simple time pressure. The enclosure would take some time to design and build and this was too long to fit into the time allowed for laser manufacture. In any case, the enclosure would be needed at the start of the manufacture period since all the laser components would have to be fitted into it. It was clear that some sort of box with considerable rigidity was required and no-one yet knew how to achieve the design requirements. Many design constraints existed for the enclosure and it had about seven roles to play in the final laser performance. We made a special agreement with the customer that this assembly would be designed and ordered before the PDS was accepted. To do this, the meetings with the customer included extra sections regarding the design of the enclosure.

Within a month of starting I had written and agreed a specification for the enclosure with the customer. The enclosure was the most expensive single subassembly of the laser and the design specification for it ran to about ten pages. Apart from its exacting mechanical requirements, it had to have a prescribed thermal, safety, protection and RF shielding performance. I also had to relearn my stress analysis and design a beam plinth to meet the deflection and weight critera. In the first days of my investigation exotic options involving such things as honeycombs of special light alloys with active temperature control looked like the only solution. If such options were pursued it would be necessary for me to orchestrate the supply of components from many different companies and fit the components together for the first time at Oxford. Needless to say, I was keen to avoid such undesirable commercial and technical arrangements. To begin with, even the best advice I'd obtained made the design prohibitively expensive.

I sent out a formal tender request with the specification of the enclosure to about ten companies whose names I had taken from trade journals and reference manuals. After a few weeks of hard work, an order was placed for an aluminium enclosure at about a fifth of the initial cost estimates with a supplier who could meet the stringent quality requirements of the customer. After collaborating with this supplier, and using some of their expertise, the design was now much simpler and more elegant. There had been a good deal of problem solving to be done but between us we had the necessary knowledge. At one point I also had to bring in a consultant from higher education for advice on the metallurgical aspects of the design.

The only technical problem we could not accurately predict was the welding of so much aluminium and the possible distortion that might result. We identified a subcontractor specializing in aluminium welding who felt confident and arranged for a shipyard press to be used to bend the aluminium beam straight if it distorted too much. This was no mean feat, since the aluminium beam was over 101 mm deep. We found a press in southern England large enough to accept the

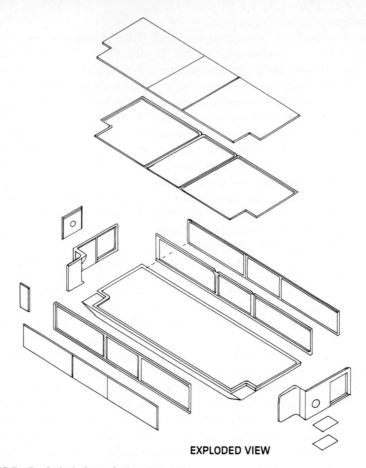

EXPLODED VIEW

Figure 13.5   Exploded view of the enclosure

entire plinth at once and we booked time on it so that, if needed, it would be available.

The commercial arrangements regarding the enclosure were now good. One supplier was responsible for ensuring that all the components fitted and met an acceptance specification rather than my trying to synchronize the production, delivery and assembly of the whole affair. In the end the aluminium did bend a little too much when welded but it was easily bent to a flatness well within the specification, and on the subsequent enclosures we adopted the procedure as standard.

Disconnection safety

One of the conditions agreed in the design report was that the laser would be safe under the simultaneous disconnection of all services. This in itself was true but it all depended on what your definition of 'safe' was.

As the design stood, personnel would be safe because the enclosure formed a safety shroud around the whole laser, even an internal fire would be contained. However, we realized that some of the laser components might not be safe under disconnection. This was because, unlike any of the other lasers built by the company, the water lines had in-line connectors that sealed themselves off when disconnected. These were included so that when the customer took a laser in or out of the rack the connectors (from which the water-cooling lines would first have to be disconnected) would not leak any of the 30 or 40 litres of water held in the cooling circuit. There was even a specification in the design report on the acceptable 'weepage' which was to be less than six drops. We started to consider what would happen under the worst conditions.

If the cooling water was stopped at the same time as the power and the laser had been at full power for a long time there would be a lot of heat stored in the plasma tube. Upon disconnection the heat from the plasma tube, at about 1400°C, would start to flow into the surrounding insulation and cooling vessels. After about half an hour without cooling (we estimated) the O-rings would probably melt, as would any plastic near the metal cooling jacket. We started to think carefully about what else might happen. The cooling circuit might also be cut if the laser was being removed for some reason. The self-sealing water connections meant if the inlet and outlet connections were broken the water in the cooling system would be contained within a trapped volume. This water would certainly boil and something would have to give.

I called the customer there and then. I asked about the imagined operation of the lasers in the future when many lasers would be in service, and the team hung on my every word. It took moments to establish that exactly the conditions we had discussed might apply. If a laser went down in service the plan was to remove the faulty unit and then forklift in a replacement. The operating crew would have their hands full for at least an hour before they could get round to servicing the faulty laser. Our previous conversation had shown that when they lifted the service panels they could expect to find melted or burnt plastics and a laser assembly that would need a complete strip down to replace all the O-rings, which is a long job. In addition, there would be a ruptured cooling circuit as the trapped cooling water boiled and eventually burst its pipes, spraying scalding water over everything, causing doubt that the control electronics in the laser would still be working. Steff spoke for all of us when she said 'All that doesn't really have the ring of "safe under disconnection" about it, does it?'

We talked, all of us. I canvassed opinion and started a discussion on the issue. None of us believed the customer would be satisfied with the present design. Ideas for a solution came thick and fast. If the heat was a problem could we get rid of enough of it to remove the problem? Could we slow down the heat transfer so that peak temperatures were low? Could we somehow direct all the heat into the massive aluminium beam that formed that base of the laser? Could we cover the laser tube in cooling fins that would reduce the peak temperature? There were many good ideas and some non-starters, in fact we had a lot of fun thinking up ideas. After about 40 minutes the team was flagging and it was clear that answers to some thermodynamic calculations were needed before any rational selection of options could proceed. Rhys was the best at that, so he went off to do the sums. We all carried on with the building programme.

Three days later we met to finalize our reaction to the simultaneous disconnection problem. After some time away from the problem the options seemed simpler. In fact, the only solution that really had any chance of success on technical grounds was surrounding the laser tube with enough water in the cooling circuit to absorb the heat without boiling. The diameter of the water cooling jacket could be increased sufficiently with a little reorganization of the internal components.

Rhys had calculated the required amount of water for safety, accounting for the duration of the process and the heat lost by convection. There were two consequences of the proposal. First, the laser tube would look rather fatter than we had imagined and we felt that we would lose the slim hi-tech appearance. In fact, it wasn't long before we nicknamed it 'the boiler'. The second consequence was an increase in the weight of the whole laser. The only real difficulty with the first consequence was that Rhys, a man skilled in the art of laser design, was anxious for his latest creation to look absolutely right and didn't much care for the aesthetics of this 'boiler'. The second consequence, however, was rather more serious.

The customer was very concerned about weight since the rack of lasers was a considerable structure in its own right and the precise alignment required for the operation's success would be upset by even a little too much extra weight. We already knew that various other parts of the laser had come out rather overweight, and although we could not yet be accurate about the final weight of the laser it was clear that it would greatly exceed the 750 kg specified in the contract. We did the sensible thing and had a technical meeting with the customer almost immediately.

There was, in fact, some spare capacity in the design of the rack and the new weight was acceptable. The rack would still work up to a weight of 1050 kg. Even with the new 'boiler' we didn't think we would get anywhere near that much. Ultimately, the final consequence of the problem was that the customer would have to buy a larger forklift truck than had been anticipated. We felt that the weight should not really have been specified in the contract, after all, we had not increased the weight intentionally. The weight of the laser would simply be a consequence of the other constraints imposed on the design. The customer had been insistent on including a weight in the specification, and in the end Oxford Lasers had been forced to agree. Paradoxically, it appeared that while Oxford Lasers was able to supply the highest specification copper lasers commercially available anywhere on the planet, estimating its weight accurately was proving a good deal more difficult!

13.1.7 Testing

We were visited by the customer for one of the many inspection meetings that occurred during the construction phase. A few days before the visit we again had the test rig running at over 100 watts for some experiments on component life and proudly showed our visitors the awesome sight of such a beam. The technical excellence and importance of this achievement was lost on the inspectors since they were only interested in the paperwork. They tested our ability to trace various components they identified in the laser to their original suppliers. The paperwork was in order and the components were traced. Various other aspects of the project's administration and some quality issues were also scrutinized. The inspectors went away happy. We were happy that they were happy but incredulous that the test rig didn't impress them.

In week 50 the whole laser was to go together and be operational. It was time to test our designs. We had tested most of the subsystems separately and had even been required to do so by the inspection and test procedures described in the design report. It was a nervous time for us and we tested several of the systems again for safety. About mid-morning there was nothing else left to delay us and we really did have to try the whole laser together.

Richard turned the whole laser on for the first time and left it going for about 5 seconds and then turned it off again very quickly. The laser had instrumentation on all the vital measuring points. John was on duty looking down the bore through a special protective attenuating plastic and reported on the behaviour of the discharge. The lights were dim so that we could quickly spot any unwanted discharges underneath the perspex covers used for testing. Richard and Rhys paced up and down alongside the laser saying technical things. Steff came to watch as well, the lights were too dim for her to work and in any case, this was an important event. Rhys and Richard finished their discussion and announced that we were satisfied and that we were going to switch it on again and, as Richard said, 'go for it'.

There is a safety protocol used for switching on a copper laser and the team automatically went into it again. 'Thyratron warm', announced Richard, 'ready?' he then asked. Two 'readies' came from myself and Rhys, followed by one from John when he was in position. Everyone looked at Steff, who wasn't used to all this laser talk, 'Yea, fine, ready, I mean', she said to all the waiting faces. 'Switching on', called Richard. The loud thump of a chain of contactors switching in confirmed we had started. 'Volts on' came next as the power controller started to deliver power into the laser tube. By now the whole machine was buzzing, 'Breakdown', called John, as the discharge started to flow properly, '14 kV'. Rhys looked at one of the oscilloscopes and paced round the laser. He and Richard exchanged glances then he looked up and smiled, 'Piece of cake' he said, joking. People always joke when they are nervous! We watched, wondering how long the system could run for without some portion of it tripping the whole laser out.

Steff looked at the laser, which to her seemed the same as normal apart from the messy test equipment and the noise, and said 'Well, that's it, then, time for them to pay us again!' In fact, the success was inconvenient for her since she was drawing one of the internal subassemblies at the moment and with the laser actually being on it wasn't really safe to climb around on top of it!

The team just kept on looking at the laser and watching it run, Rhys and Richard were all for putting copper in and getting light out that day but we refrained and decided not to get carried away with our success.

About half an hour later John went to the toilet at the back of the unit and something strange happened. The moment he flushed it, the laser tripped out. We soon found that the laser had shut itself down because it detected a drop in water pressure. The laser was interlocked to the water pressure so that if the supply slowed down and there was a risk of overheating it switched itself off. The water supply to the unit was not enormous and it turned out that there was not enough for the laser and the loo. The laser obviously took priority and so the loo was closed, and John is the last person ever to have used that particular convenience.

Over the next fortnight we became more and more confident with the laser and planned to demonstrate it at the next milestone meeting with copper in the tube as Richard and Rhys wanted, ahead of the plan. The laser wouldn't just be discharging, it would be lasing.

In week 52 the second milestone of a laser continuously running went ahead on time, a year to the day after the project started. We ran the laser at about three-quarters of full power and the beam of green light given at this power looked rigid enough to sit on. Our customer was delighted with the sight. A scientific delegation had been sent this time, including their most senior scientist, a man who appreciated what he was looking at on occasions like this. The delegation accepted without reservation that if we were able to achieve this demonstration ahead of schedule we wouldn't have too much difficulty with the few outstanding circuit boards and control systems that needed completion. In the team afterwards we joked that there was a month in the plan between this milestone and the next of demonstrating full power, yet we were confident now that all we had to do was turn up the power control knob the full power and we would be fine. 'Not a bad month's work' we said.

In reality, of course, we used the month of time we had clawed back to help in the problem areas where we were behind, particularly the alignment system, which John and I were able to sort out completely in just two weeks now that we didn't have interruptions taking us away from it every hour or so.

The customer arrived in a delegation for a demonstration of full power, and, as usual, our jeans and trainers were swapped for suits and ties. We were getting very practised at hosting visitations from our customer and the initial detailed planing and effort that had gone into preparation for the early meetings had given way to a more relaxed and informal feel. However, each meeting still contained a formal recognition of the state of the project and minutes were always taken recording actions and decisions.

The demonstration of the laser running at full power went off without incident, although behind the scenes there had been more problems than expected. Laser output power had not been as steady as we would have liked, and in the week just before the test Richard began to suspect that the plasma tube had cracked. We discussed it and, on his recommendation, decided to fit a replacement. This was not a popular decision, since there is a great deal of work involved and Richard and John particularly would have a lot of overtime to do in order to meet the deadline. In the event, the tube was cracked, copper had leaked out and a new tube was the right decision. On the day of the test the tube was hardly cool from all the installation processes and running that accompanies the fitting of a new tube before we were running it again for the milestone.

The major block of work for the team, however, was now to complete the characterization tests of the new product. A laser needs an extensive set of tests to determine the operating envelope and these would take many days of running time to complete. In addition, there was a need for a continuous test of 100 hours of unbroken running. One stoppage for only a few moments would negate the whole test and it would have to be started again. The other daunting task was the continuing construction of the second laser. In week 70 that was due to be discharging stably. Work continued with the regular weekly meetings forming the basis for planning. John constructed the second laser, with Richard and Rhys finishing the testing and construction of the first one. Steff continued drawing while I wrote the manual and other support documentation. I also arranged shipping, including insurance and all the other details that come with freighting out a large laser by surface to the customer's site.

After the full-power milestone meeting we regrouped and listed all the 'little jobs' that needed to be completed before the pre-shipment inspection. The laser was due for delivery in week 68 and this was our next and fourth milestone. The tasks required a large amount of time finalizing the first laser. It's not just the tasks themselves, in a sense that's the easy bit, it's the documentation that goes with them so that you can make many more of the same. Steff's list of drawings to prepare was as long as ever, many of which were associated with quite small changes made recently to components that had been modified. Many such changes involved modifications to the general assembly drawings higher up the bill along with the relevant component drawings. The drawing set was now up to about 250 and still had some way to go before it would form the minimum statement of the designs with which we would be happy. There were also other drawings required for the operation and service manual that had to be prepared.

I wrote the manual, a task which took many weeks and which involved further consultation with and approval from the customer. There are technical manual writing services but we had reasonable provision for reprographics and decided to keep the task entirely in-house. Besides, it would have been necessary to agree the use of subcontractors for this purpose with the customer and we didn't want to have to do that.

There was also a great deal of financial information to straighten out. The data was all there, how much had been spent on each subassembly, how long each had taken to make. I started a review and produced a cost for the whole project. Overall the costs were fine, which was not a surprise since we had been monitoring them during the weekly meetings, but some subassemblies were up to three time more expensive than we had expected while others were much cheaper.

Particularly inaccurate were the estimates of time taken to build. Some subassemblies required vast amounts of testing and setting up. The financial data provided an interesting guide to what design changes should be made. We naturally had out own 'gut feeling' ideas about what could be done better and in some cases the financial data supported these. However, there were other areas that we regarded as very straightforward and technically easy but that consumed more than their fair share of the cost.

One week later the first laser was due for commissioning. The commissioning of an instrument such as a laser requires planning and careful attention to detail. The plan was to forklift the laser onto a lorry at Oxford, drive to a public weighbridge to provide an accurate weight reading and continue to the customer site, where another crane would be waiting. The laser would be forklifted off and then lifted by crane into a second-floor laboratory. It is a heartstopping spectacle to see the culmination of such a large proportion of one's time and effort covered with polythene and dangling from a crane jib in the breeze and rain. We all breathed again when it was safely inside the laboratory. The whole transport process had to be insured, the route and arrival time agreed, and clearance papers for all personnel to enter and leave the customer's site were required. There were details of the tests to be carried out on-site to be agreed, costs for consumables during the test, and a van full of spare parts would be taken in case of accidents during transportation or difficulties during commissioning. It is always difficult to control arrangements at a distance but by having agreed itineraries and schedules between Oxford Lasers and the customer it was all achieved, and in the event the commissioning went very smoothly and the laser survived its journey well.

Rhys and Richard stayed to commission the laser while John and I returned to Oxford to continue with the next one. When we got back Steff was really bored. She had been stuck in the industrial unit on her own for two days getting loads of work done but being very bored. I promised her the installation trip next time, but, as it happened, she wouldn't be going on that trip either.

13.1.8 Epilogue

After the delivery of the second laser Rhys and I had one of our reflective sessions on project progress. We had spent about a third of the project time thinking about what we were going to do, about a third actually doing it, and about a third measuring what we did and confirming that it worked. This was a ratio that served us well, and we planned to repeat it when we were next hired to build a 200 watt laser! Our minds were turning more and more to the future. It was clear that things would change. There wasn't enough work for the whole group between now and the arrival of an order for lasers to fill the rest of the customer's rack. Even if we were successful, it was some way off. In addition, much as we would have loved it, there was no 200 watt copper laser programme anywhere in the world for which we could tender.

The group finalized a new contract for the production of an oscillator, a special variant on the laser design to be used by the customer to inject light at the start of each chain of lasers in the rack. We had anticipated this order but it was only for one machine, and it involved much less work than one of the big amplifying lasers. It would, however, provide work for a few more months. It was becoming more certain that there would be a considerable time lag between the customer receiving the first two lasers and the first oscillator and then testing the lasers in the rack to prove, in principle, that their proposed system would work. Only then would they be in a position to order the future amplifiers and oscillators. Such an order would clearly be a very attractive proposition to Oxford Lasers and stood to be their biggest order to date, probably for quite some time to come.

Sadly, Steff resigned from Oxford Lasers and returned to her original employers. I was concerned by this development, the team had worked well and I had thought everyone was happy. We had a chat and I asked why she wanted to leave. She explained that there was little of the electronics drafting that she had liked so much in her previous job. There had also been a number a changes at her previous employer and the reasons for her leaving in the first place no longer applied. The company had contacted Steff and had offered new terms and conditions that were difficult to refuse. She had considerable expertise in electronics drafting and we had both expected there to be more of this work on the Cu100 than in fact there was. We chatted some more, partly about Oxford Lasers, partly about our own lives and partly about the rest of the group. She would be sad to leave the team but she felt the move was best. She was right: things were moving on at Oxford Lasers too.

The Cu100 project had generated a considerable profit and it was re-employed by the board to develop the company. Oxford Lasers was undertaking a second major product development programme to completely redesign the existing range of lasers and bring them into new markets. I had been asked to assist in this venture and after discussion with my superiors an appointment was made. The job was a more senior post than the one I currently held and was very attractive. The following Monday I left the Cu100 group to run a new project as Head of Engineering. The project was to redesign the company's flagship laser and bring its ageing technology up to date.

Much work had already been done on this project since money had come in already from the Cu100 in stage payments. The design was in fact virtually complete. Lasers were promised for delivery later that year and there was no way out of this. In hindsight, there were many mistakes made at this time. The project was not well focused, the project plans had already gone wrong several times, the documentation was poor and a large number of vital systems had only received cursory testing. However, I joined the project and after much expenditure, hard work, frustration and a great deal of help, the lasers were shipped on time. The commissioning went disastrously, the project was not a success technically or financially. As a company, much was learnt from it. The hard way.

Later that year I met with Rhys for a drink to mark the passage of the project. The oscillator was delivered. Commissioning was over and the project was completed. There was nothing in the AVLIS market on the horizon until the order for the lasers to fill the rack was received, and this was still six months off at least.

Richard's particular skills were needed on another product development programme to develop a new design of copper laser altogether using compounds of copper, instead of elemental, which would operate at much lower temperatures but had the disadvantage of only producing very low powers. However, there was the possibility of commercial exploitation, and Oxford Lasers decided to pursue a research project into it. Richard started this work straightaway and spent some time working in both groups before leaving the Cu100 group altogether.

A year later, when Oxford Lasers did receive an order for more amplifiers and oscillators to fill the rack, only John and Rhys remained from the original team of five. The receipt of the order was a massive commercial achievement and vindication of all our original time and effort.

In the much longer term, however, it was not AVLIS that sustained Oxford Lasers as had been imagined. The commercial situation regarding the supply of uranium did not develop as expected and there are no plans to build a full-scale separation facility in the UK. To begin with, there is a large oversupply of enriched uranium in the world, due in part to a much smaller increase in demand than was projected and also to the collapse of the former USSR, which liberated an enormous amount of uranium onto the free market. Second, developments in separation technology have eroded the price advantage, though if demand were greater it would still be advantageous. Oxford Lasers was therefore faced with considerable difficulty when the order for lasers to fill the rack had been fulfilled. The company had been developed in style and position to deal with a type of customer whose requirements no longer existed. New markets had to be found. It so happened that the last AVLIS business was completed in the depths of an economic recession and nearly half of the staff had to be laid off. After a long and difficult search new markets are being developed, particularly in materials processing, where the very high power density of the copper laser pulse is an advantage. New markets are also being developed in the area of flow visualization, an extension of high-speed photography in which the very short and bright pulse of the copper laser is used to illuminate rapidly moving gases in, for example, an internal combustion engine, and so obtain time-resolved data describing combustion, flame propagation and gas flow over the valves.

Even if the AVLIS opportunity did not unfold as had been hoped, the project to develop the first two lasers was a success. The lasers were delivered on time and within budget. It was certainly one of our better projects.

Today, Oxford Lasers is continuing to actively seek large corporate partners for whom it can be a high-technology provider. At the time of writing, a multiphase investigation and development project has just been signed in the area of materials processing with a large multinational, the successful outcome of which will make the AVLIS business appear very small.

In the end, the lessons of the Cu100 project and the experience of working with a large customer have remained as strengths of the company.

13.2 The Cu100 project debrief

In the case study many management issues were raised. We will now consider these following the structure of the book.

Nature of organizations

In Chapter 2 we considered the establishment of organizations, the role of corporate objectives and the strategies needed to ensure a company's survival. From the case study we see the growth of a new company following from the development of a novel product. Even at a very early stage, the strategies of the company, in their marketing and product development, were focused towards their overall objective, which was to be able to supply the AVLIS market. From this primary goal we see that the company also set itself other objectives, in terms of growth, and that the objectives were reviewed annually. Without this type of focus and review it would be easy for a company to spread itself too thinly, trying to achieve too much and not achieving anything. In addition, without review of objectives there is no way to plan for change and ensure that decisions taken previously are still the most appropriate given the changing market and business environment.

The case study highlights the difficulty that companies can face when they try to meet particular customer or market needs. If Oxford Lasers had not pursued other markets and customers away from AVLIS then the company would have surely folded when that market was denied. As it was, the company was damaged but survived.

The company has a strong research and development focus, and at one time half of the workforce were employed in R&D activities. This shows the way in which the nature of the product and the market affects the company organization. With a proven product or a stable market there would not be the requirement for such a large R&D activity with the associated high level of speculative investment.

Functions of organizations

In Chapter 3 we looked at the different functions that were required in an engineering organization. Although the Cu100 team was multifunctional, we still see many references in the case to the discrete functions. Parts were ordered by the team to specifications set by them. We even saw an example of James visiting suppliers, working with them and using their specialist knowledge (the distortion due to welding problem). The company had established ordering and goods receipt procedures and these were used by the team, while retaining responsibility for inspection and storage, as required by the customer. This is a prime example of how it is important to draw on resources that are available when they meet the criteria. It would have been expensive and time consuming to operate a separate procedure.

In this project we see little interaction of the team with the manufacturing function, apart from recruiting one of their technicians. However, throughout the project the plan was always that the Cu100 laser would go into production by receiving the order for more lasers. Because of this, the team were developing the product with manufacture in mind, defining bills of material and providing a complete drawing set.

James and Rhys were involved in the 'sales' function, in negotiating the contract with the customer. Throughout the project they also retained close contact with the customer and were finally responsible for installation and commissioning on the customer's premises. The company's finance department were used for the processing of orders, although it is clear from the case that responsibility for managing the project budget remained within the team.

The whole case describes the product development process for the Cu100. The project team were wholly responsible for all aspects of the design, from producing the design report to building the laser. Drawings were originally subcontracted but part-way through the project the team had its own draughtswoman.

Finally, we consider the quality function. It was clear in the case that there was a stringent quality system requirement from the customer. References implying this relate to the checking of inspection schedules, the requirement to trace parts to source, and the specific reference to having to use an enclosure supplier who could meet the quality requirements of the customer.

Finance

In Chapter 4 we saw a general need for monetary control and then looked at some specific accounting techniques, including the preparation of balance sheets, product costing, and the use of budgets. We see in the case that the project has its own budget and also that the customer makes stage payments when milestones are achieved. The first of these means that there is a difficult financial task for the team, the second, that the team has responsibility for producing significant cash in-flows to the company.

Managing a budget means that you have to plan recruitment, purchases and equipment hire, and allow for the costs of operating an industrial unit. James wouldn't have needed to know about balance sheets and profit and loss accounts for this job, but he must have had to cost the lasers and ensure that there was sufficient cash flow to cover necessary expenditure.

Product development

The whole project was a product development exercise and thus we see that all aspects of management are required to manage product development successfully. However, in relation to the specific issues raised in Chapter 5 we can see that there was a conscious decision by Oxford Lasers to use a project organization to run this project, even though the company was functionalized. For this project there were clearly a number of advantages to this. For example, there were no problems with functional interfaces when the project moved from 'design' to 'manufacture'.

We saw the importance of defining customer requirements and working with the customer, particularly when there were technical difficulties such as the disconnection problem, which required that the design team have technical information from the customer. We also saw the importance of reviews and the role of documentation in the design process.

Operations management

This case study has very little to offer on operations management, as it is very much a design-and-build product. However, there was always the clear intention that the lasers would go into manufacture and the effect of this is noted by the documentation requirements to ensure that manufacture could take place at a later stage.

Quality management

The case study refers to the quality requirements of the customer, particularly in relation to documentation, traceability and acceptance testing. However, the achievement of quality and the means of that permeate the case from defining the customer's requirements, and agreeing them through the design report, to having trained personnel and ensuring the safe delivery of the laser.

Project planning and management

In Chapter 8 we looked at the requirements for project management and at some specific planning techniques. This was one large project and it was obviously well managed. All the deadlines were met. The customer was so pleased that more orders were placed.

The project constraints, the milestones and deadlines were all defined by the customer. However, the way in which these requirements were met was determined by James and Rhys. Plans were developed during the first eleven weeks of the project and must have been costed as they led to a fixed project budget. Further detailed plans, in the form of quality plans, were agreed with the customer in the design report.

Monitoring of the project by Rhys and James seems to have been performed on a fairly informal basis, although there was a much more formal review with the customer. The day-to-day management of the project and scheduling of tasks was done by James during the weekly meeting. There is an indication that there may have also been a very informal review of progress at the monthly company meeting.

Personnel management

In Chapter 9 we considered some of the issues related to employing people. These included legislation governing employment and recruitment, motivation, appraisal, training and development, and job design. There are many examples of the first four of these topics, and we shall consider these in turn.

In the case we see an example of the recruitment and selection process from the candidate's point of view when James applied for the job. We also saw the other side of this when James and Rhys were recruiting for other members of the team. You notice from the case that they had a clear idea of not only the skills they were looking for but also the type of person they needed for the team. There was also an example of the problems that can arise if you are not able to make a satisfactory appointment (James having to prepare some drawings, subcontractors doing others).

Unfortunately, we also saw another aspect of employing people, 'poaching' of staff (when Steff's old company went directly to her with a new job offer). Following this, Steff gave notice to terminate her employment with Oxford Lasers.

In the case the high level of motivation of the team is obvious. Not so obvious are the many techniques employed to keep that motivation high. Working in a small group usually works well, but there were other things like the weekly meeting. The team spirit was engendered by being different from the rest of the company, in their own industrial unit, and socializing together (beer and curry). In particular, there was the total involvement of everyone with the whole project and with the customer, which only broke down when Steff wasn't asked to go when the laser was delivered.

We saw that Oxford Lasers operate a two-tier appraisal system for everyone in the company. We also saw that it took a considerable amount of James's and Rhys's time. But as James said, he needed to establish how things really were going and he didn't normally get time for that. The appraisal system provided that opportunity.

Throughout the text we see many references to training and development. These start with James learning about laser physics and having to relearn his stress analysis. This is not unusual: again, as engineering is so vast and is constantly changing we can't be experts in everything. What we have as engineers is a general knowledge, such as management, thermodynamic principles, etc., that is widely applicable. However, we also have the ability to acquire more knowledge, in particular we have the ability to define what knowledge we need. In the case we also have a reference to John who was trained by the Production Department. It is likely that as a technician John acquired much of his formal training while working in the company, although this may have been coupled with some form of tertiary education.

Team working and creativity

In Chapter 10 we looked at the selection and use of teams and the process of planning innovation, as exemplified by the Cu100 product development. This whole case is about team working and no further discussion of that is necessary. However, we saw an example of problem solving by team work and the way in which the whole group contributed ideas to the brain storming session to solve the disconnection problem.

Personal management

In Chapter 11 we considered some techniques that can be used to achieve good personal management. In the case we see few references that we could point to as examples of personal management. However, it is clear that in the early weeks of the project James was very much on his own, designing the enclosure and learning about laser physics. He would not have been able to do these tasks and meet the deadline if he had not had an effective time management system, a well-organized filing system and generally good personal organization.

Communication skills

In Chapter 12 we saw that the job of an engineer is very largely concerned with gathering and using information, and we looked at some techniques for doing these things. Throughout the case it is obvious that information control was of the utmost importance. This was a big project and they could not afford to repeat mistakes or lose information. Experiments were recorded and design files were kept. All the drawings were cross-referenced via the general assembly drawings. There must have been a document control procedure because the customer was insisting on quality plans and on traceability. In addition, there were inspection and test schedules that had to be met.

Information gathering was done in a variety of ways. James talks of receiving instruction from his boss, Rhys, and from his subordinate, Richard, who had specific technical expertise. The customer provided information, particularly about the system into which the lasers would fit, and then later at technical meetings. The suppliers of the enclosure also provided technical expertise and further specialist help was provided by an institution of higher education. Without all these inputs to the product development process it is unlikely that the target would have been met. We now live in a world of such complexity that no one person can have an adequate grasp of all the knowledge required to do a job like this. The effective engineer uses all the resources available to provide the extra knowledge required.

In Chapter 12 we also looked at the requirement for engineers to be able to communicate well both in written work and orally. We considered some specific techniques that would be of use here and at methods of communicating in the workplace. Communication is such a big issue in engineering that the case study is full of examples; throughout there are constant references to communication issues. The major written communications were James's job application, the design report, and the operating manuals for the lasers. However, all the design was documented with experiments being recorded. There was obviously a significant requirement for all the team members to communicate in writing.

Examples of oral communications range from the very formal communi-cations with the customer to the weekly meetings with the team. These are equally important. You have to be able to identify the customer need but you can only meet that need if you can then translate it into tasks and activities by communicating with the people who work with you and for you. Another aspect of communication was that of wider communication with the other people in the company, and in fact we saw that Oxford Lasers put so much emphasis on this that a monthly company meeting was held.

The vocation of engineering management

In this chapter we have given you a case study, a real engineer working on a real engineering project. The project would not have been achievable without that engineer's management skill. Every single management issue we have described as a separate issue within the body of the book can be seen to be in use in this case study. The work described in the case study is clearly an engineering undertaking, and yet, as we have seen, the work is filled with management issues.

In the introduction to this book we asked whether is was worthwhile for an engineer to study the contents and so develop their management skills. In this case study we have conclusively shown that engineers use management and technical skills; it is clear that engineers are capable of being excellent at both of these. We have shown that management can indeed be done well when it is done by engineers.

Index